Lecture Notes in Mathematics

Edited by A. Dold and B. Eckmann

Subseries: Institute for Mathematics and its Applications, Minneapolis
Advisers: H. Weinberger and G. R. Sell

1035

The Mathematics and Physics of Disordered Media:

Percolation, Random Walk, Modeling, and Simulation

Proceedings of a Workshop held at the IMA,
University of Minnesota, Minneapolis
February 13–19, 1983

Edited by B. D. Hughes and B. W. Ninham

Springer-Verlag
Berlin Heidelberg New York Tokyo 1983

Editors

B. D. Hughes
B. W. Ninham
Department of Applied Mathematics, Institute for Advanced Studies
Australian, National University
Canberra, A.C.T. 2600, Australia

AMS Subject Classifications (1980): 60 J15, 60 J20, 76 S05, 82 A42

ISBN 3-540-12707-0 Springer-Verlag Berlin Heidelberg New York Tokyo
ISBN 0-387-12707-0 Springer-Verlag New York Heidelberg Berlin Tokyo

Printing and binding: Beltz Offsetdruck, Hemsbach/Bergstr.
2146/3140-543210

The Mathematics and Physics of Disordered Media

PREFACE

The successes of the new physics in the 17th and 18th centuries were inextricably interwoven with the discovery of the calculus. Following Newton, Leibniz, and Laplace, the scientific rationalists, in a burst of enthusiasm over the omnipotence of deductive reasoning, made mathematics Big Bird of the Sciences. The orderliness of that God who made this best of all possible deterministic worlds stood revealed. In the words of Descartes, *Cum Deus calculat, fit mundus*!

Then came Gauss and Riemann with non-Euclidean geometries, Cantor with infinite sets and a host of logicians with the axiomatization of mathematics. All of these developments challenged preconceived notions, and many researchers went deeper and deeper into the foundations of mathematics, trying to find The Perch for Big Bird. These hopes, along with the dreams of the rationalists, ended suddenly when Godel showed that any sufficiently rich system of axioms contains undecidable propositions.*

This loss of faith, like the dread acciditas, that dry soul-withering wind which afflicted the good monks of Egypt so many centuries ago, resulted in a crisis of confidence. Coincident with this internal attack on the foundations, one also finds many applied scientists viewing mathematics as irrelevant to the real world. These developments are in no small measure responsible for the present separation between pure mathematics, applied mathematics, and the sciences generally. Unlike their counterparts in nuclear and particle physics, astronomy and space sciences, biology, and computer and earth sciences, who trumpeted their triumphs real or imagined, mathematicians sat mute and musing.

Awareness of the increasing gap between science and mathematics prompted the National Science Foundation to fund the Institute for Mathematics and its Applications (IMA) at Minnesota along with a sister Institute at Berkeley in 1982.

---

*These developments have been documented in the challenging and extraordinarily lucid writings of Morris Kline in Mathematics, the Loss of Certainty, Oxford, (1981).

The purpose of the IMA is to facilitate the flow of problems and ideas between scientists and mathematicians.

An exciting glimpse of one bold new shape for the mathematics of the 1980's emerged from a workshop held at the IMA in February 1983. The workshop was devoted to the Mathematics and Physics of Disordered Media. Our attempt to define a charter read as follows:

> One of the fundamental questions of the 1980's facing both mathematicians and scientists is the mathematical characterisation of disorder. Until recently it had been impossible to conceptualize or even to contemplate the possibility that a tractable calculus might emerge. A classic example is that of liquids. This state of matter is intermediate between gases, where characteristic distribution functions are uniform to lowest order, and crystals, where the unperturbed distribution functions are periodic. There is no rigorous mathematical description appropriate to the intermediate irreducibly disordered state. Beyond the simple statement of Lindemann's law there is no theory of melting. The situation is analogous in the field of irregular porous media, ubiquitous in areas as diverse as earth sciences and the food industry.
>
> The last decade has seen the beginnings of a unity of methods and approaches in statistical mechanics, transport in amorphous and disordered materials, properties of heterogeneous polymers and composite materials, turbulent flows, phase nucleation, and interfacial science. All have an underlying structure characterized in some sense by chaos, self-avoiding random walk, percolation and fractals.
>
> Some real progress has been made in understanding random walks and percolation processes on the one hand, and through mean field or effective medium approximations and simulation of liquids and porous media on the other. The subject is directly connected with the statistics of extreme events and important pragmatic areas like fracture of solids, comminution of particulate materials, and flow through porous media.

The key words are fractals, percolation, random walks, and chaos. The realization, so eloquently expounded by Mandelbrot, that the Hausdorff dimension of the length of a real (linear) coast line is not one, is an astonishing one. Things are not what they seem. They depend on how one looks at them. This should not be kept secret, and unquestionably progress in the topics of this Volume will throw new light on renormalization groups, particle physics, and phase transitions.

The meeting drew together mathematicians, pure and applied, chemists, chemical engineers, physicists, computer scientists, materials and polymer scientists and statisticians from industry and academia. An honored guest was John M. Hammersley, an Oxford Mathematician who invented the subject of percolation theory thirty years ago. Unlike most mathematical terms, percolation needs no explanation. It means exactly what it means in good old-fashioned English. Another pioneer, Elliott Montroll, was unable to attend because of illness, but nonetheless has co-authored two of our papers.

This very diverse group of scientists found themselves speaking the same language. Progress does seem to be in the air on the difficult problem of developing a calculus to describe chaotic and random systems. Most things in nature are chaotic and random. Mathematical constructs such as Cantor sets, and continuous non-differentiable functions, hitherto considered to be highly abstract and far from the real world, loom large in these recent developments, but in a nice comprehensible way. Perhaps after all and in the longer view, mathematicians <u>are</u> on the right track. Have a good read.

Barry Hughes George Sell
Barry Ninham Hans Weinberger

June 1983

ACKNOWLEDGEMENTS

The organizers of the Workshop thank the staff of IMA, especially Susan Anderson, Debbie Bradley, and Pat Kurth for their assistance with local arrangements. We especially thank Debbie Bradley for her careful typing of this volume.

THE MATHEMATICS AND PHYSICS OF DISORDERED MEDIA

TABLE OF CONTENTS

Preface III

I. INTRODUCTORY LECTURES

B.D. Hughes and S. Prager, "Random processes and random systems: an introduction" 1

II. CONFERENCE LECTURES

E.W. Montroll and M.F. Shlesinger, "On the wedding of certain dynamical processes in disordered complex materials to the theory of stable (Lévy) distribution functions" 109

M.F. Shlesinger and E.W. Montroll, "Fractal stochastic processes: clusters and intermittencies" 138

R.F. Voss, R.B. Laibowitz and E.I. Alessandrini, "Percolation and fractal properties of thin gold films" 153

J. Koplik, D. Wilkinson and J.F. Willemsen, "Percolation and capillary fluid displacement" 169

S. Redner, "Recent progress and current puzzles in percolation" 184

J.M. Hammersley and G. Mazzarino, "Markov fields, correlated percolation, and the Ising model" 201

S. Redner, "Directionality effects in percolation" 246

J.W. Halley and C. Dasgupta, "Percolation and related systems in equilibrium statistical mechanics" 260

S.G. Whittington, "Branched polymers, polymer gels, and percolation theory" 283

J.C. Wierman, "Critical probabilities in percolation models" 300

M. Sahimi, "Critical exponents and thresholds for percolation and conduction" 314

G. Stell, "Models of disordered media: some new results including some new connections between composite-media, fluid-state, and random-flight theories" 347

G.H. Malone, S.Y. Suh, and S. Prager, "Variational bounds on the diffusive and hydrodynamic permeabilities of randomly perforated sheets" 370

G.C. Papanicolaou, "Diffusions and random walks in random media" 391

M.J. Stephen, "Waves in disordered media" 400

W.J. Pardee, W.L. Morris and B.N. Cox, "Microscopic origins of stochastic crack growth" 414

RANDOM PROCESSES AND RANDOM SYSTEMS:

AN INTRODUCTION

Barry D. Hughes* and Stephen Prager+

*Department of Chemical Engineering and Materials Science
+Department of Chemistry
University of Minnesota
Minneapolis, Minnesota 55455

ABSTRACT

We introduce and review a number of topics drawn from the theories of random processes and random systems. In particular we address the following subjects: random walks in continuous spaces and on lattices; continuum limits of random walks and stable distributions; master equations, generalized master equations and continuous-time random walks; self-avoiding walks on lattices; percolation theory; steady-state and transient transport in random lattices; and diffusion and conduction in heterogeneous continua.

Table of Contents

Introduction

Part A: Discrete Models (by Barry D. Hughes)

1. Random walks and random flights 6
2. Random walks on lattices 9
3. Continuum limits of random walk processes 17
4. Master equations, generalized master equations, and continuous-time random walks 22
5. Self-avoiding walks 29
6. Percolation theory 38
7. Equilibrium transport in random lattices 48
8. Transient motion in random lattices 55

References for Part A 70

Part B: Diffusion in disordered media: Variational bounds (by Stephen Prager)

1. Statement of problem and variational formulation 87
2. Bounds independent of phase geometry 96
3. Bounds involving spatial correlations of $g(\vec{r})$ 99
4. Bounds for beds of spherical particles 100
5. Bounds involving knowledge of other bulk properties 102
6. Bounds for many-body diffusion problems 103

References for Part B 106

Introduction

The present article has been written with several purposes in mind. The first is that it should serve as a self-contained introduction, suitable for a wide audience, to some of the topics discussed from a more specialized point of view in other articles in this volume. The second is that it should be useful as a review, and a guide to the primary literature. However, the main thrust of the article is the development of two distinct themes, random processes and random systems, which are gaining increasing importance in the physical sciences. In Part A of the article, we emphasize models in which random processes and random systems have discrete structure. Part B is devoted to transport problems in disordered continua.

Overview of Part A

The notion of a random process, as we see it, grows from an attempt to describe microscopically complex processes by statistical equations of evolution. A prototypical example is the erratic Brownian motion of dust particles or pollen grains in solution due to collisions with solvent molecules. Although above the quantum mechanical level, the process is entirely deterministic, the motion of each grain is sufficiently erratic that it may be taken as random (with the simplest model being that of a "random walk", as discussed in sections 1 to 4). The random processes which have been most extensively and successfully studied are those which possess no memory effect, or very simple memory effects. Much harder and less well understood are problems with strong memory effects, exemplified by the problem of a self-avoiding walk (section 5).

By modelling a physical phenomenon as a random process, we usually are adopting the view of a natural phenomenon as a drama played out on a fairly simple and uniform stage, but with a random script. The direct antithesis of this view is what we call the random system, in which the script is written out and orderly, but the stage setting is chaotic. The simplest physical example of a random system is an irregular porous medium. If fluid flows steadily through the voids in the medium, the streamlines are fixed in space, but tortuous due to the spatial variation in local geometry and topology. Percolation theory (section 6) gives a precisely formulated mathematical model, in the context of which random geometry and topology can be investigated quantitatively. It can be generalized to predict the hydraulic or electrical resistance, or other steady-state transport properties of a random system, as outlined for lattice systems in section 7; transport in random continua is deferred to Part B. Many fundamental outstanding questions in this area remain to be resolved.

While the notions of random process and random system, as we have outlined them, are apparent opposites, it is now known that for successful modelling of important physical phenomena the two concepts must be fused together. For example, in the dispersion (speading) of a blob of dye convected through a porous medium, the effects of tortuousness of the streamlines (random system) compete with diffusion between streamlines (random process). We discuss simple models for random processes in random systems in section 8.

The topics discussed in Part A represent a somewhat arbitrary selection from an enormous body of work contained in the mathematical, physical, chemical, and engineering literatures. We have avoided wherever possible the discussion of problems which require a knowledge of the deeper concepts and technicalities of contemporary probability theory. Extensive references are given, with a distinct bias towards the applied literature and no attempt to place the topics discussed in the broader contexts of Markov processes and their derivatives, Markov random fields, and so on. Even within the applied literature, we make no claim to bibliographic completeness. A recent random walk bibliography [L.H. Liyange, C.M. Gulati and J.M. Hill, "A bibliography of applications of random walks in theoretical chemistry and physics", <u>Advances in Molecular Relaxation and Interaction Processes 22</u> (1982), 53-72] lists almost 300 references, and yet represents but the tip of the iceberg. It is hoped that the references supplied here will prove sufficient to guide the reader into those parts of the literature which arouse his or her interest.

We draw to the reader's attention the existence of published proceedings of a number of recent conferences on random processes and random systems*. A clear introduction to some of the concepts and applications of random walk theory has been given by G.H. Weiss ["Random walks and their applications", Amer. Scientist 71 (1983) 65-71], and a delightful account of the historical antecedents of random walk theory, entitled "A wonderful world of random walks", has been compiled by E.W. Montroll and M.F. Shlesinger, and appears in a volume dedicated to M. Lax, edited by H. Falk and published by the Physics Department of City College of the City University of New York. A compendium of papers edited by N. Wax, Selected Paper in Noise and Stochastic Processes (New York, Dover, 1954) remains valuable, but gives no idea of the wealth of developments which were shortly to follow. Some aspects of the theory of homogeneously disordered systems, from the perspective of solid state physics, form the subject of a major book by J.M. Ziman entitled Models of Disorder (Cambridge University Press, 1979). Three important collections of papers on modern theoretical and numerical approaches to thermodynamic critical phenomena contain survey articles which are relevant to the problems of self-avoiding walk, percolation, and conduction in random systems: Numerical Methods in the Study of Critical Phenomena, ed., J. Della Dora, J. Demongeot and B. Lacolle (Berlin, Springer-Verlag, 1981); Monte Carlo Methods in Statistical Physics, ed. K. Binder (Berlin, Springer-Verlag, 1979); and Real-Space Renormalization, ed. T.W. Burkhardt and J.M.J. van Leeuwen (Berlin, Springer-Verlag, 1982).

---

* (a) The proceedings of the Symposium on Random Walks and Their Application to the Physical and Biological Sciences (National Bureau of Standards, Gaithersburg, Maryland, 1982), published as a special issue of the Journal of Statistical Physics (Volume 30, No. 2, 1983); some additional papers presented at the symposium will be appearing in an American Institute of Physics Conference Proceedings Volume edited by M.F. Shlesinger and B.J. West.

(b) "Percolation Processes and Structures", Annals of the Israel Physical Society, Vol. 5, ed. G. Deutscher, R. Zallen and J. Adler, (Bristol, Adam Hilger, 1983); this volume contains articles by several contributors to the present volume (Hammersley, Halley and Redner) and many others.

(c) Electrical Transport and Optical Properties of Inhomogeneous Media, ed. J.C. Garland and D.B. Tanner, AIP Conference Proceedings, Vol. 40 (New York, American Institute of Physics, 1978).

(d) Disordered Systems and Localization, ed. C. Castellani, C. Di Castro and L. Peliti, Lecture Notes in Physics Vol. 149 (Berlin, Springer-Verlag, 1982).

(e) Macroscopic Properties of Disordered Media, ed. R. Burridge, S. Childress and G. Papanicolaou, Lecture Notes in Physics, Vol. 154 (Berlin, Springer-Verlag, 1982).

(f) Ill-condensed Matter, ed. R. Balian, R. Maynard and G. Toulouse (Amsterdam, North-Holland, 1979).

Overview of Part B

In Part B we turn to problems of transport and conduction in random continua. Our discussion is phrased in terms of the diffusion problem, though a number of other problems, including thermal and electrical conduction, electric permittivity and magnetic permeability are mathematically equivalent to the diffusion problem. In the main we focus on the application of variational methods to obtain rigorous bounds on overall diffusion rates. The presentation relies heavily on the reader's physical intuition and common sense - we take it for granted throughout that the passage to samples of infinite size will not lead to grief, and that the details of the boundary conditions at the sample surfaces do not matter so long as a given overall flux or concentration gradient in the diffusing species is maintained. The literature cited is a sampling rather than a list of the work that has been published; the intent is to convey some feeling for the different ways in which information about the structure of the material may be given, and for the methods by which such knowledge may be converted into calculable bounds.

In section 1 we define the effective diffusion coefficient and state the basic variational inequalities. Section 2 introduces the remarkable bounds discovered by Hashin and Shtrikman, which are known to be the best obtainable for an isotropic two-phase material if only the volume fraction of the phases is given. The use of spatial correlations to obtain better bounds is the subject of Section 3, and Section 4 discusses bounds for systems of particles, especially the overlapping spheres model of Weissberg. Bounds derived from information on other bulk properties of the material are described in section 5. We conclude with a brief discussion of Brownian motion in large groups of interacting particles, and obtain an upper bound on the relative velocity of two solute species subject to uniform external forces.

One topic is absent from Part B: percolation. Even far from any percolation threshold, as portions of the material in question become impermeable to the diffusing species, the lower bound on the effective diffusion coefficient goes to zero; even though the upper bound remains useful, this precludes any rigorous statement about percolation thresholds or how they are approached.

Acknowledgements

The authors have been privileged to work on topics reviewed in this article with H.T. Davis, E.W. Montroll, M. Sahimi, M.F. Shlesinger and L.E. Scriven. To these gentlemen, and to B.W. Ninham and H.F. Weinberger, who have greatly influenced our vision and understanding of fundamental problems, we express our sincere thanks. The preparation of this article was supported by the U.S. Department of Energy.

PART A: DISCRETE MODELS

1. Random Walks and Random Flights

The problem of "random walk" was first posed, but not solved, in a letter written to Nature in 1905 by Karl Pearson [1]:

> "A man starts from a point O and walks ℓ yards in a straight line; he then turns through any angle whatever and walks another ℓ yards in a second straight line. He repeats this process n times. I require the probability that after these n stretches he is at a distance between r and $r + \delta r$ from his starting point O.
>
> The problem is one of considerable interest....".

A rather informal asymptotic solution of this problem in the limit of large n had been given twenty-five years earlier by Lord Rayleigh [2] in a paper on the addition of sound waves of equal amplitude but random phase, as Rayleigh pointed out in his own letter to Nature [3], duly acknowledged by Pearson [4]. An integral representation of the solution valid for arbitrary values of n was provided almost immediately by Kluyver [5], and contributions to the necessary mathematics were also made by Markoff [6]. Lord Rayleigh returned to the problem in 1919 [7], expanding on earlier analyses of Pearson's problem, and developing the three-dimensional generalization, which he called "random flight".

To present-day scientists much of the analysis associated with the Pearson and Rayleigh problems, as we present it below, is straightforward, while probabilists would readily identify it as a particularly simple case of the well-studied general problem of addition of independent random variables, and perhaps assign it very little importance. However the random walk or random flight concept has been most fruitful in a wide variety of different fields, as may be gauged from major reviews of the field [8,9,10], and the terminology transcends linguistic barriers - Problem des Irrwanderns [11] to the Teutonic, marche aléatoire to the Francophiles. (For the lattice analogue of the Rayleigh-Pearson problem, Pólya, used herumwandernde Punkte and Irrfahrt [12], and promenade au hasard [13].)

Let $P_n(\vec{x})$ denote the probability density function for the position $\vec{x}$ of a random walker (or aviator) in the E-dimensional continuum after n steps (displacements) have been made. The steps are taken to be independent random variables and we write $p_n(\vec{x})$ for the probability density function for the nth step. Then the evolution of the random walk is governed by the equation

$$P_{n+1}(\vec{x}) = \int p_{n+1}(\vec{x} - \vec{x}')\, P_n(\vec{x}')\, d^E\vec{x}'. \qquad (1.1)$$

The implicit assumption of translational invariance, embodied in the simple convolution in Eq. (1.1), ensures that the formal solution of the problem is easily constructed using Fourier transforms (characteristic functions [14] to probabilists). Let

$$\tilde{P}_n(\vec{q}) = \int e^{i\vec{q}\cdot\vec{x}}\, P_n(\vec{x})\, d^E\vec{x} \qquad (1.2)$$

and

$$\tilde{p}_n(\vec{q}) = \int e^{i\vec{q}\cdot\vec{x}}\, p_n(\vec{x})\, d^E\vec{x}\ . \qquad (1.3)$$

Taking the Fourier transform of Eq. (1.1) and using the convolution theorem for the Fourier transform, we deduce that

$$\tilde{P}_{n+1}(q) = \tilde{p}_{n+1}(\vec{q})\, \tilde{P}_n(\vec{q})\ , \qquad (1.4)$$

whence

$$P_n(\vec{x}) = \frac{1}{(2\pi)^E} \int e^{-i\vec{q}\cdot\vec{x}}\, \tilde{P}_n(\vec{q})\, d^E\vec{q} \qquad (1.5)$$

$$= \frac{1}{(2\pi)^E} \int e^{-i\vec{q}\cdot\vec{x}}\, \tilde{P}_o(\vec{q}) \prod_{j=1}^{n} \tilde{p}_j(\vec{q})\, d^E\vec{q}\ . \qquad (1.6)$$

For the problems of Pearson and Rayleigh, the directions of allowed steps are isotropically distributed, and all steps have the same length ℓ , so that

$$p_j(\vec{x}) = \ell^{-E+1} A_E^{-1} \delta(|\vec{x}| - \ell) \tag{1.7}$$

where A_E is the surface area of the hypersphere of unit radius in E dimensions. The radial symmetry enables all of the Fourier transforms to be reduced to single integrals involving Bessel functions if E is even, and trigonometric functions if E is odd and greater than unity (see, e.g., Watson [15] or Bochner and Chandrasekharan [16]). One finds that

$$P_n(\vec{x}) = \frac{1}{2\pi} \int_0^\infty J_0(\rho|\vec{x}|) \{J_0(\rho\ell)\}^n \rho d\rho \tag{1.8}$$

in two dimensions, while in three dimensions

$$P_n(\vec{x}) = \frac{1}{2\pi^2|\vec{x}|} \int_0^\infty \sin(\rho|\vec{x}|) \{ \frac{\sin(\rho\ell)}{\rho\ell} \}^n \rho d\rho . \tag{1.9}$$

For large values of n , $P_n(\vec{x})$ converges to a Gaussian or normal distribution, reflecting the central limit theorem of probability theory:

$$P_n(\vec{x}) \sim \{ \frac{En}{2\pi\ell^2} \}^{E/2} \exp(- \frac{En|\vec{x}|^2}{2\ell^2}) ; \tag{1.10}$$

an analysis of higher order terms is possible. The explicit evaluation of $P_n(\vec{x})$ for modest values of n is a somewhat harder problem. In <u>three dimensions</u>, Rayleigh [7] gave explicit solutions for $n \leq 4$, Chandrasekhar [17] for $n \leq 6$ and Vincenz and Bruckshaw [18] for $n \leq 8$. The folklore arose, and perhaps persists in some quarters, that the general problem of finding closed-form representations of $P_n(\vec{x})$ in three dimensions for arbitrary $n > 2$ is unsolved. However, Treloar [19] was able to derive a general solution almost forty years ago, using methods drawn from sampling theory. (A particularly elegant discussion of the problem along these lines can be found in Feller [20], pp. 32-33.) A direct derivation of Treloar's result from Eq. (1.9) has been given by Dvořák [21]. In the region $|\vec{x}| < n\ell$ where $P_n(\vec{x}) > 0$, it may be written in a variety of equivalent forms, with perhaps the simplest being [21]

$$P_n(\vec{x}) = \frac{1}{4\pi\ell^2|\vec{x}|2^n(n-2)!}\sum_{\nu=0}^{n}(-1)^\nu\binom{n}{\nu}\left[n-2\nu-\frac{|\vec{x}|}{\ell}\right]^{n-2}\operatorname{sgn}\left[n-2\nu-\frac{|\vec{x}|}{\ell}\right] .$$

(For computational purposes, a recurrence relation satisfied by $P_n(\vec{x})$ is more efficient when $n > 10$ [21].) The determination of a closed form expression for $P_n(\vec{x})$ in two dimensions is much harder (for reasons discussed in [20], p. 33) and an expression for arbitrary n is apparently not available at present.

Many generalizations and extensions of the preceding analyses have been given, and we list but a few here. Barakat has considered the case when the lengths of the steps are randomly distributed [22], and the case when the number of steps taken is a random variable [23]; Nossal and Weiss [24] have examined the case when the distribution of step directions in Pearson's walk is anisotropic; Montroll and West [9] and Hughes, Montroll and Shlesinger [25] have considered generalizations of Pearsons's and Rayleigh's problems in which the lengths of the individual steps are random variables with infinite variances.

2. Random Walks on Lattices

A considerable conceptual simplification in random walk problems is achieved if the walk is confined to a lattice, or discrete space. The simplest example of a discrete space is the E-dimensional hypercubic lattice, which consists of sites having coordinates $\vec{\ell} = (\ell_1, \ell_2, \ldots, \ell_E)$ [ℓ_j integral] with each site connected to its nearest-neighbour sites by a bond. The idea of confining a random walk to a hypercubic lattice occurred to Pólya in 1921 [12], and he asked the following specific question: Is a walker who steps at random between nearest-neighbour sites (with all allowed steps equally likely) certain to return to his starting site? He was able to answer this question: YES if $E = 1$ or $E = 2$, NO if $E \geq 3$, i.e. in sufficiently low dimensions the walk is recurrent or persistent, while in three or more dimensions eventual escape is certain and the walk can be called transient. Pólya's question is but one of a number of questions which are easy to pose and answer for lattice walks, but difficult to discuss for walks in continuous spaces without measure-theoretic analysis. A thorough treatment of many aspects of the theory of random walks on lattices, directed to a mathematical audience, is contained in a book by Spitzer [26]. More applications-oriented discussions may be found in the book by Barber and Ninham [8] and a recent review by Weiss and Rubin [10]. A number of particularly elegant and influential contributions to the field have been made by Montroll and co-workers (e.g. [9] and [27-30]).

Let $P_n(\vec{\ell})$ denote the probability that the walker is at site $\vec{\ell}$ after n steps, and assume without loss of generality that the walker starts at the origin of coordinates, i.e. $P_0(\vec{\ell}) = \delta_{\vec{\ell}\vec{0}}$. Also let $p(\vec{\ell})$ be the probability that any step consists of a vector displacement $\vec{\ell}$. Then

$$P_{n+1}(\vec{\ell}) = \sum_{\vec{\ell}'} p(\vec{\ell} - \vec{\ell}')\, P_n(\vec{\ell}') . \tag{2.1}$$

Introducing discrete Fourier transforms

$$\tilde{P}_n(\vec{\theta}) = \sum_{\vec{\ell}} e^{i\vec{\ell}\cdot\vec{\theta}}\, P_n(\vec{\ell}) , \tag{2.2}$$

$$\lambda(\vec{\theta}) = \sum_{\vec{\ell}} e^{i\vec{\ell}\cdot\vec{\theta}}\, p(\vec{\ell}) , \tag{2.3}$$

we find that

$$\tilde{P}_{n+1}(\vec{\theta}) = \lambda(\vec{\theta})\tilde{P}_n(\vec{\theta}) \quad \text{i.e.,} \quad \tilde{P}_n(\vec{\theta}) = \lambda(\vec{\theta})^n , \tag{2.4}$$

and so

$$P_n(\vec{\ell}) = \frac{1}{(2\pi)^E} \int_B e^{-i\vec{\ell}\cdot\vec{\theta}}\, \tilde{P}_n(\vec{\theta})\, d^E\vec{\theta} \tag{2.5}$$

$$= \frac{1}{(2\pi)^E} \int_B e^{-i\vec{\ell}\cdot\vec{\theta}}\, \lambda(\vec{\theta})^n\, d^E\vec{\theta} . \tag{2.6}$$

The integral is taken over the first Brillouin zone $B = [-\pi,\pi]^E$. Equation (2.6) gives the formal solution of the random walk problem.

The function $\lambda(\vec{\theta})$ is usually called the <u>structure function</u> by physicists, the terminology being borrowed from the theory of lattice dynamics. For a <u>Pólya walker</u> (which we define to be a walker who steps between nearest-neighbour sites, with all allowed steps equally likely) $\lambda(\vec{\theta})$ reflects the connectivity structure and dimensionality of the lattice. For the E-dimensional hypercubic lattice, with $\vec{\theta} = (\theta_1, \theta_2, \ldots, \theta_E)$,

$$\lambda(\vec{\theta}) = (1/E) \sum_{j=1}^{E} \cos\theta_j \ . \tag{2.7}$$

Pólya walks on Bravais lattices other than the simple cubic lattice can be defined in terms of walks on the simple cubic lattice with some nearest-neighbour steps forbidden, and some next-nearest-neighbour steps allowed, so that only a subset of the simple cubic lattice is visited. (The body-centred cubic and face-centred cubic lattices have been analysed by Montroll [27]; for the two-dimensional triangular lattice see Montroll [30].) For lattices which are periodic, but not of the Bravais type, i.e. which have more than one site per unit cell, the formulation of the random walk problem is more complicated. (See [31] for the diamond and hexagonal close packed lattices, [32] for several other three-dimensional lattices, [33] for the two-dimensional hexagonal or honeycomb lattice, and [34-35] for tree-like pseudo-lattices.)

The requirement that $p(\vec{\ell}) \geq 0$ and $\sum_{\vec{\ell}} p(\vec{\ell}) = 1$ imposes restrictions on the structure of $\lambda(\vec{\theta})$ near $\vec{\theta} = \vec{0}$. If the mean-squared displacement per step

$$\langle \ell^2 \rangle = \sum_{\vec{\ell}} \vec{\ell}^{\,2} \, p(\vec{\ell}) \tag{2.8}$$

is <u>finite</u>, it is easily seen that

$$\lambda(\vec{\theta}) = 1 + i\vec{\mu}\cdot\vec{\theta} - \tfrac{1}{2}\vec{\theta}\cdot A\cdot\vec{\theta} + \ldots \tag{2.9}$$

where

$$\vec{\mu} = \sum \vec{\ell}\, p(\vec{\ell}) \quad \text{and} \quad A = \sum \vec{\ell}\,\vec{\ell}\, p(\vec{\ell}) \ . \tag{2.10}$$

The tensor **A** is symmetric, so there exists a rotation operator which diagonalizes it. For walks which are symmetric, in the sense that $p(\vec{\ell}) = p(-\vec{\ell})$, i.e. the mean displacement per step $\vec{\mu}$ is zero, this implies that in terms of components θ_j' of $\vec{\theta}$ relative to principal axes,

$$1 - \lambda(\vec{\theta}) \sim \sum_{j=1}^{E} c_j (\theta_j')^2 \quad \text{as} \quad |\vec{\theta}| \to 0 \ ; \tag{2.11}$$

the constants c_j are non-negative (since $|\lambda(\vec{\theta})| \leq 1$ for all $\vec{\theta}$).

We now derive Pólya's theorem, and analyse various statistical properties of random walks on lattices.* It is necessary to introduce an auxillary quantity, the <u>first-passage time</u> distribution for transitions from site $\vec{0}$ to site $\vec{\ell}$. We denote by $F_n(\vec{\ell})$ the probability that a walker starting from site $\vec{0}$ first arrives at site $\vec{\ell}$ when exactly n steps have been taken, and we adopt the convention that $F_0(\vec{\ell}) = 0$. Partitioning over the times of first arrival at site $\vec{\ell}$, we see that

$$P_n(\vec{\ell}) = \delta_{n0}\,\delta_{\vec{\ell}\vec{0}} + \sum_{m=1}^{n} F_m(\vec{\ell})\,P_{n-m}(\vec{0}) \,. \tag{2.12}$$

The sum can be decoupled by use of generating functions. Let

$$P(\vec{\ell};\xi) = \sum_{n=0}^{\infty} P_n(\vec{\ell})\,\xi^n \,, \quad 0 < \xi < 1 \,, \tag{2.13}$$

and

$$F(\vec{\ell};\xi) = \sum_{n=1}^{\infty} F_n(\vec{\ell})\,\xi^n \,, \quad 0 < \xi < 1 \,. \tag{2.14}$$

We multiply both sides of Eq. (2.12) by ξ^n and sum over n, giving

$$P(\vec{\ell};\xi) = \delta_{\vec{\ell}\vec{0}} + F(\vec{\ell};\xi)\,P(\vec{0};\xi) \,. \tag{2.15}$$

Thus if one of the generating functions $P(\vec{\ell};\xi)$ or $F(\vec{\ell};\xi)$ is known, the other is easily found. Of the two, $P(\vec{\ell};\xi)$ can be written most simply, using Eq. (2.6):

$$P(\vec{\ell};\xi) = \frac{1}{(2\pi)^E} \int_B \frac{e^{-i\vec{\ell}\cdot\vec{\theta}}\,d^E\vec{\theta}}{1 - \xi\lambda(\vec{\theta})} \,. \tag{2.16}$$

For walks of Pólya type, $P(\vec{\ell};\xi)$ can be evaluated in terms of elementary functions in one dimension and complete elliptic integrals in two dimensions. In three dimensions, expressions for $P(\vec{\ell};\xi)$ in terms of generalizations of the hypergeometric function are possible, but the analysis involved in any systematic discussion of the problem is extraordinarily difficult [36,37].

---

* Very clear accounts of this material may be found in the papers of Montroll and co-workers [27-30]. For some statistical properties of lattice walks not considered here, see [29] and [10]. Much of the analysis here is not restricted to periodic lattices, but a few subtleties arise in generalizing it, as discussed in [35].

To derive Pólya's theorem we note that the probability of eventual return to the starting point is

$$R = \sum_{n=1}^{\infty} F_n(\vec{0}) = \lim_{\xi \to 1^-} F(\vec{0};\xi) = 1 - \lim_{\xi \to 1^-} P(\vec{0};\xi)^{-1} . \qquad (2.17)$$

Thus the walk is recurrent (return certain: $R = 1$) if and only if $P(\vec{0};\xi) \to \infty$ as $\xi \to 1$ from below. For all lattice walks, $1 - \lambda(\vec{\theta})$ vanishes at $\vec{\theta} = \vec{0}$. Divergence of $P(\vec{0};\xi)$ as $\xi \to 1$ from below is therefore possible if $1 - \lambda(\vec{\theta})$ vanishes sufficiently rapidly as $\vec{\theta} \to \vec{0}$. We restrict our attention here to symmetric walks ($\vec{\mu} = \vec{0}$) and exclude the case when the walk is confined to a subspace of the lattice having dimension less than E by assuming that all the c_j in (2.11) are non-zero. Let $e(\delta)$ denote the hyperellipsoid $\sum_j c_j(\theta_j')^2 < \delta^2$. The convergence or divergence of the integral (2.16) at $\xi = 1$ is the same as that of

$$\int_{e(\delta)} \frac{d^E\vec{\theta}'}{\sum c_j \theta_j'^2} = (c_1 \dots c_E)^{-1/2} \int_{|\vec{\phi}| < \delta} \frac{d^E\vec{\phi}}{\phi_1^2 + \dots + \phi_E^2} \quad \text{(rescaling the axes)}$$

$$= \text{constant} \cdot \int_0^\delta \frac{\phi^{E-1}\, d\phi}{\phi^2} = \infty \quad \text{if} \quad E \le 2 .$$

This is part of Pólya's theorem: a symmetric lattice random walker with a finite mean-squared displacement per step is certain to return to the starting site in one or two dimensions.*

If the condition that the mean-squared displacement per step be finite is relaxed, Pólya's theorem is modified, and in a sense the effective dimension of the random walk is changed [39]. If, for a symmetric walk,

---

* So long as $\vec{\theta} = \vec{0}$ is the only zero of $1 - \lambda(\vec{\theta})$ inside or on the boundary of the Brillouin zone B, this analysis also establishes that $R < 1$ in three or more dimensions, since in this case $P(\vec{0},1) < \infty$. For Pólya's walk on hypercubic lattices, $\vec{\theta} = \vec{0}$ is indeed the unique zero of $1 - \lambda(\vec{\theta})$, but for the body-centred and face-centred cubic lattices $1 - \lambda(\vec{\theta})$ also vanishes at isolated points on the boundary of B. However, $1 - \lambda(\vec{\theta})$ vanishes quadratically at these points, just as it does at the origin, and $P(\vec{0},1)$ is therefore finite. (Conditions under which $\vec{\theta} = \vec{0}$ is the unique zero of $1 - \lambda(\vec{\theta})$ in B are discussed in an Appendix to [38]; the implications of the additional zeros on the structure of $P(\vec{0};\xi)$ are discussed in [37], footnote 57.)

$$p(\vec{\ell}) \propto |\vec{\ell}|^{-E-K} \quad \text{as } |\vec{\ell}| \to \infty \tag{2.18}$$

with $0 < K < 2$, so that $\sum \vec{\ell}^2 p(\vec{\ell}) = \infty$ and

$$1 - \lambda(\vec{\theta}) \propto |\vec{\theta}|^K \quad \text{as } \vec{\theta} \to \vec{0} , \tag{2.19}$$

then the walk is recurrent if $E \leq K$ and transient if $E > K$. (This may be seen by a modest extension of the preceding derivation of Pólya's theorem.)

When $R < 1$, its value can be calculated using (2.16) and (2.17), but in practice this is a very difficult problem. For Pólya walks on the simple cubic, body-centred cubic and face-centred cubic lattices one is required to evaluate the three integrals

$$I_{SC} = \frac{1}{\pi^3} \iiint_0^\pi \frac{du\,dv\,dw}{3 - \cos u - \cos v - \cos w} , \tag{2.20}$$

$$I_{BCC} = \frac{1}{\pi^3} \iiint_0^\pi \frac{du\,dv\,dw}{1 - \cos u \cos v \cos w} , \tag{2.21}$$

$$I_{FCC} = \frac{1}{\pi^3} \iiint_0^\pi \frac{du\,dv\,dw}{3 - \cos v \cos w - \cos w \cos u - \cos u \cos v} , \tag{2.22}$$

respectively. These integrals were first expressed in terms of known special functions by Watson [40] and are consequently called <u>Watson integrals</u>. Watson was able to express all three integrals in terms of products of complete elliptic integrals, and was further able to express I_{BCC} and I_{FCC} (but not I_{SC}) in terms of products of gamma functions. The reduction of I_{SC} to gamma functions was finally effected by Glasser and Zucker [41]. The results are:

$$I_{SC} = \frac{\sqrt{6}}{96\pi^3} \Gamma(1/24)\Gamma(5/24)\Gamma(7/24)\Gamma(11/24) ,$$

$$I_{BCC} = \frac{1}{4\pi^3} \Gamma(1/4)^4 , \quad I_{FCC} = \frac{3\Gamma(1/3)^6}{2^{14/3}\pi^4} . \tag{2.23}$$

Ishioka and Koiwa [31] related $P(\vec{0},\xi)$ for the diamond (D) and hexagonal close-packed (HCP) lattice to $P(\vec{0},\xi)$ for the FCC lattice. The numerical values of R given in Table 2 are deduced from the Watson-Glasser-Zucker-Ishioka-Koiwa results:

TABLE 1

Lattice	z	R
D	4	0.442239
SC	6	0.340537
BCC	8	0.282230
FCC	12	0.256318
HCP	12	0.256318

Return probabilities R (correct to six significant figures) for Pólya walks on three dimensional lattices; z, the coordination number, is the number of nearest neighbours of any site. The FCC and HCP values are identical [31].

That the return probabilities in three dimension could be evaluated using the Watson integrals was apparently first noted by Montroll [27], who also gave an asymptotic expansion of R for hypercubic lattices of dimension $E \gg 1$.

Although the return probability R is the most important random walk statistic, many other properties of the walk can be inferred from the behaviour of $F(\vec{0};\xi)$ near $\xi = 1$. For example, the conditional mean recurrence time, i.e. the mean time to return for any walker who does return is

$$\tau_c = R^{-1} \sum_{n=1}^{\infty} nF_n(\vec{0}) = R^{-1} \frac{\partial}{\partial\xi} F(\vec{0};\xi) \Big|_{\xi=1} . \qquad (2.24)$$

For symmetric Pólya walkers, it can easily be shown that $\tau_c = \infty$ in one or two dimensions, i.e. return, though certain, takes forever on the average. The higher dimensional cases are very interesting. If $E < 4$, $\tau_c = \infty$, but if $E > 4$, τ_c is finite [42]. One may summarize these results by saying that in low dimensional systems, many of the recurrent paths are very long, while in high dimensional systems, most recurrent paths are very short. Indeed, $\tau_c \to 2$ as $E \to \infty$.

Another random walk statistic of interest is the expected number of distinct sites visited in the first n steps of a random walk, denoted by S_n. It can be shown [29] that for periodic lattices with all sites probabilistically equivalent,

$$\sum_{n=0}^{\infty} S_n \, \xi^n = (1-\xi)^{-2} P(\vec{0};\xi)^{-1} . \qquad (2.25)$$

If the leading order behaviour of $P(\vec{0};\xi)$ near $\xi = 1$ is known, the asymptotic form of S_n as $n \to \infty$ can be inferred from (2.25) using Tauberian theorems [20] (which relate the asymptotic behaviour of the coefficients in a power series to the singularity structure of its analytic continuation). For <u>transient</u> walks

$$S_n \sim n/P(\vec{0},1) = \{1 - R\}\, n\ . \tag{2.26}$$

For Pólya walks, the results $S_n \sim (8n/\pi)^{1/2}$ for one dimension and $S_n \sim \pi n/\ell n\, n$ for the square lattice were given by Pólya in 1938 [13] (in a footnote, and using slightly different terminology). The behaviour of S_n for walks with infinite mean-squared displacement per step differs from that for Pólya walks. Gillis and Weiss [43] have analysed some examples. (For a one-dimensional walk with $p(\ell) \propto |\ell|^{-2}$ as $|\ell| \to \infty$, they find that $S_n \propto n/\ell n(n)$: a one-dimensional walk with long-ranged transition probabilities gives the same asymptotic dependence of S_n as the two-dimensional Pólya walk.)

The results concerning lattice walks presented here represent but a small fraction of a vast body of work, into which the interested reader may be guided by a number of monographs and reviews [8,9,10,26,28]; many applications of random walk ideas are also discussed extensively in several of these sources [8,9,10]. However, we indicate briefly here a few of the interesting problems for lattice walks which have been considered in the physics literature. Walks on infinite lattices with one or more boundaries, or possessing a small number of defective sites, can often be treated expeditiously by a "defect" technique introduced by Montroll [28]. The case in which the defective sites are traps is of particular relevance in solid state physics, and in the theory of photosynthesis [30], and some very elegant contributions to the theory have been made recently by den Hollander and Kasteleyn [44]. Much of the formalism developed for infinite lattices is relevant to finite lattices with periodic boundary conditions; a considerable reduction of computational difficulties associated with finite systems has been achieved by Hatlee, Kozak and Walsh [45]. Some random walk processes in Euclidean space related to fractal sets have been examined by Hughes, Montroll and Shlesinger [25,38,39,46], and are discussed in the article by Shlesinger and Montroll in the present volume; Rammal and Toulouse [47] have examined Pólya random walks on fractal lattices, i.e. lattices which have dilational rather than translational invariance. The basic random walk formalism can be adapted to multistate random walk processes [48] and this generalization can be adopted to describe motion of clusters on surfaces [48], diffusion in multiphase systems [49] and other problems. Cummings and Stell [50] have considered analogies between random walk problems and the Ornstein-Zernike equation of liquid state physics, as discussed briefly in the article by Stell in the present volume. Joyce [51] has shown how the spherical model, one of the few exactly solved problems in equilibrium statistical mechanics, is intimately related to random walks, and Montroll and Reiss [52] have examined random walk analogies in a non-equilibrium critical phenomenon modelled via a dynamic Ising model. A remarkable paper of Gillis [53] gives the exact solution of a one-dimensional random walk problem in the presence of an external field (in the sense that the individual step transition probability law is not translationally invariant, and changes according to distance from the origin in a prescribed manner).

3. Continuum limits of random walk processes

The general problem of the behaviour of random walk processes after a large number of steps is the province of the central limit theorem, and related theorems, of probability theory. However an instructive, if not entirely rigorous, analysis of the problem can be given without straying far outside random walk terminology and concepts. We illustrate this for one-dimensional lattice walks. Our end result is the production of differential equations describing the 'continuum limit' of the walk in a certain sense. (Work along these lines was instigated by Rayleigh [2] in his early analyses of a problem equivalent to Pearson's walk discussed in Section 1, and has been pursued intermittently since then.)

We introduce two parameters, which we shall ultimately allow to approach zero in a suitably constrained manner to generate a continuum limit: Δ, the lattice spacing, and τ, the time between successive steps. We work in Fourier space, rather than real space, to avoid the necessity of assuming that the mean-squared displacement per step is finite. From Eq. (2.4), we have

$$\frac{1}{\tau}\{\tilde{P}_{n+1}(\theta) - \tilde{P}_n(\theta)\} = -\frac{1}{\tau}[1 - \lambda(\theta)]\tilde{P}_n(\theta) . \quad (3.1)$$

If we write $t = n\tau$ and $q = \theta/\Delta$ and define

$$\tilde{P}_n(\theta) = \tilde{p}(\Delta^{-1}\theta, n\tau) = \tilde{p}(q,t) \quad (3.2)$$

and let Δ and $\tau \to 0$ simultaneously, then (3.1) implies that

$$\frac{\partial}{\partial t}\tilde{p}(q,t) = -a(q)\,\tilde{p}(q,t) , \quad (3.3)$$

provided that the limit is taken in such a manner that

$$a(q) = \lim_{\Delta,\tau\to 0} \frac{1}{\tau}[1 - \lambda(\Delta q)] \quad (3.4)$$

exists, a question to which we return shortly. The notation $\tilde{p}(q,t)$ has been employed in anticipation of the fact that when Δ and $\tau \to 0$, $\tilde{p}(q,t)$ is the Fourier transform of a continuum probability density function $p(x,t)$. To establish this, we define

$$p(x,t) = \Delta^{-1}P_n(\ell) \quad (3.5)$$

if $[\ell - 1/2]\Delta < x < [\ell + 1/2]\Delta$. Then from Eq. (2.5),

$$p(x,t) = \frac{1}{2\pi\Delta}\int_{-\pi}^{\pi} e^{i\ell\theta}\,\tilde{P}_n(\theta)d\theta \quad (3.6)$$

$$= \frac{1}{2\pi}\int_{-\pi/\Delta}^{\pi/\Delta} e^{-ixq + i\eta q}\,\tilde{p}(q,t)dq \quad (3.7)$$

where $|\eta| < \Delta/2$. Formally letting $\Delta \to 0$ we deduce that

$$p(x,t) = \frac{1}{2\pi}\int_{-\infty}^{\infty} e^{-iqx}\,\tilde{p}(q,t)dq . \quad (3.8)$$

For a symmetric walk with finite mean-squared displacement (measured in lattice spacings) σ^2 per step, we know from Eq. (2.9) that $1 - \lambda(\theta) \sim 1/2\sigma^2\theta^2$ as $\theta \to 0$. Consequently, so long as we insist that $\Delta^2 \sim C\tau$ (with C constant) as $\Delta,\tau \to 0$, the function $a(q)$ defined by (3.4) exists, and is simply $1/2\sigma^2Cq^2$. If we write $D = 1/2\sigma^2C$, we find the equation

$$\frac{\partial}{\partial t}\ \tilde{p}(q,t) = -Dq^2\tilde{p}(q,t)\ , \qquad (3.9)$$

which we recognize as the Fourier transform of the diffusion equation

$$\frac{\partial}{\partial t}\ p(x,t) = D\frac{\partial}{\partial x^2}\ p(x,t)\ . \qquad (3.10)$$

If the mean-squared displacement per step is infinite, the diffusion equation no longer necessarily obtains as the continuum limit. For example, if $p(\ell) \propto |\ell|^{-1-\mu}$ as $|\ell| \to \infty$, with $0 < \mu < 2$, so that $1 - \lambda(\theta) \propto |\theta|^{\mu}$ as $\theta \to 0$, we find that the constraint on Δ and τ required for a sensible continuum limit is that $\Delta^{\mu} \propto \tau$. In this case, we find that

$$\frac{\partial}{\partial t}\tilde{p}(q,t) = -A|q|^{\mu}\ \tilde{p}(q,t) \qquad (3.11)$$

with A a positive constant. Except for some rational values of μ [54], Eq. (3.11) cannot be converted to a partial differential equation for $p(x,t)$. However, for a random walk commencing at $\ell = 0$, for which $p(x,0) = \delta(x)$ (so that $\tilde{p}(q,0) = 1$) Eq. (3.11) has the solution

$$\tilde{p}(q,t) = \exp\{-A|q|^{\mu}t\}\ . \qquad (3.12)$$

A spatial probability density function $f(x)$ having the Fourier transform

$$\tilde{f}(q) = \int_{-\infty}^{\infty} e^{iqx}\ f(x)dx = \exp\{i\gamma q - c|q|^{\mu}[1 + i\nu\omega(q,\mu)\mathrm{sgn}(q)]\}\ , \qquad (3.13)$$

with $0 < \mu < 2$, $-1 < \nu < 1$, $c > 0$, γ real and

$$\omega(q,\mu) = \begin{cases} \tan\left(\frac{1}{2}\pi\mu\right) & \mu \neq 1 \\ (2/\pi)\ \ln|q| & \mu = 1 \end{cases}\ , \qquad (3.14)$$

is called a (one dimensional) stable or Lévy distribution of order μ. The first systematic investigation of these distributions was made by Lévy [55], who established their significence as limiting laws for the distributions of sums of independent random variables; a very detailed discussion of stable distributions is given by Gnedenko and Kolmogorov [56].

The content of Eq. (3.12) is that the continuum limit of a symmetric random walk generated by a suitably long-ranged probability density function is a time-evolving Lévy distribution or Lévy flight. (It is not difficult to derive the more general non-symmetric Lévy distribution via a suitably defined continuum limit of a biased random walk. If $\mu < 1$ and $\nu = 1$ or -1, then the stable distribution defined by (3.13) is one-sided, in the sense that $f(x) = 0$ for $x < \gamma$ or $x > \gamma$ respectively: see [20] pp. 569, 571-572.) In the case $\mu = 2$, the Lévy distribution is simply the Gauss or normal distribution, and the continuum limit $p(x,t)$ of the random walk satisfies the diffusion equation. When $\mu < 2$, the Lévy distribution has infinite variance: the mean-squared displacement at any time t, in the continuum limit, is infinite.

The analysis sketched here for one-dimensional lattice walks can be rephrased for walks on the one-dimensional continuum. It can also be generalized to higher dimensions, leading to E-dimensional isotropic Lévy distributions, with

$$\tilde{p}(\vec{q},t) = \int e^{i\vec{q}\cdot\vec{x}} p(\vec{x},t) d^E x = \exp(-A|\vec{q}|^{\mu} t) . \tag{3.15}$$

A variety of interesting attributes of continuous-time stochastic processes characterized by (3.15) has been given by Mandelbrot [57,58]; discrete time analogues of (3.15) have also been investigated ([25,38,39,46]; see also the articles by Shlesinger and Montroll, and Montroll and Shlesinger in the present volume).

In the preceding informal analysis, we have seen how Lévy distributions arise from continuum limits of random walk processes. A second characterization of Lévy distributions (or more precisely a <u>class</u> of distributions of which the Lévy distributions are the most important members) is possible. Limiting our attention to one spatial dimension we may ask which continuous space and time probability distributions $p(x,t)$ possess (i) temporal homogeneity, and (ii) translational invariance. These two requirements are embodied in the equation

$$p(x,t) \equiv \int_{-\infty}^{\infty} p(x-y,t-\tau)\, p(y,\tau) dy \ , \quad 0 < \tau < t . \tag{3.16}$$

Forms of this equation are variously attributed to Bachelier, Smoluchowski, Chapman and Kolmogorov. Montroll and West [9] have assigned it the usefully short and descriptive name of "the chain equation" and examined some of its implications. In Fourier space, (3.16) becomes

$$\tilde{p}(q,t) = \tilde{p}(q,t-\tau)\tilde{p}(q,\tau) . \quad (3.17)$$

If $\tilde{p}(q,t)$ is assumed to be a continuous function of t, by setting $t = \tau = 0$ in Eq. (3.17) we deduce that

$$\tilde{p}(q,0)[1-\tilde{p}(q,0)] = 0 . \quad (3.18)$$

Since $\tilde{p}(q,t)$ is the Fourier transform of a probability density function, it is necessarily a continuous function of q (see [59], p. 141), and $\tilde{p}(0,t) = 1$ for all t. It follows therefore from Eq. (3.18) that $\tilde{p}(q,0) \equiv 1$, and so $p(x,0) = \delta(x)$: $p(x,t)$ describes the evolution of an initially localized distribution. Further suppose that $\tilde{p}(q,t)$ is continuously differentiable with respect to time. Then differentiating Eq. (3.17) with respect to τ and letting $\tau \to 0$ yields a differential equation, the only solutions of which (subject to the condition $\tilde{p}(q,0) = 1$) have the form

$$\tilde{p}(q,t) = \exp\{-a(q)t\} . \quad (3.19)$$

Certain restrictions on $a(q)$ have to be imposed if Eq. (3.19) is to define a valid probability density function for all times t, and what these restrictions are is by no means immediately apparent. Referring back to Eq. (3.3), we see that distributions of this form arise as continuum limits of random walks in discrete space and time. In particular the Lévy or stable distributions are solutions of the chain equation, and indeed the most important solutions. They are, however, not the only solutions. Let $\phi(x)$ be any distribution which is infinitely divisible, in the sense that for each n the random variable which it describes can be represented as the sum of n independent random variables with common distribution $\phi_n(x)$ (i.e. $\tilde{\phi}(q) = \tilde{\phi}_n(q)^n$); every stable distribution is infinitely divisible, but the stable distributions are only a small subclass of the infinitely divisible distributions. Then it can be shown ([20] p. 557) that for each real, positive t, $\tilde{\phi}(q)^t$ is the Fourier transform of a valid probability distribution. Hence

$$\tilde{p}(q,t) = \tilde{\phi}(q)^t = \exp\{t\ln\tilde{\phi}(q)\} \tag{3.20}$$

is a valid solution of the chain equation (3.16).

4. Master equations, generalized master equations, and continuous-time random walks

In the preceding section we showed how discrete time random walk processes can be used to generate continuous-time processes by taking a continuum limit in both space and time. However, it is possible to generate continuous-time processes without the loss of lattice structure by the use of linear rate equations or, as they are usually known in physics, master equations [60]. The simplest of these equations has the form

$$\frac{d}{dt}\,p(\vec{\ell},t) = \sum_{\vec{\ell}'} W(\vec{\ell},\vec{\ell}')\{p(\vec{\ell}',t) - p(\vec{\ell},t)\} . \tag{4.1}$$

The coefficients $W(\vec{\ell},\vec{\ell}')$ are the rates at which transitions occur between sites $\vec{\ell}$ and $\vec{\ell}'$: the walker's probability of being at site $\vec{\ell}$ decays in proportion to the probability of being at site $\vec{\ell}$, and accumulates in proportion to the probability of being at neighbouring sites. It has been assumed that the $W(\vec{\ell},\vec{\ell}')$ are symmetric, i.e. $W(\vec{\ell},\vec{\ell}') = W(\vec{\ell}',\vec{\ell})$, for simplicity. In a later section (section 8) we shall examine a case in which the $W(\vec{\ell},\vec{\ell}')$ depend on the absolute location of the sites $\vec{\ell}$ and $\vec{\ell}'$, but for the present we assume that there is translational invariance:

$$W(\vec{\ell},\vec{\ell}') = W(\vec{\ell} - \vec{\ell}') . \tag{4.2}$$

Given that (4.2) holds, master equations can be analysed in exactly the same manner as random walk processes with the generating function (2.13) replaced by the Laplace transform

$$\hat{p}(\vec{\ell},u) = \int_0^\infty e^{-ut}\,p(\vec{\ell},t)dt . \tag{4.3}$$

Indeed, some of the analysis is simplified, since the theory of differential equations is more familiar than the theory of difference equations. We are no longer able to pose questions concerning individual steps or visits to sites, but we can easily derive formulae for the probability of visiting a given site in a given time interval, and so on. Exactly as in the ordinary random walk problem, we obtain diffusive behaviour, in the sense that the mean-squared displacement at time t,

$$R^2(t) = \sum_{\vec{\ell}} \vec{\ell}^2 \, p(\vec{\ell},t) \ , \tag{4.4}$$

grows linearly with time at large times and a meaningful diffusion constant can be defined, provided that long-ranged transitions are suitably restricted, i.e.

$$\sum_{\vec{\ell}} \vec{\ell}^2 \, W(\ell) < \infty \ . \tag{4.5}$$

When this constraint is violated, lattice analogues of the Lévy distributions arise.

We record here for later reference the solution of (4.1) subject to the initial condition $p(\vec{\ell},0) = \delta_{\vec{\ell},\vec{0}}$ in the case of unbiased motion involving only nearest-neighbour transitions, on a translationally invariant periodic lattice. Writing

$$W(\vec{\ell} - \vec{\ell}') = \begin{cases} W_0 & \vec{\ell}' \in \{\vec{\ell}\} \\ 0 & \text{otherwise} \end{cases} \ , \tag{4.6}$$

where $\{\vec{\ell}\}$ denotes the set of nearest-neighbour sites of $\vec{\ell}$, and using a discrete Fourier transform and the Laplace transform (4.3), we readily show that

$$\hat{p}(\vec{\ell},u) = \frac{1}{(2\pi)^E} \int_B \frac{\exp(-i\vec{\ell}\cdot\vec{\phi}) \, d^E\vec{\phi}}{u - zW_0[1 - \lambda(\vec{\phi})]} \ . \tag{4.7}$$

Here, as before, $B = [-\pi,\pi]^E$ and the coordination number z is the number of nearest neighbours of any site. In the time domain Eq. (4.7) becomes

$$p(\vec{\ell},t) = \frac{1}{(2\pi)^E} \int_B \exp\{-i\vec{\ell}\cdot\vec{\phi} - zW_0[1 - \lambda(\vec{\phi})]t\} d^E\vec{\phi} \ . \tag{4.8}$$

It is straightforward to show that for a symmetric walk $[\lambda(\vec{\phi}) = \lambda(-\vec{\phi})]$

$$R^2(t) = zW_0\{-\nabla^2_{\vec{\phi}}\lambda(\vec{\phi})|_{\vec{\phi}=\vec{0}}\}t \tag{4.9}$$

and that

$$p(\vec{0},t) \propto t^{-E/2} \quad \text{as} \quad t \to \infty , \tag{4.10}$$

i.e. the initial 'disturbance' decays more rapidly as the dimensionality increases. For an E-dimensional simple hypercubic lattice the integral (4.8) can be expressed in terms of modified Bessel functions:

$$p(\vec{\ell},t) = \prod_{j=1}^{E} \{ I_{\ell_j}(W_0 t)\, e^{-W_0 t} \} , \tag{4.11}$$

using the integral representation $I_n(z) = (1/\pi) \int_0^\pi \exp(z \cos\theta) \cos(n\theta)\, d\theta$.

In adopting a master equation, rather than a discrete-time random walk formalism, the class of possible stochastic behaviours has not been broadened: one still has diffusive behaviour as the norm, with Lévy flight behaviour (we propose the term superdiffusive behaviour) if the individual transitions are very long ranged. The question arises as to whether we can generate subdiffusive behaviour, in the sense that the mean squared displacement defined by (4.4) grows with time more slowly than linearly. It transpires that the answer is yes, if we are prepared to allow a memory effect in the governing equation, as we discuss below, or if we disorder the rate coefficients $W(\vec{\ell},\vec{\ell}')$ in a suitable manner, as discussed later (section 8).

The equation

$$\frac{d}{dt} p(\vec{\ell},t) = \sum_{\vec{\ell}'} \int_0^t d\tau\, W(\vec{\ell} - \vec{\ell}', t - \tau)\, \{p(\vec{\ell}',\tau) - p(\vec{\ell},\tau)\} \tag{4.12}$$

is often called a generalized master equation, and the function $W(\vec{\ell},t)$ its memory kernel. Such equations are particularly esteemed by physicists since they arise from a coarse-graining of the exact equations of motion of systems with many degrees of freedom [61], and are useful in modelling a number of phenomena, including exciton transport in solids [62]. Since the memory effect in (4.12) is of convolution form, a Laplace transform effectively removes it:

$$u\hat{p}(\vec{\ell},u) - p(\vec{\ell},0) = \sum_{\ell'} \hat{W}(\vec{\ell} - \vec{\ell}',u)\, [\hat{p}(\vec{\ell}',u) - \hat{p}(\vec{\ell},u)] . \tag{4.13}$$

One can therefore treat the generalized master equation as easily as the simpler memoryless master equation.

A conceptual drawback with generalized master equations is the difficulty in visualizing for a particular memory kernel what the qualitative behaviour of the random transport process will be. For example, does it matter if $W(\vec{\ell},t)$ changes sign? We suspend discussion of these matters for a short while, and analyse a different continuous-time random process, in which well-defined individual steps take place at well-spaced times. Specifically, we consider here a random walk on a lattice, for which the times t_j between successive steps are independent, identically distributed random variables having a probability density function $\psi(t_j)$, which may be called the waiting-time or pausing-time density. The introduction of this random process into modern physics, under the name continuous-time random walk, is due to Montroll and Weiss [29]. From a mathematical standpoint, it is a particularly useful example of the semi-Markov process of Lévy [63], Feller [64], Smith [65] and Pyke [66]. Feller ([20], pp. 479-483), and Connolly [67] have made some detailed investigations of a one-dimensional process they call randomized random walk, which is the Montroll-Weiss continuous time random walk in the special case of an exponential waiting-time density

$$\psi(t) = \alpha e^{-\alpha t} \quad (\alpha > 0) , \tag{4.14}$$

but as we shall see below, working with a general $\psi(t)$ leads to very informative results.

Let $\psi_n(t)$ denote the probability density function for the time of occurrence of the nth step. By convention $\psi_0(t) = \delta_+(t)$. Then as

$$\psi_n(t) = \int_0^t \psi(t)\, \psi_{n-1}(t - \tau)\, d\tau , \tag{4.15}$$

the introduction of Laplace transforms

$$\hat{\psi}_n(u) = \int_0^\infty e^{-ut}\, \psi_n(t)dt , \quad \hat{\psi}(u) = \int_0^\infty e^{-ut}\, \psi(t)dt \tag{4.16}$$

leads to the conclusion that

$$\hat{\psi}_n(u) = \hat{\psi}(u)^n . \tag{4.17}$$

The temporal probability density function describing first passage from site $\vec{0}$

to site $\vec{\ell}$ is

$$f(\vec{\ell},t) = \sum_{n=1}^{\infty} F_n(\vec{\ell})\, \psi_n(t) , \tag{4.18}$$

with $F_n(\vec{\ell})$ the first passage time distribution for discrete-time random walks. Its Laplace transform is

$$\hat{f}(\vec{\ell},u) = \sum_{n=1}^{\infty} F_n(\vec{\ell})\, \hat{\psi}(u)^n = F(\vec{\ell};\hat{\psi}(u)) , \tag{4.19}$$

where $F(\vec{\ell};\xi)$ is the first passage time generating function for a discrete time walk, defined by (2.14). Many statistical properties of the walk at long times can be extracted from Eq. (4.19) using Tauberian theorems for the Laplace transform [20]. We give here but one example. Return to the origin is certain if and only if $\int_0^\infty f(\vec{0},t)dt = \hat{f}(\vec{0},0) = 1$. So long as the waiting time density is <u>non-defective</u>, i.e. $\int_0^\infty \psi(t)dt = \hat{\psi}(0) = 1$, so that motion cannot freeze, we see that the value of $\hat{f}(\vec{0},0)$ is insensitive to the nature of $\psi(t)$, and depends only on the individual step length distribution $p(\vec{\ell})$; in particular Pólya's theorem holds for such continuous-time random walks.

If we require only the probability density $p(\vec{\ell},t)$ for occupancy of site $\vec{\ell}$ at time t, given that the walk started at site $\vec{0}$, we may calculate it as follows [29]. Let $Q(\ell,t)$ be the probability density function for arrival at site $\vec{\ell}$ <u>exactly</u> at time t, and let $\Psi(t)$ be the probability that a pause between successive steps is of duration <u>at least</u> t, i.e.

$$\Psi(t) = 1 - \int_0^t \psi(\tau)d\tau . \tag{4.20}$$

Then

$$p(\vec{\ell},t) = \int_0^t Q(\vec{\ell},\tau)\, \Psi(t-\tau)d\tau , \tag{4.21}$$

whence it follows that

$$\hat{p}(\vec{\ell},u) = \hat{Q}(\vec{\ell},u)\hat{\Psi}(u) = \hat{Q}(\vec{\ell},u)\, u^{-1}\, [1 - \hat{\psi}(u)] . \tag{4.22}$$

However,

$$Q(\vec{\ell},t) = \sum_{n=0}^{\infty} P_n(\vec{\ell})\,\psi_n(t) , \tag{4.23}$$

with $P_n(\vec{\ell})$ the solution of the discrete-time random walk problem, and so

$$\hat{Q}(\vec{\ell},u) = \sum_{n=0}^{\infty} P_n(\vec{\ell})\,\hat{\psi}(u)^n = P(\ell;\hat{\psi}(u)) , \tag{4.24}$$

with $P(\vec{\ell};\xi)$ the generating function for the discrete-time random walk defined by (2.13). It follows that

$$\hat{p}(\vec{\ell},u) = u^{-1}[1 - \hat{\psi}(u)]\; P(\vec{\ell},\hat{\psi}(u)) \tag{4.25}$$

and the properties of the continuous-time random walk are readily deduced using Tauberian theorems for the Laplace transform, if $P(\vec{\ell};\xi)$ is known. Using Eq. (2.8) we see that $\hat{p}(\vec{\ell},u)$ can be expressed solely in terms of $\hat{\psi}(u)$ and the structure function $\lambda(\vec{\phi})$ of the associated discrete-time random walk:

$$\hat{p}(\vec{\ell},u) = \frac{1 - \hat{\psi}(u)}{u(2\pi)^E} \int_B \frac{e^{-i\vec{\ell}\cdot\vec{\phi}}\, d^E\vec{\phi}}{1-\hat{\psi}(u)\lambda(\vec{\phi})} . \tag{4.26}$$

An example now suffices to establish that the continuous-time random walk describes a richer class of random processes than either the ordinary discrete-time random walk or the master equation. It is easily shown that the mean-squared displacement at time t corresponding to (4.26) has Laplace transform

$$\mathcal{L}\{R^2(t)\,;t \to u\} = - \nabla^2_{\vec{\phi}} \left\{ \frac{1 - \hat{\psi}(u)}{u[1-\hat{\psi}(u)\lambda(\vec{\phi})]} \right\}_{\vec{\phi}=\vec{0}} \tag{4.27}$$

and so for a symmetric one-dimensional walk with finite mean-squared displacement per step σ^2 we find that

$$\mathcal{L}\{R^2(t);t \to u\} = \frac{\sigma^2}{u[1-\hat{\psi}(u)]} . \tag{4.28}$$

So long as the <u>mean waiting time between steps</u>

$$\tau \equiv \int_0^{\infty} dt\; t\psi(t) \tag{4.29}$$

is finite,

$$1 - \hat{\psi}(u) \sim \tau u \quad \text{as} \quad u \to 0 \tag{4.30}$$

and we infer from Eq. (4.28), via a Tauberian theorem [20], that

$$\langle R^2(t)\rangle \sim (\sigma^2/\tau)t \tag{4.31}$$

as $t \to \infty$, i.e. motion is diffusive in character. On the other hand, if $\tau = \infty$, as will be the case if

$$\psi(t) \propto t^{-1-\alpha} \quad \text{as} \quad t \to \infty \quad , 0 < \alpha < 1 \; , \tag{4.32}$$

we have

$$1 - \hat{\psi}(u) \propto u^{\alpha} \tag{4.33}$$

and we find that

$$R^2(t) \propto t^{\alpha} \; . \tag{4.34}$$

Continuous-time random walks thus generate subdiffusive motion, in the sense that the mean-squared displacement grows with time more slowly than linearly, if the mean time between steps is infinite.

The above ideas can be adapted to continuous time random walks on the continuum, walks with spatial defects, multistate processes [48] and walks for which the spatial and temporal behaviour is coupled, in the sense that the probability $p(\vec{\ell},t)$ of a displacement $\vec{\ell}$ a time t after the previous displacement is not of the form $p(\vec{\ell})\psi(t)$ [68] . An analysis of possible long-time behaviours of such coupled walks, and of walks in which either the mean waiting-time between steps, or the mean-squared displacement per step, or both, are infinite, has been given by Shlesinger, Klafter and Wong [69].

We conclude this section with a very elegant result due to Kenkre, Montroll and Shlesinger [70]. We consider a continuous-time random walk, with waiting-time density $\psi(t)$ and single-step distribution $p(\vec{\ell})$ and a generalized master equation with memory kernel

$$W(\vec{\ell} - \vec{\ell}',t) = p(\vec{\ell} - \vec{\ell}')\phi(t) \; . \tag{4.35}$$

If both processes are subject to the same initial condition $(p(\vec{\ell},0) = \delta_{\vec{\ell},\vec{0}}$, say) the solution of the continuous-time random walk is given by Eq. (4.26), while that of the generalized master equation is given by

$$\hat{p}(\vec{\ell},u) = \frac{1}{(2\pi)^E} \int_B \frac{e^{-i\vec{\ell}\cdot\vec{\theta}} d^E\vec{\theta}}{u-\hat{\phi}(u)[\lambda(\vec{\theta})-1]} , \qquad (4.36)$$

with $\lambda(\vec{\theta})$ the usual structure function. The solutions are seen to be identical if, and only if,

$$\hat{\phi}(u) = u\hat{\psi}(u)/[1 - \hat{\psi}(u)] . \qquad (4.37)$$

There is thus a unique mapping between the two processes! Moreover, (4.37) enables us to determine when, if ever, a continuous-time random walk is exactly equivalent to a master equation. The generalized master equation degenerates to a master equation when $\phi(t) = \alpha\delta_+(t)$, with $\alpha > 0$, and (4.37) then implies that $\hat{\psi}(u) = \alpha/(u + \alpha)$, i.e. $\psi(t) = \alpha e^{-\alpha t}$. The "randomized random walk" of Feller [20] and Connolly [67] is the only continuous-time random walk precisely equivalent to a master equation. (Some conditions under which the solution of a master equation corresponds asymptotically to that of a continuous-time random walk have been analysed by Bedeaux et al. [71]. Modifications of the continuous-time random walk to allow for the first step of the random walk to have a different waiting time density from that for the subsequent steps, and the consequent modification in the relation to generalized master equations, are discussed in [10].)

5. Self-Avoiding Walks

In the preceding sections we have reviewed some random processes which are Markovian in the sense that they have no memory effect, or very weakly non-Markovian, with a convolution-type memory which is no obstacle to calculation. For random walks on lattices in discrete time, one may ask what kind of memory effect is needed to destroy the elegant simplicity of the theory developed above. It is evident that a finite range memory, in which the probability distribution for the nth step depends on the position of the walker for the previous k steps $(k < \infty)$, leads to nothing new, since the problem can be reformulated as an ordinary random walk with internal states, or a Markov chain. (For examples of finite memory calculations, see [8], pp. 46-59.)

An extreme example, in which the walker remembers all of his previous steps, can be defined as follows: a random walker starts at site $\vec{0}$, and executes a random walk in which he is forbidden to visit any site more than once. At each step, only motion to any one of the 'allowed' nearest-neighbour sites of the current site is permitted; all such sites have equal probability of being visited next. Such a self-avoiding walker's probability distribution and statistical attributes must differ radically from those of an ordinary Pólya walker. For example, in a two dimensional walk, the possibility arises that a configuration similar to that shown in Fig. 1 may occur, and the walk is obliged to terminate. To avoid such a phenomenon, but retain the non-Markovian nature of the problem one may assign an 'energy of interaction' when the walker revisits sites previously visited, i.e. there is an energy penalty (or a bonus, in the case of a "self-attracting" walk [72]) associated with multiple visits. The problem arises as to exactly how the interaction energy is to be related to the probability of evolution of the walk. A logical way to proceed is to assign each path or configuration $\{\vec{\ell}_j\}$ in an n-step walk (defined as the set of sites visited, in the order in which they are visited) an interaction energy $U(\{\vec{\ell}_j\})$ and take as the probability of the configuration

$$P(\{\vec{\ell}_j\}) = Z^{-1} \exp[-\beta U(\{\vec{\ell}_j\})] , \tag{5.1}$$

with β a positive constant. (In applications $\beta = (kT)^{-1}$, k being Boltzmann's constant and T the absolute temperature.) The quantity

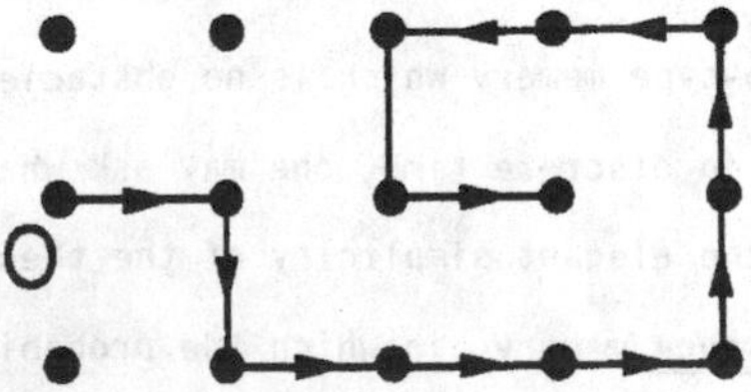

Fig. 1: Termination of a self-avoiding walk on the square lattice after a finite number of steps:

$$Z = \sum_{\substack{\text{all configurations} \\ \{\vec{\ell}_j\}}} \exp[-\beta U(\{\vec{\ell}_j\})] \qquad (5.2)$$

is a normalization constant. If $U(\{\vec{\ell}_j\})$ depends upon the order in which the sites $\vec{\ell}_j$ are visited, then the walk generated by (5.1) is a valid, time-evolving but non-Markovian process. Amit et. al. [73] have proposed that such a system should be called the true self-avoiding walk problem. If the $U(\{\vec{\ell}_j\})$ depend only upon the elements of the set $\{\vec{\ell}_j\}$, without regard to their order, the statistical properties of the walk are different [73]. This latter model is related to the excluded volume problem in polymer theory [74-75], and its limit in the case when the interaction energy is infinite, i.e. overlap is forbidden, is what is usually referred to as the self-avoiding walk. We discuss here only self-avoiding walks in this latter sense: we consider possible configurations of a non-overlapping chain, n bonds in length, confined to an infinite periodic lattice. The most interesting properties of such a system are

(i) c_n, the number of such chain configurations,

(ii) $P_n(r)$, the probability distribution function for the distance r between the ends of the chain.

A quantity derived from $P_n(r)$, and somewhat easier to analyse, is its second moment,

(iii) $$R_n^2 \equiv \int_0^\infty r^2 P_n(r)\, dr \; . \qquad (5.3)$$

Another quantity often analysed is

(iv) u_n, the number of closed polygons with n sides which have no self-intersections;

this is related to the probability that the two ends of a self-avoiding walk are nearest neighbours, and u_n/c_n is sometimes called the ring closure probability.

We limit our discussion in the main here to the large n asymptotic forms of

c_n , u_n , and R_n^2 , and to some connections between self-avoiding walks and statistical mechanics. Much fuller accounts of these matters, and many related topics can be found in Barber and Ninham [8], de Gennes [75], Domb [76], Domb [77], McKenzie [78] and Whittington [79]. We draw the reader's attention to two problems of very recent interest not discussed here: self-avoiding walks in the presence of boundaries [80,81], and relations between self-avoiding walks and fractals [82].

In an early paper on Monte Carlo methods, Hammersley and Morton [83] established what is believed to be the first nontrivial rigorous result concerning self-avoiding walks, in the particular case of a tetrahedral (diamond) lattice, namely that if we write $c_n = \exp(n\,K(n))$ then as $n \to \infty$, $K(n) \to \kappa$, a finite, positive constant. This result was later extended by Broadbent and Hammersley [84] and Hammersley [85] to a very wide class of lattices. It includes all infinite lattices with the properties that (i) all sites are topologically equivalent and have finite coordination number and (ii) at least one self-avoiding walk of infinite length can be started from any site (but the result proved in [84] and [85] holds for much more general lattices than these).

The quantity κ is not a universal constant, but rather a property of the lattice, and was called the connective constant by Broadbent and Hammersley [84]. It is now more common to write

$$\mu = e^{\kappa} \tag{5.4}$$

and call μ the <u>connective constant</u>; we adopt this latter convention here. The Broadbent-Hammersley-Morton result can be rewritten as

$$c_n = \mu^n \exp\{o(n)\} , \quad n \to \infty , \tag{5.5}$$

with no estimate available of the $o(n)$ term. Several years later, Hammersley [86] established, for d-dimensional simple hypercubic lattices, the companion result

$$u_n = \mu^n \exp\{o(n)\} , \text{ n even}, \quad n \to \infty . \tag{5.6}$$

The restriction that n be even is necessary, since on a simple hypercubic lattice, all closed polygons have an even number of sides. We see, therefore, that the same constant μ governs the growth both of c_n and u_n. It is a kind of effective coordination number, since for ordinary Pólya walks on a lattice with coordination number z, $c_n = z^n$.

Hammersley and Welsh [87] established the stronger result that for hypercubic lattices

$$c_n = \mu^n \exp\{O(n^{1/2})\}, \tag{5.7}$$

and some additional rigorous results were furnished by Kesten [88], but numerical evidence suggests that in two and three dimensions,

$$c_n \sim Cn^g\mu^n, \quad u_n \sim Un^{-h}\mu^n \tag{5.8}$$

with C, U lattice-dependent constants, and the exponents g and h <u>universal</u> in the sense that they depend only on dimensionality, and not on lattice structure [76]. Numerical evidence also suggests that in 2 or 3 dimensions [76]

$$R_n^2 \sim Kn^{2\nu}, \tag{5.9}$$

with the exponent ν universal, and K a lattice-dependent constant. The reader is warned that the notation for critical exponents is by no means standard. Other notations include: γ or $\gamma - 1$ for g; $-\alpha$, $\alpha - 2$, $1 - h$ or $-h$ for h; and $\frac{1}{2}\nu$, γ or $\frac{1}{2}\theta$ for ν.

Fisher [89] noted that generalizing a mean-field type argument of Flory leads one to predict that for self-avoiding walks in E dimensions,

$$\nu = \begin{cases} 3/(E+2) & E = 1,2,3 \\ 1/2 & E > 4 \end{cases}. \tag{5.10}$$

The significance of this result is that in more than four dimensions the self-avoiding efect becomes negligible and $\langle R_n^2\rangle$ grows in the same way as in ordinary Pólya walks. In the language of critical phenomena, $E = 4$ is the <u>(upper)</u> <u>critical dimensionality</u> for the self-avoiding walk problem. With judicious hindsight, one may remark that its existence is not surprising. The non-Markovian nature of the self-avoiding walk resides in long-ranged memory. However it is known that the conditional mean recurrence time τ_c for an ordinary Pólya walker [defined by Eq. (2.24)] is finite if $E > 4$, so that, loosely speaking, revisits to a site occur soon, or almost not at all. Thus if $E > 4$ the memory effect is sufficiently weak that behaviour qualitatively similar to a Pólya walk ensues.

The evaluation of the connective constant μ for a given lattice and the rigorous determination of the values of the critical exponents g,h and ν are problems of difficulty commensurate with their importance. Twenty years ago, Hammersley [90] derived the rigorous inequality

$$2E - 1 - \log(2E - 1) \leq \mu \leq 2E - 1 \qquad (5.11)$$

for the connective constant of an E-dimensional simple hypercubic lattice. In the intervening years, considerable efforts have been made to determine μ,g,h and ν numerically by Monte Carlo methods, or by series expansion techniques. In the latter method, as many terms as practicable in the series $\sum c_n\xi^n$, $\sum u_n\xi^n$ and $\sum R_n^2\xi^n$ are evaluated exactly by careful enumeration of all walks with n steps. Values of μ,g,h and ν are then deduced via an extrapolation technique which attempts to determine the location and the leading order behaviour of the analytic continuation of the series near its singularity closest to the origin, and which assumes that the asymptotic forms (5.8) and (5.9) hold. We give sample results in Tables 2 and 3. (See Domb [76] and Watts [91]; the latter paper is the source of the data tabulated here. A more extensive table, including additional critical exponents, is given by McKenzie [78].) The series predictions of ν coincide exactly with the Fisher-Flory expression [Eq. (5.10)].

A new era in the self-avoiding walk problem began with a brief paper of de Gennes [92], in which a relation was established between this conformational problem and a degenerate limit of a problem of equilibrium statistical mechanics. The idea was explored at greater length by Bowers and McKerrell [93] and des Cloizeaux [94], and given a very clear presentation by Sarma [95]; it is discussed in some detail in the article by Halley in the present volume. Briefly, one con-

Table 2: Connective constant μ for periodic lattices of dimension E = 2,3, derived by Padé approximant analysis of series expansions [91]; z is the coordination number of the lattice. Error 'estimates' are subjective (though reasonable) confidence limits, not rigorous bounds.

	lattice	z	μ
E = 2	honeycomb	3	1.8478 ± 0.0002
	square	4	2.6385 ± 0.0003
	triangular	6	4.1520 ± 0.001
E = 3	diamond	4	2.8792 ± 0.0005
	simple cubic	6	4.6838 ± 0.001
	body-centred cubic	8	6.5295 ± 0.002
	face-centred cubic	12	10.035 ± 0.005

Table 3: Critical exponents g and h [Eq. (5.8)] and ν [Eq. (5.9)] estimated by analysis of series expansions [91] for dimension E = 2,3 (but note the discussion following Eq. (5.15) below, and [99]).

E	g	h	ν
2	1/3	3/2	3/4
3	1/6	7/4	3/5

---

siders the <u>O(n) or n-vector model</u>, in which each site $\vec{\ell}$ of an E-dimensional lattice carries an n-dimensional vector spin $S(\vec{\ell})$ of length $\sqrt{n}$. The spins on nearest-neighbour sites $\vec{\ell}$ and $\vec{\ell}'$ have an energy of interaction $-KS(\vec{\ell})\cdot S(\vec{\ell}')$. Although n is subject to the constraint that it be a positive integer, the correspondence to self-avoiding walks comes about by treating n as a continuous parameter and taking the formal limit $n \to 0$ at an appropriate point in the analysis. Generating functions for c_n and other statistical properties of self-avoiding walks correspond to various thermodynamic properties of the n-vector model in the $n \to 0$ limit. For example, if $G(N,\vec{\ell},\vec{\ell}')$ is the number of self-avoiding walks of n steps beginning at site $\vec{\ell}$ and ending at site $\vec{\ell}'$, and $C(\vec{\ell},\vec{\ell}')$ is the spin correlation between sites $\vec{\ell}$ and $\vec{\ell}'$ in the n-vector model [i.e. the expected value of $S_1(\vec{\ell})\, S_1(\vec{\ell}')$], then

$$\lim_{n\to 0} C(\vec{\ell},\vec{\ell}') = \sum_{N=0}^{\infty} (\beta K)^N G(N,\vec{\ell},\vec{\ell}') \,. \tag{5.12}$$

(Here $\beta = (kT)^{-1}$, with k Boltzmann's constant and T the absolute temperature.) The addition of a non-zero external magnetic field to the n-vector model enables one to consider more complicated problems than the self-avoiding walk, but some subtleties arise, as discussed by Gujrati [96] and Wheeler and Pfeuty [97].

The discovery of the connection of self-avoiding walks to statistical mechanics via the n-vector model has had significant consequences. The first of these is the development of scaling theories (along the same lines as the arguments of Kadanoff and Widom [98] for statistical mechanical models). Extensive discussions of these ideas are given in de Gennes' book [75] and McKenzie's review [78]; similar ideas apply to percolation problems, as discussed in section 6. A key result of scaling arguments is a formal connection between the exponent ν describing growth in mean-squared displacement [Eq. (5.9)], the exponent h in the expression for the number of self-avoiding polygons [Eq. (5.8)] and the dimension E (for $E < 4$) :

$$E\nu = h \; . \tag{5.13}$$

The best numerical estimates of ν and h in two dimensions ($\nu = 3/4$, $h = 3/2$) satisfy this relation identically, but if the relation holds in three dimensions also, then one or both of the (accepted) values $h = 7/4$ and $\nu = 3/5$ must be in error, and there is room for additional precise numerical work [99].

A second consequence of the statistical mechanical analogy is the ability to use renormalization and related tools as alternatives to Monte Carlo and series expansion methods for numerical work. While a number of investigations along these lines have been made [100], perhaps the most successful is the analysis by Derrida [101] of the problem of self-avoiding walks on the square lattice, using a transfer matrix method. Derrida finds $\mu^{-1} = 0.37905 \pm 0.00003$ (i.e. $\mu \simeq 2.6382$) and $\nu = 0.7503 \pm 0.0002$, results very close to 'accepted' values. The error estimates are not rigorous, but are reasonable estimates of the possible error.

Perhaps the most important consequence of the statistical mechanical analogy is the formal determination of the connective constant and several exponents for self-avoiding walks on the hexagonal or honeycomb lattice. The solution is due to

Nienhuis [102], who obtained explicit formulae for critical exponents and the critical temperature for n-vector models when $-2 < n < 2$. Translated into self-avoiding walk terminology [77], Nienhuis' results are:

$$\mu = (2 + \sqrt{2})^{1/2} \approx 1.847759 \tag{5.14}$$

$$h = 3/2 \quad , \quad \nu = 3/4 \; . \tag{5.15}$$

Since the exponents h and ν should be universal, these results confirm for two dimensions the Fisher-Flory prediction of ν , the accepted value of h and the scaling relation [Eq. (5.13)]. The closeness of the series prediction $\mu = 1.8478 \pm 0.0002$ [91] to Nienhuis' value is remarkable. Although Nienhuis' analysis does not lead to an unequivocal determination of the exponent g characterizing the growth of the number of self-avoiding walks of n steps, it strongly suggests [77] that $g = 11/32 \approx 0.344$, in place of the 'accepted' value $g = 1/3$.

We draw the reader's attention to some other interesting non-Markovian problems. Kasteleyn [103] has enumerated exactly the number of self-avoiding walks which pass through every site on a finite square lattice with periodic boundary conditions (i.e. wrapped about a torus), and Manhattan oriented bonds: all bonds conduct in one direction only, like one-way streets, and any two adjacent, parallel streets are oppositely directed. Several authors have analysed directed self-avoiding walks and some exact results are available [104]. For example, in the case of a square lattice with all bonds parallel to one axis directed, $c_n \sim \text{constant} \cdot \mu^n$ as $n \to \infty$, with $\mu = 1 + \sqrt{2}$. This result supports the thesis of an article by Redner in this volume: directedness changes critical exponents. A two-dimensional continuum self-avoiding walk has been analysed by Weigel [105], who argues that in a certain sense this problem is "asymptotically Markovian", although an explicit solution cannot be exhibited.

With a view to describing the properties of branched polymer chains, many authors have considered the configurational statistics of 'lattice animals', which are connected clusters of sites on a lattice. (See, for example, the articles by Whittington and by Sahimi in this volume.) Of considerable recent interest are two models in which clusters are grown in a random manner. In the Eden model [106], a cluster grows outward from a single seed site. Each stage of the growth process consists of the 'infection' of a randomly chosen site adjacent to the cluster. In the Witten-Sander model [107], the cluster growth is diffusion mediated. Random walkers are released one at a time some distance from the cluster, and if they arrive at a site adjacent to the cluster, they remain there. The Eden and Witten-Sander models have strikingly different qualitative features.

6. Percolation Theory

We turn now from the theory of random processes to the theory of random systems or structures. We consider for the moment random structures which are created by the removal at random of a fraction of the sites (vertices) or bonds (edges) of a network (graph); in the main we shall focus on periodic lattices. The study of such discrete random structures is the province of percolation theory. (At the end of this section we shall have a little to say about random continua, and the problem of "continuum percolation".)

Although antecedents of the percolation idea undoubtedly exist, and one day are certain to be analysed by zealous historians of the mathematical sciences, the basic concept of percolation theory was first suggested by Broadbent [108] in the published discussion of a paper of Hammersley and Morton [83] on Monte Carlo methods. As a mathematical discipline, percolation theory was founded in landmark papers by Broadbent and Hammersley [84], and Hammersley [85], papers which the reader may recall are also milestones in the self-avoiding walk literature (see Section 5).

Classical percolation theory centres around two problems called the bond problem and the site problem. In the bond problem, the bonds of the lattice are assigned at random (and independently of each other) the probability p of being occupied, and the probability $1 - p$ of being vacant. For an infinite lattice, this assignment is equivalent to removing a fraction $1 - p$ of all bonds at random. Two sites are called connected if there exists at least one path between them consisting solely of occupied bonds and in each realization of the bond problem, the resulting 'bond-disordered' lattice consists of connected clusters of sites, linked together by occupied bonds, and bounded by vacant bonds. For a finite lattice, for example an $N \times N$ square lattice, so long as $p > 0$ there is a finite probability that a connected cluster touches each boundary of the lattice. If occupied bonds are regarded as open channels, this means that there is a non-zero probability that fluid introduced at one boundary can infiltrate the lattice and reach the other boundaries, or that fluid introduced at one arbitrary site in the lattice can percolate to reach the boundaries. One may ask what happens as the size of the lattice increases in some prescribed manner ($N \to \infty$ in the $N \times N$ square lattice used for illustrative purposes), or if we realize bond percolation on an infinite lattice. If p is sufficiently small, the size of any connected cluster is likely to be small, and connected clusters should appear as islands. On the other hand, if p is close to 1, the lattice should be entirely connected, apart from occasional small holes. At some value of p, there must be a transition in the topological structure of the random lattices; this value is called the bond percolation threshold, p_c^B. It may be defined rigorously as follows (assuming that the underlying lattice is infinite and that all of its sites are equivalent for simplicity here): Let $P_B(p)$ denote the (bond) percolation probability, i.e. the probability that, when the fraction of occupied bonds is p, a given site belongs to a connected cluster of infinite extent. (In the mathematical literature $P_B(p)$ is sometimes denoted by $\theta(p)$.) The bond percolation threshold is defined as the dividing point between values of p for which $P_B(p) > 0$ and values of p for which $P_B(p) = 0$:

$$p_c^B = \inf \{p : P_B(p) > 0\} . \qquad (6.1)$$

In the preceding paragraph we have defined the bond problem. In the site (or atom) problem, sites of the lattice are occupied with probability p and vacant with probability $1 - p$. Two nearest-neighbour sites are called connected if they are both occupied, and connected clusters on the lattice are again defined in the obvious manner. Every bond percolation problem can be mapped onto a site problem on a closely related lattice (see Wierman's article in the present volume). A site percolation threshold p_c^S may be anticipated, and it can be defined in exactly the same manner as the bond percolation threshold via the appropriate (site) percolation probability $P_S(p)$ for the site problem:

$$p_c^S = \inf \{p : P_S(p) > 0\} . \tag{6.2}$$

The definitions (6.1) and (6.2) are nontrivial, because the values of the critical probabilities or thresholds p_c^B and p_c^S so defined are non-zero. The establishment of this fact, and the deduction of inequalities satisfied by p_c^B and p_c^S, was the key achievement of the founding papers of percolation theory [84,85], and two other papers of Hammersley [109,110]. We shall not give the strongest form of their results here, but rather restrict attention to the case of an infinite lattice with no disconnected components, and all sites equivalent to each other. (This class includes all crystallographic lattices.) The restrictions made are sufficient to ensure the existence of the connective constant μ , defined in the context of self-avoiding walks in the preceding section, and the key result is the inequality [109,110]

$$\mu^{-1} \leq p_c^B \leq p_c^S . \tag{6.3}$$

Upper bounds are known under slightly more restrictive hypotheses than those for which (6.3) holds, with the most useful rigorous result, due to Hammersley [111], being

$$\mu^{-1} \leq p_c^B \leq p_c^S \leq 1 - \mu^{-1} ; \tag{6.4}$$

this inequality does not hold for the linear chain ($\mu = 1$) , the honeycomb lattice ($\mu \simeq 1.85$) , or any lattice so sparsely branched that $\mu < 2$. In a sense it is the best possible inequality because for a Bethe lattice or Cayley tree (lattice lacking closed loops) of coordination number 3 , equality holds everywhere [112]. An alternative bound has been suggested from numerical evidence, but not proven, by Ishioka and Koiwa [31]:

$$\mu^{-1} < p_c^B < p_c^S < R \,, \qquad (6.5)$$

with R the probability of return to the origin of a Pólya random walker (section 3). The upper bound (6.5) improves with increasing dimensions, whereas Hammersley's rigorous upper bound deteriorates. For a Bethe lattice of coordination number $z \geq 3$, equality holds everywhere in Eq. (6.5) [43].

Table 4

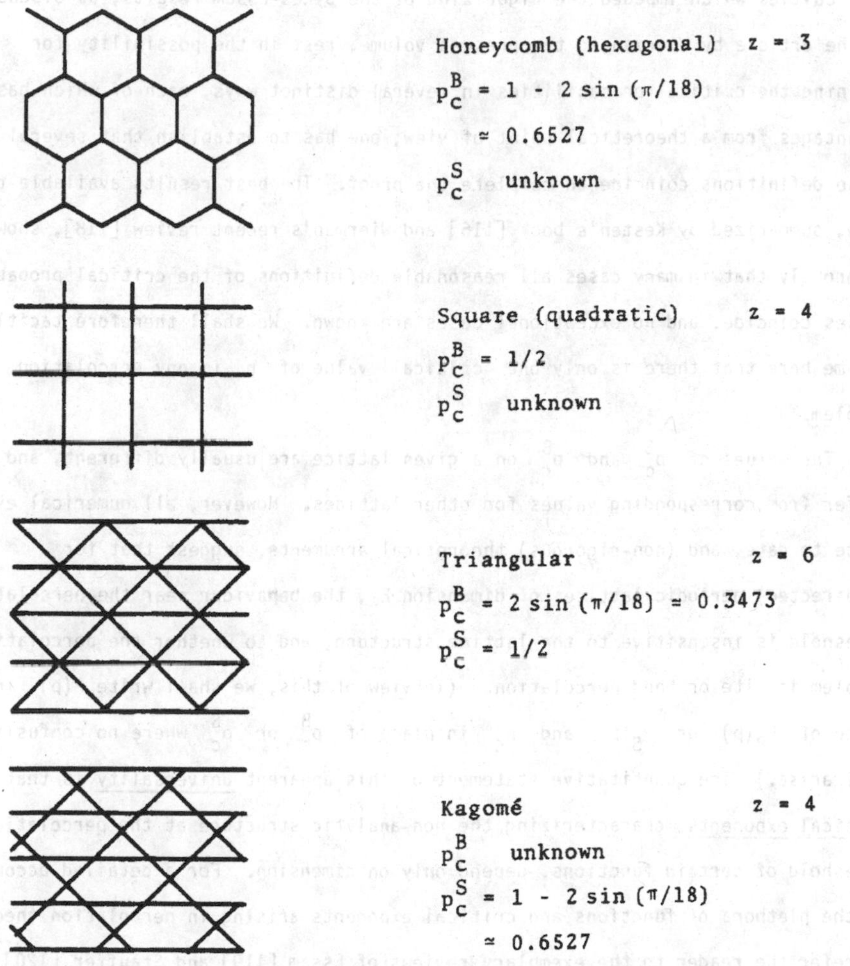

Lattice	z	p_c^B	p_c^S
Honeycomb (hexagonal)	$z = 3$	$p_c^B = 1 - 2\sin(\pi/18) \simeq 0.6527$	unknown
Square (quadratic)	$z = 4$	$p_c^B = 1/2$	unknown
Triangular	$z = 6$	$p_c^B = 2\sin(\pi/18) = 0.3473$	$p_c^S = 1/2$
Kagomé	$z = 4$	unknown	$p_c^S = 1 - 2\sin(\pi/18) \simeq 0.6527$

Rigorously established values of percolation thresholds [114-117]; z = coordination number. For conjectures on, and numerical estimates of, missing values see Sahimi's article in this volume.

A survey of recent progress in the numerical determination of p_c^B and p_c^S is given in an article by Sahimi in the present volume. The exact derivation of values of critical probabilities is to date possible only for certain lattices related to the Bethe lattice [112,113], and for a few two-dimensional lattices. The two dimensional results are summarized in Table 4. They were first obtained via ingeneous arguments by Sykes and Essam [114], but the analysis has only recently been made entirely rigorous by Kesten [115,116] and Wierman [117]. The difficulties which impeded the rigorizing of the Sykes-Essam results, as discussed in the article by Wierman in the present volume, rest in the possibility for defining the critical probabilities in several distinct ways, each of which has advantages from a theoretical point of view; one has to establish that several of these definitions coincide to complete the proof. The best results available to date, summarized by Kesten's book [116] and Wierman's recent review [118], show rigorously that in many cases all reasonable definitions of the critical probabilities coincide, and no exceptional cases are known. We shall therefore tacitly assume here that there is only one 'critical' value of p in any percolation problem.

The values of p_c^S and p_c^B on a given lattice are usually different, and differ from corresponding values for other lattices. However, all numerical evidence to date, and (non-rigorous) theoretical arguments, suggest that for (undirected) periodic lattices of dimension E, the behaviour near the percolation threshold is insensitive to the lattice structure, and to whether the percolation problem is site or bond percolation. (In view of this, we shall write $P(p)$ in place of $P_B(p)$ or $P_S(p)$ and p_c in place of p_c^B or p_c^S where no confusion will arise.) The quantitative statement of this apparent <u>universality</u> is that <u>critical exponents</u>, characterizing the non-analytic structure at the percolation threshold of certain functions, depend only on dimension. For a detailed account of the plethora of functions and critical exponents arising in percolation theory, we refer the reader to the exemplary reviews of Essam [119] and Stauffer [120] from which much of the following material is drawn (a short up-date of the latter review is available [121]). We shall introduce briefly the five most important exponents: $\alpha, \beta, \gamma, \delta$ and ν. Following Stauffer [120], we phrase our definitions in terms of site percolation, and emphasise the importance of the function

$$n_s(p) = \text{average number (per lattice site) of clusters of size } s\ . \tag{6.6}$$

The probability that a given site of the lattice belongs to a finite cluster of size s is given by sn_s, while there is a probability $P(p)$ that it belongs to an infinite cluster, and a probability $1 - p$ that it is unoccupied. Thus

$$1 - p + P(p) + \sum_s sn_s(p) = 1\ , \tag{6.7}$$

so that $P(p)$ is given in terms of the first moment of $n_s(p)$. Other important moments of n_s are the zeroth moment

$$K(p) \equiv \sum_s n_s(p)\ , \tag{6.8}$$

which is the expected <u>number of finite clusters per site</u> of the lattice, and the second moment $\sum s^2 n_s(p)$. Since $sn_s(p)$ is the probability of a site belonging to the finite cluster of size s, the sum $\sum s^2 n_s(p)$ is essentially the mean cluster size, counting only finite clusters, and the quantity

$$S(p) = \frac{\sum_s s^2 n_s(p)}{\sum_s sn_s(p)} \tag{6.9}$$

is sometimes (inadvisedly [120]) called the <u>mean cluster size.</u> (For variants on this definition see the article by Halley and Dasgupta in this volume.) Kesten [116,122] has obtained weak bounds on the behaviour of these three functions near the percolation threshold p_c, but the conventional wisdom in the physics literature is that up to a multiplicative constant:

$$P(p) \sim (p - p_c)^{\beta} \qquad \text{as } p \to p_c^+\ , \tag{6.10}$$

$$\frac{d^3K}{dp^3} \sim \begin{cases} (p_c - p)^{-1-\alpha} & \text{as } p \to p_c^- \\ (p - p_c)^{-1-\alpha'} & \text{as } p \to p_c^+ \end{cases}\ , \tag{6.11}$$

$$S(p) \sim \begin{cases} (p_c - p)^{-\gamma} & \text{as } p \to p_c^- \\ , & \\ (p - p_c)^{-\gamma'} & \text{as } p \to p_c^+ \end{cases} \qquad (6.12)$$

with α and α' negative, and β, γ and γ' positive. Moreover it is generally believed that $\alpha = \alpha'$ and $\gamma = \gamma'$ and we assume this here.

The notation used for these critical exponents has been chosen by physicists to reflect formal connections between percolation theory and equilibrium statistical mechanics. Two such connections have been made. The first, discussed in this volume by Redner, and by Halley and Dasgupta, is due to Kasteleyn and Fortuin [123]. It relates percolation theory to the formal limit $q \to 1$ of the q-state Potts model, a generalization of the Ising model of ferromagnetism, in which each site of a lattice carries a spin capable of occupying one of q states. The second, due to Dunn et al. [124] (see also Essam [119]), relates percolation to a model of dilute ferrogmagnets. The critical probability p_c corresponds to the Curie temperature T_c of the magnetic analogue, and the parameter regimes $0 \leq p < p_c$ and $p_c < p \leq 1$ correspond respectively to the disordered phase $(T > T_c)$ and the ordered phase $(T < T_c)$ of the magnet [125]. (The use of primes to distinguish between critical exponents for the same physical quantity defined on different sides of p_c follows the convention in the magnetic model literature that primed exponents correspond to the ordered phase.)

In Table 5 we collect the key functions of percolation theory, together with their analogues in magnetic models. The magnetic analogy suggests the usefulness of defining an additional function, $\sum_s sn_s(p_c)e^{-hs}$ $(h > 0)$, corresponding to the mean magnetic moment at the critical temperature. The small h behaviour of this function defines a new critical exponent δ (which in the magnetic analogue gives the shape of the critical isotherm for small external field):

$$\sum sn_s(p_c) - \sum sn_s(p_c)\, e^{-hs} \sim h^{1/\delta} \quad \text{as } h \to 0+ . \qquad (6.13)$$

An additional important function of percolation theory which cannot be related directly to $n_s(p)$ is the <u>correlation or coherence length</u> $\xi(p)$. For a defini-

Table 5

Key functions of percolation theory (cf. Stauffer [120,121] and Essam [119]); several other definitions of the mean cluster size $S(p)$ are in use, but give the same critical exponent γ. The key function $n_s(p)$ is the number of clusters of finite size s per lattice site, for site percolation with a fraction p of all sites present.

Function (and magnetic analogue)	Definition	Asymptotic form	Critical exponent in E dimensions
mean number of clusters per site (free energy per site)	$K(p) = \sum_s n_s(p)$	$\frac{\partial^3 K}{\partial p^3} \sim \lvert p-p_c\rvert^{-1-\alpha}$ as $p \to p_c$	$E = 2$ $\alpha = -2/3$ $E = 3$ $\alpha \simeq -0.5$ $E > 6$ $\alpha = -1$
percolation probability (spontaneous magnetization)	$P(p) = p - \sum_s sn_s(p)$	$P(p) \sim (p - p_c)^{\beta}$ as $p \to p_c^+$	$E = 2$ $\beta = 5/36$ $E = 3$ $\beta \simeq 0.4$ $E > 6$ $\beta = 1$
mean size of finite clusters (susceptibility)	$S(p) = \frac{\sum_s s^2 n_s(p)}{\sum_s sn_s(p)}$	$S(p) \sim \lvert p - p_c\rvert^{-\gamma}$ as $p \to p_c$	$E = 2$ $\gamma = 43/18$ $E = 3$ $\gamma \simeq 1.7$ $E > 6$ $\gamma = 1$
not named (mean magnetic moment on the critical isotherm)	$\mu(p_c,h)=\sum sn_s(p_c)e^{-hs}$	$\mu(p_c,0) - \mu(p_c,h) \sim h^{1/\delta}$ as $h \to 0^+$	$E = 2$ $\delta = 91/5$ $E = 3$ $\delta \simeq 5.$ $E > 6$ $\delta = 2$
correlation or coherence length (correlation length)	see Redner's first article in this volume	$\xi(p) \sim \lvert p - p_c\rvert^{-\nu}$ as $p \to p_c$	$E = 2$ $\nu = 4/3$ $E = 3$ $\nu \simeq 0.8$ $E > 6$ $\nu = 1/2$

tion and discussion of the sigificance of $\xi(p)$, we refer the reader to the article by Halley and Dasgupta and the first article by Redner in the present volume. Essentially $\xi(p)$ measures the size of connected clusters for $p < p_c$, and the length scale over which the random lattice appears homogeneous for $p > p_c$. The correlation length diverges as $p \to p_c$, leading to two new critical exponents ν and ν' :

$$\xi(p) \sim \begin{cases} (p_c - p)^{-\nu} & p \to p_c^- \\ (p - p_c)^{-\nu'} & p \to p_c^+ . \end{cases} \qquad (6.14)$$

It is believed that $\nu = \nu'$.

Although we have now introduced five critical exponents (α, β, γ, δ and ν, assuming that $\alpha = \alpha'$, $\gamma = \gamma'$ and $\nu = \nu'$), scaling arguments [120] may be used to deduce relations between them. Using the ansatz of Stauffer [120] $n_s(p) \sim s^{-\tau} f([p - p_c]s^{\sigma})$ for $s >> 1$ and $|p - p_c| << 1$, with $f(z)$ finite at $z = 0$, one may infer that for a periodic lattice of dimension $E \geq 2$,

$$2 - \alpha = \gamma + 2\beta = \beta(\delta + 1) . \tag{6.15}$$

Any two of the exponents α, β, γ and δ uniquely determine the other two, if the scaling argument is correct. Scaling laws of the form (6.15) are well-known in the theory of critical phenomena [125], where they are derived using a similar scaling ansatz. The relation $\alpha + 2\beta + \gamma = 2$ is called Rushbrooke's scaling law and the relation $\gamma = \beta(\delta - 1)$ is called Widom's scaling law. Thermodynamic arguments can be used to establish the inequalities $\alpha' + 2\beta + \gamma' \geq 2$ [126] and $(1 + \delta)\beta \geq 2 - \alpha'$ [127], but no rigorous proof that equality holds is yet available.

A noteworthy feature of Eq. (6.15) is the absence of any explicit dependence on the dimension E . An additional scaling law, called Josephson's scaling law or the hyperscaling law, relates the correlation length exponent ν to the other exponents, but does involve E explicitly:

$$2 - \alpha = E\nu . \tag{6.16}$$

Since the correlation length is not defined in terms of $n_s(p)$, its exponent cannot be related to the other critical exponents using the scaling ansatz for $n_s(p)$ described above. Additional assumptions on the radii of clusters have to be introduced [120].

Unlike Eq. (6.15), the hyperscaling relation holds only for a restricted range of lattice dimensions, namely $2 \leq E \leq E_c = 6$, and $E_c = 6$ is called the upper critical dimension [128] (see Redner's first article in this volume). For $E > E_c = 6$, the values of the exponents are independent of E , and coincide with values predicted by "mean field theories" [129,130]:

$$\alpha = -1 , \quad \beta = 1 , \gamma = 1 , \quad \delta = 2 , \quad \nu = 1/2 . \qquad (6.17)$$

The mean field values of α, β, γ and δ are the same as those for the Bethe lattice [112]; the correct definition of the correlation length for a Bethe lattice is a matter of contention, and different definitions give different values for ν [131].

At $E = E_c = 6$, the critical exponents are believed to take their mean field values, but the asymptotic forms of the functions defining them must be corrected by the insertion of prefactors which are powers of $\ell n|p - p_c|$. When $E < 6$, the hypothesis of local homogeneity upon which mean field theory is based is grossly invalid, and becomes successively worse as E decreases. However _renormalization group_ calculations can be used to determine the values of the critical exponents. Such calculations fall into two classes: "momentum-space" and "real-space" calculations. The former approach leads to formal asymptotic expansions for the exponents in terms of the variable $\varepsilon = 6 - E$, treated as a small, continuous parameter [129,132]. The latter approach is discussed in Redner's first article in this volume.

For the dimensionalities of most physical interest ($E = 2$ or $E = 3$) some progress has been made in the determination of the critical exponents, but no rigorous results are yet at hand. In Table 5 we list values of the exponents in two dimensions, inferred from a number of conjectures on the q-state Potts model (due to den Nijs [133], Nienhuis _et al._ [134] and Pearson [135]), which are generally believed to be exact. We also list approximate values of the exponents in three dimensions [121]. For a full account of recent advances in the determination of values of the critical exponents, the reader may refer to Sahimi's article in the present volume. (However, we draw to the reader's attention an interesting paper of Pearson [136] in which number-theoretic arguments are applied to the calculation of critical exponents for the q-state Potts model in two dimensions, for $q = 2,3,4$.)

It is not possible here to outline the rich variety of applications of percolation theory, or to describe interesting generalizations of the percolation model and recent theoretical advances. (Some idea of the interest and progress in the field may be gained from the papers collected in the present volume.) We restrict ourselves here to noting a few of the more important reviews and expository articles not already cited in this section [137-141].

Our discussion of percolation theory has been confined to percolation problems on periodic lattices and tree-like structures. It is possible to consider percolation problems on topologically disordered lattices such as that generated by the Voronoi random tessellation of space [142], and percolation problems in continua [143,144]. The latter problem is considerably more difficult to analyze than lattice problems, but numerical evidence to date suggests that the critical exponents are the same as for percolation on a periodic lattice [144]. Most of the work on continuum percolation to date is based on a micromorphology in which space is partitioned into 'occupied' and 'empty' volumes as follows. Initially all space is 'empty'. Small 'elementary regions' of space, typically identical spheres, are the marked 'occupied' at random until a macroscopic volume fraction ϕ of occupied space is attained. The value of the critical volume fraction ϕ_c above which the 'occupied' region percolates depends upon the nature of the 'elementary regions' (monodisperse, i.e. identical spheres, or polydisperse spheres; spheres or ellipsoids, etc.) and the correlations allowed between then (e.g. the extent to which individual spheres may overlap). The significance of microstructure in random continua is brought out in Part B of the present article, but we refer the reader here to an excellent article by Zallen [145] on the general subject of 'stochastic geometry', and a monograph of Roach [146].

7. Equilibrium transport in random lattices

Percolation theory, as outlined in the preceding section, attempts to answer only geometrical and topological questions about random lattices, such as: is there macroscopic connectivity? However it is possible to generalize percolation theory so that the random lattice supports a physical process. Some examples of such generalizations are discussed in the article by Halley and Dasgupta in the present volume, but two other conceptually simple extensions of percolation theory are the subject of this section, and the following section. We address in this section problems which are static in time, while in section 8 we examine systems which evolve with time.

We begin with a problem of great interest to physicists over the last decade [147,148], which owes its popularity to a seminal paper of Kirkpatrick [147]: the problem of the random resistor network (or the percolation conduction problem). We consider bond percolation on a periodic lattice in E dimensions, with each vacant bond being an insulator, and each occupied bond a conducting element (resistor). Thus, the jth bond of the lattice has a conductance g_j, where $\{g_j\}$ is a set of independent random variables, with common probability density function

$$f(g) = (1 - p)\delta_+(g) + ph(g) . \tag{7.1}$$

The function $h(g)$ is the probability density function for the conductance of a bond, given that its conductance is non-zero. A particularly simple choice is $h(g) = \delta(g - g_0)$, so that all conducting bonds have the same conductance, and $f(g)$ has the binary distribution

$$f(g) = (1 - p)\delta_+(g) + p\delta(g - g_0) . \tag{7.2}$$

We are interested in knowing when a macroscopic fragment of the random lattice will conduct electricity if it is subjected to a potential difference, and in determining the value of the macroscopic conductivity of the lattice. Evidently, if p does not exceed the (bond) percolation threshold p_c of the lattice, a sufficiently large sample should be non-conducting [149].

Before addressing lattices of dimensionality $E > 1$, we look briefly at the problem of a one-dimensional random resistor network. If n bonds, with random conductivities g_j governed by the probability density function (7.1), are placed in series, the conductance of the composite is simply $(\sum_j g_j^{-1})^{-1}$. Thus the composite system is equivalent to a uniform system of conductance σ for each bond, where

$$\sigma^{-1} = \frac{1}{n} \sum_{j=1}^{n} g_j^{-1} . \tag{7.3}$$

The strong law of large numbers [20] tells us that if the random variables g^{-1} have finite expectation $\langle g^{-1}\rangle$, then σ^{-1} converges to this value with probability one. Therefore the conductance per bond, or conductivity, of an infinite linear chain is

$$\sigma = \langle g^{-1}\rangle^{-1} \quad , \quad \text{where} \quad \langle g^{-1}\rangle = \int dg \, g^{-1} f(g) \, . \tag{7.4}$$

When $p < 1$, $\sigma = 0$ (one cut suffices to disconnect a linear chain).

In view of Jensen's inequality on the expectation of a convex function of a random variable [20], we see that for the one-dimensional random resistor network,

$$\langle g^{-1}\rangle^{-1} = \sigma \leq \langle g\rangle \, , \tag{7.5}$$

with $\langle g\rangle$ the expectation (mean) of g. If $f(g)$ if other than a single delta function, the inequality is strict. With a correct definition of σ for systems in two or more dimensions, which we outline in a moment, the relation (7.5) generalizes to the inequality

$$\langle g^{-1}\rangle^{-1} \leq \sigma \leq \langle g\rangle \, . \tag{7.6}$$

Since the continuum analogue of Eq. (7.6) is established in Part B, we shall not prove Eq. (7.6) here. Moreover, Eq. (7.6) is not a particularly sharp inequality for percolation conduction problems, since if we use the binary distribution (7.2) for $f(g)$ we obtain the bounds

$$0 \leq \sigma \leq pg_0 \, , \tag{7.7}$$

which give no information on the value of the percolation threshold p_c.

We now consider random resistor networks of dimension $E > 1$, and for simplicity of exposition we restrict our attention initially to the E-dimensional simple cubic lattice, with individual bond conductances distributed according to the binary distribution (7.2). Since the macroscopic conductivity σ of a solid is defined as the current per unit area which flows in response to a unit potential gradient, an appropriate definition of the effective conductivity σ of the random resistor network is as follows. Consider a finite hypercube containing

L^E sites, defined by the inequalities $1 \leq \ell_j \leq L$ for $j = 1,\ldots,E$. The faces $\ell_1 = 1$ and $\ell_1 = L$ of the hypercube are subjected to a potential difference V (and so there is an average potential gradient $V/(L-1) \approx V/L$ if $L \gg 1$). The boundary conditions on the other faces should be relatively unimportant when $L \gg 1$, but in numerical work periodic boundary conditions are preferred (i.e. the faces $\ell_j = 1$ and $\ell_j = L$, $j = 2,\ldots,E$, are joined). A total current I flows in response to the potential gradient; the current per unit area is I/L^{E-1}. We therefore define the conductivity of the hypercube to be

$$\sigma_L = \langle L^{2-E} I/V \rangle \tag{7.8}$$

with the angular brackets denoting an ensemble average over all possible realizations of the hypercube, and we call

$$\sigma = \lim_{L\to\infty} \sigma_L \tag{7.9}$$

the <u>(effective)</u> <u>conductivity</u> of the random lattice. In the case when $p = 1$, this definition returns the value $\sigma = q_0$.

The determination of σ by numerical simulation (see [147,148] and Sahimi's article in the present volume) is rather expensive, since it requires the solution of Kirchhoff's equation

$$\sum_{\vec{j}\in\{\vec{\ell}\}} q(\vec{j},\vec{\ell})\,\{V(\vec{j}) - V(\vec{\ell})\} = 0\,, \tag{7.10}$$

i.e. the numerical inversion of a large matrix. (Here $V(\vec{\ell})$ is the potential at site $\vec{\ell}$, $\{\vec{\ell}\}$ denotes the set of nearest-neighbours of site $\vec{\ell}$, and $q(\vec{j},\vec{\ell})$ is the conductivity of the bond joining nearest-neighbour sites $\vec{j}$ and $\vec{\ell}$.) It is not at present feasible, for example, to exceed $L = 20$ in three dimensions.

Numerical simulations and other theoretical approaches are consistent with the following picture: up to a constant factor [147,148],

$$\sigma \sim (p - p_c)^t \quad \text{as} \quad p \to p_c^+\,, \tag{7.11}$$

where the <u>conductivity</u> <u>critical</u> <u>exponent</u> t is universal, i.e. the same for all lattices of a given dimension E , and an increasing function of E for $2 < E < E_c = 6$. The best numerical estimate of t for $E = 2$ is $t \approx 1.28$ [150]. For $E > E_c = 6$, it is now generally accepted that $t = 3$ [151-152]. The only exact solution of the random resistor problem available to date is Stinchcombe's solution for a Bethe lattice or Cayley tree [153]. For the Bethe lattice, certain ambiguities arise in the definition of the conductivity σ , but Straley [154] has proposed a reasonable definition, which leads to the critical exponent $t = 3$. (Stinchcombe's original definition gave $t = 2$, but his analysis is easily adapted to Straley's definition.)

The theoretical prediction of t , and its relation to the geometrical and topological critical exponents α, β, γ, δ and ν of section 6, are problems of great current interest, as discussed in Redner's first article in this volume. The difficulty lies in the need to know the topological structure of the backbone (current-carrying portion of the lattice) as $p \to p_c^+$ in order to relate t to other exponents [155]. However, a qualitative understanding of the random resistor problem can be obtained through an approximation technique called the "effective medium approximation". The technique has quite a long history for conduction problems in random continua [156], where it is sometimes called the "self-consistent method", and has also been used in several areas of condensed matter physics under the name "coherent potential approximation" [157]. The application of the technique to the random resistor problem is due to Kirkpatrick [158,147]. Since in section 8 we show in detail how the effective medium approximation can be applied to a problem which encompasses the random resistor problem as a special case, we shall only state the basic results for the random resistor problem here. For an arbitrary lattice in which all sites have coordination number z , the effective medium approximation predicts that the bond percolation threshold is

$$p^* = 2/z \; . \tag{7.12}$$

This is a good approximation in two dimensions (compare Eq. (7.12) with Table 4 of Section 6, and the discussion of 'dimensional invariants' in the article by Halley and Dasgupta in this volume); it is a poor approximation in higher dimensions. The effective medium approximation further predicts that the conductivity is

$$\sigma^* = \begin{cases} [\frac{p - p^*}{1 - p^*}]g_0 & p > p^* \\ 0 & p < p^* \end{cases} ; \qquad (7.13)$$

the predicted conductivity is linear in p, and so it is predicted that $t = 1$. Despite its failure to yield a good approximation for p_c and t in three or more dimensions, the effective medium approximation has proved useful. It is numerically quite accurate outside of a small region (the "critical region") and it does predict a percolation threshold, a virtue which is not shared by other approximations based on finite expansions in powers of p or $1 - p$. (Moreover it has recently been shown by Sahimi et al. [159] that a combination of the effective medium approximation with simple real-space renormalization methods leads to an excellent numerical estimate of σ, even quite close to the percolation threshold.)

The effective medium approximation in its simplest form is highly inaccurate for the Bethe lattice, but an ingeneous modification of it due to Essam et al. [160] enables the correct critical exponent to be predicted. Bethe lattice exact solutions and effective medium approximations for ordinary lattices have proved a a useful tool for investigating more complicated, and perhaps more subtle, variants of the random resistor problem. For example, Kogut and Straley [161] have argued along these lines that if the conductances of occupied bonds are distributed (i.e. the conductances are drawn from (7.1) rather than (7.2)), the critical exponent t is independent of the subdistribution $h(g)$ so long as

$$h_{-1} = \int dg\, g^{-1}\, h(g) \qquad (7.14)$$

is finite. If $h_{-1} = \infty$, so that very weakly conducting bonds are relatively abundant, the exponent t appears to depend upon the asymptotic behaviour of $h(g)$ near $g = 0$.

Although we have concentrated on the random resistor problem generated by bond percolation, it is possible to formulate the problem for site percolation, or for a more general correlated model [147,162]. It is also possible to consider a a random superconductor, in which a fraction $1-p$ of the bonds have a finite conductance g_0, while the remaining fraction p are superconductors:

$$f(g) = (1-p)\,\delta(g-g_0) + p\delta(g-\infty)\ . \tag{7.15}$$

If $p > p_c$, the network is superconducting, while for $p < p_c$ the conductivity is finite, but diverges as $p \to p_c^-$:

$$\sigma \sim (p_c - p)^{-s}\ . \tag{7.16}$$

Straley [163] has developed a scaling theory for the conductivity for the general problem

$$f(g) = (1-p)\,\delta(g-g_1) + p\delta(g-g_2)\ , \tag{7.17}$$

with $0 < g_1 < g_2 < \infty$, and has also established by a duality argument for the square lattice the remarkable result [164] that

$$s = t \quad \text{in two dimensions.} \tag{7.18}$$

It appears that s decreases with dimensionality and that $s = 0$ for $E > E_c = 6$ (i.e. if $E > E_c$, the conductivity diverges very weakly as the percolation threshold is approached from below).

Despite the challenging mathematical questions raised by the random resistor and superconductor problems, they have received little attention to date in the mathematical literature, Kesten's monograph [116] being the major exception; a synthesis of recent progress by physicists has been given by Straley [165]. The problem is one of considerable technical importance, and applications in condensed matter physics have been in the minds of most of the physicists who have taken up the problem. However, hydrodynamic problems in porous media can be modelled in the first approximation by network models equivalent to the random resistor problem [166], and are but one example of the many potential "real-life" applications of the basic ideas of percolation and conduction in random lattices. Recent simulations of Jerauld et al. [167] of conduction on topologically disordered lattice structures suggest that there is no need, in the first instance, to go beyond percolation on periodic lattices: average coordination number and dimensionality appear to be the key parameters.

8. Transient motion in random lattices

In the preceding sections we have pursued two different themes: random processes, and random systems. To describe phenomena in nature it is often useful to emphasize one of these themes and suppress the other. However, problems do arise in which it is absolutely essential that the two themes be blended harmoniously if a useful model is to be constructed. Areas in which good theories of random processes in random media are needed include diffusion in polycrystalline solids [168], electrical transport in amorphous semiconductors [169], and dispersion in flow through porous media [170]. We indicate here the kinds of models that are of current interest for such problems, the limits of achievements to date, and the pressing need for further work, by examining two simple systems: discrete time random walk in a random environment, and master equation mediated transport in a random environment. In both cases the 'random environment' is a periodic lattice with the disorder or randomness built into local transition probabilities or transition rates. For a sample of recent progress on random motion in disordered continua, the reader may consult Papanicolaou's article in the present volume.

A discrete-time random walk in a random environment can be defined by generalizing the Pólya lattice random walk discussed in section 2, with the transition probability $p(\vec{\ell} - \vec{\ell}')$ for a transition from site $\vec{\ell}'$ to site $\vec{\ell}$ in Eq. (2.1) replaced by a function $p(\vec{\ell}|\vec{\ell}')$. For each $\vec{\ell}'$, the function $p(\vec{\ell}|\vec{\ell}')$ is drawn at random from a space of allowed functions of $\vec{\ell}$. The simplest example, and the one which has been most studied by mathematicians, is the one-dimensional walk with

$$p(\ell|\ell') = \alpha_{\ell'}\,\delta_{\ell,\ell'+1} + \{1 - \alpha_{\ell'}\}\delta_{\ell,\ell'-1}\;; \tag{8.1}$$

here $\{\alpha_\ell\}$ is a set of independent, identically distributed random variables assuming values in the interval $[0,1]$. A realization ω of the lattice consists in the assignment of the parameters $\{\alpha_\ell\}$, which remain constant throughout any subsequent random walk process supported by the lattice.

It is neither possible nor relevant to obtain an explicit solution $P_n(\ell|\omega)$

of the random walk equations

$$P_{n+1}(\ell) = \sum_{\ell'} p(\ell|\ell')\, P_n(\ell') \quad , \quad P_o(\ell) = \delta_{\ell 0} \ , \tag{8.2}$$

for an arbitrary realization ω of the lattice. Rather one seeks properties of the system which correspond to some macroscopic or observable quantity. The simplest questions to ask concern ensemble-averaged properties of the transport which correspond to macroscopic observable quantities. For example, if the angular brackets $\langle\ \rangle_\omega$ denote the average with respect to realizations ω, one may seek the values of $\langle P_n(\ell|\omega)\rangle_\omega$ and $\langle R_n^2\rangle_\omega \equiv \langle \Sigma_\ell\, \ell^2\, P_n(\ell|\omega)\rangle_\omega$. Much harder questions, of greater interest to mathematicians (though perhaps of less importance in applications) concern the qualitative behaviour of walks a.e. (almost everywhere), i.e. properties of almost all random walks in almost all realizations of the lattice (with 'almost all' meaning apart from a set of measure zero in an appropriately defined probability space).

We first summarize some known rigorous results which bring out subtleties arising from the interplay of random motion with random environment. (These results cannot be proved here due to lack of space, and the proofs invoke concepts and terminology which we have not used elsewhere in this article.) The first rigorous investigation of the problem defined by Eqs. (8.1) and (8.2) seems to have been made by Kozlov [171], and a fuller study (leading to the results we quote here) was made by Solomon [172].

It is natural to anticipate that the qualitative properties of the walk should depend on whether the walk is in some sense statistically biased towards $+\infty$ or $-\infty$, or whether it is unbiased. We write $\beta_\ell = 1 - \alpha_\ell$, and (α,β) for any pair (α_ℓ,β_ℓ). The average with respect to the distribution of α is denoted by angular brackets $\langle\cdot\rangle$. It is found [172] that an appropriate measure of the bias of the walk is not in terms of $\langle\alpha - \beta\rangle$, but rather $\langle\sigma\rangle$ and $\langle\sigma^{-1}\rangle$, where $\sigma = \beta/\alpha = (1 - \alpha)/\alpha$. In view of Jensen's inequality [20] for the expectation of a convex function of a random variable, possible values of $\langle\sigma\rangle$ and $\langle\sigma^{-1}\rangle$ are constrained to satisfy the inequality $\langle\sigma\rangle\langle\sigma^{-1}\rangle \geq 1$. The cases $\langle\sigma\rangle < 1$, $\langle\sigma^{-1}\rangle < 1$ and $\langle\sigma\rangle^{-1} \leq 1 \leq \langle\sigma^{-1}\rangle$ therefore exhaust all possibilities.

Results of Solomon [172] characterize the behaviour of X_n, the position of the walker after n steps, and T_ℓ, the first passage time to site ℓ, given that the walker started from the origin of coordinates. If $\langle\sigma\rangle < 1$ or $\langle\sigma^{-1}\rangle < 1$, the walk is sufficiently strongly biased that X_n/n converges to a non-zero value which may be interpreted as a <u>mean drift velocity</u>:

$$\lim_{n\to\infty} \frac{X_n}{n} = V_+ \text{ a.e. and } \lim_{\ell\to\infty} \frac{T_\ell}{\ell} = \frac{1}{V_+} \text{ a.e. if } \langle\sigma\rangle < 1 ; \tag{8.3}$$

$$\lim_{n\to\infty} \frac{X_n}{n} = -V_- \text{ a.e. and } \lim_{\ell\to\infty} \frac{T_{-\ell}}{\ell} = \frac{1}{V_-} \text{ a.e. if } \langle\sigma^{-1}\rangle < 1 ; \tag{8.4}$$

where

$$V_+ = \frac{1-\langle\sigma\rangle}{1+\langle\sigma\rangle} \quad\text{and}\quad V_- = \frac{1-\langle\sigma^{-1}\rangle}{1+\langle\sigma^{-1}\rangle} . \tag{8.5}$$

In the remaining case $(\langle\sigma\rangle^{-1} \leqslant 1 \leqslant \langle\sigma^{-1}\rangle)$, the mean drift velocity is zero, even though the walk may still be biased and have the mean value of X_n growing with time:

$$\lim_{n\to\infty} \frac{X_n}{n} = 0 \text{ a.e. and } \lim_{\ell\to\infty} \frac{T_\ell}{\ell} = \lim_{\ell\to\infty} \frac{T_{-\ell}}{\ell} = \infty \quad \text{a.e.} \quad . \tag{8.6}$$

An illustrative example is given by assigning to α the probability density function

$$\phi(\alpha) = (1-p)\delta(\alpha - [1-a]) + p\delta(\alpha - a) , \quad \tfrac{1}{2} < a < 1 . \tag{8.7}$$

If $p > a$ the walker drifts towards $+\infty$ with mean velocity

$$V_+ = (2a-1)(p-a)/\{2a(1-p) + p - a\} ,$$

while if $p < 1 - a$ he drifts towards $-\infty$ with mean velocity

$$-V_- = -(2a-1)(1-a-p)/\{2ap + 1 - a - p\} .$$

The 'exceptional' case corresponding to (8.6) arises when $1 - a \leqslant p \leqslant a$: the mean drift velocity is zero in an <u>interval</u>, at all but one point of which the walk is biased. Such behaviour was observed in simulations over a decade ago by Temkin

[173]. It arises when the microstructure fluctuations are sufficiently severe that (loosely speaking) the walker does not spend on the average equal intervals of time in spatial regions of equal size, but instead tends to linger in some regions, and pass rapidly through others.

Much remains to be determined on the behaviour of X_n/n in the case where the mean drift velocity vanishes, and on the spread of X_n about its mean value, but we note the following major advances. Sinai has obtained some remarkable results [174] under the hypothesis that $\langle \log \sigma \rangle = 0$. It is easily established using Jensen's inequality [20] that his hypothesis implies that $\langle \sigma \rangle > 1$ and $\langle \sigma^{-1} \rangle > 1$, so that his results apply to systems with zero mean drift velocity, characterized by Eq. (8.6). Roughly speaking, Sinai's result is that $X_n/\log^2 n$ approaches a (realization-dependent) limit as $n \to \infty$, so that X_n is $O(\log^2 n)$. In the example (8.7), Sinai's hypotheses are satisfied when $p = \frac{1}{2}$ and $\frac{1}{2} < a < 1$, and his results should be compared with those for Pólya's walk ($p = \frac{1}{2}$, $a = \frac{1}{2}$), for which X_n is $O(n^{1/2})$. What might seem to be quite mild randomness in the environment can have very drastic consequences on the evolution of the random walk.

Kesten et al. [175] have obtained a number of results for systems for which $-\infty < \langle \log \sigma \rangle < 0$, subject to the additional assumptions that (i) $\log \sigma$ has a non-arithmetic distribution [176], a requirement which excludes the example (8.7) from consideration, and (ii) there exists a positive number κ such that $\langle \sigma^\kappa \rangle = 1$ and $\langle (\sigma^\kappa \log \sigma) \rangle < \infty$. They characterize the distributions of the position X_n and the first passage time T_ℓ in terms of stable distributions. Their results imply that when $\kappa < 1$, the position X_n is $O(n^\kappa)$ as $n \to \infty$ and the first passage time T_ℓ to a site $\ell > 0$ bonds remote from the starting site is $O(\ell^{1/\kappa})$.

A more physically motivated discussion of the problem of one-dimensional random walk in a random environment has been given by Derrida and Pomeau [177]. They consider periodic replications of a random lattice of length N, and calculate the ensemble-averaged <u>inverse</u> velocity $\langle V_N^{-1} \rangle_\omega$ in the limit $N \to \infty$. They find that $\langle V_N^{-1}\rangle_\omega = [1 + \langle\sigma\rangle]/[1 - \langle\sigma\rangle]$ if $\langle\sigma\rangle < 1$ and $\langle V_N^{-1}\rangle_\omega = -[1 + \langle\sigma^{-1}\rangle]/[1 - \langle\sigma^{-1}\rangle]$ if $\langle\sigma^{-1}\rangle < 1$. Since they are able to show that fluctuations in V_N are $O(N^{-1/2})$, they can use the relation $\langle V_N^{-1}\rangle_\omega^{-1} \sim \langle V_N\rangle_\omega$ as $N \to \infty$ to confirm Solomon's results. Recently Derrida [178] has extended the periodic replication approach, and obtained some very beautiful results, including an explicit expression for the diffusion or dispersion coefficient

$$D \equiv \lim_{n \to \infty} \frac{1}{2n} \{\overline{X_n^2} - (\overline{X}_n)^2\} \tag{8.8}$$

(with the bar denoting the average over all walks in a fixed realization ω of the environment), on the assumption that $\langle\sigma^2\rangle < 1$:

$$D = -\frac{1}{2}\frac{(1 - \langle\sigma\rangle)^2}{\langle 1/\alpha\rangle^2} + \frac{1 - \langle\sigma\rangle^2}{1 - \langle\sigma^2\rangle}\frac{[\langle 1/\alpha\rangle\langle\sigma/\alpha\rangle + \frac{1}{2}\langle 1/\alpha^2\rangle(1 - \langle\sigma\rangle)]}{\langle 1/\alpha\rangle^3} \tag{8.9}$$

for almost all realizations of the random environment. A consequence of Derrida's analysis is that a random walk in a random environment may possess a well-defined positive mean drift velocity, but have initially localized distributions (pulses) broadening much more rapidly than in uniform systems.

In the preceding discussion we have focussed on one-dimensional discrete time random walks in a static, random environment. For generalizations of these problems to lattices in two or more dimensions, very few rigorous exact results are known. Kalikow [179] has established some results on the transience or persistence of random walks on higher dimensional random lattices, but the proofs of his intuitively reasonable results are quite intricate. Of possibly greater interest is a paper of Anshelevich et al. [180] in which a multi-dimensional random walk in a random environment with a certain symmetry property at each site is examined. On the assumption that the perturbation of the transition properties about their average values is bounded strongly enough to exclude bonds with arbitrarily small transition probabilities, they establish the existence of a well-defined diffusion coefficient D, and present an explicit (though computationally intractable) formula for D. Their hypothesis excludes the extreme case (unfortunately that of most current interest to physicists) in which some of the bonds of a lattice are cut at random, with the random walker then performing a

Pólya walk on the resulting disordered lattice, the statistical properties of which are described by percolation theory. The random motion is that of a drunkard in a maze, or (following de Gennes [181], who proposed the problem as a probe of structure near the percolation threshold) the <u>ant in the labyrinth</u> [148,181-186]. We shall comment further on this problem shortly, after a discussion of one continuous-time analogue of discrete time random walk in a random environment, which has recently been studied quite intensively by physicists.

We consider a <u>randomized master equation</u>,

$$\frac{d}{dt}\,p(\vec{\ell},t) = \sum_{\vec{j}\in\{\vec{\ell}\}} \{W(\vec{j},\vec{\ell})\,p(\vec{j},t) - W(\vec{\ell},\vec{j})\,p(\vec{\ell},t)\} \tag{8.10}$$

where $\{\vec{\ell}\}$ denotes the set of nearest neighbours of site $\vec{\ell}$, and the coefficients $W(\vec{j},\vec{\ell})$ are random variables. The simplest case, and the only one which we discuss in detail here, is that in which the rates are symmetric, i.e. $W(\vec{j},\vec{\ell}) = W(\vec{\ell},\vec{j})$, but otherwise independent and identically distributed, with a probability density function $f(w)$. The randomized master equation thus becomes

$$\frac{d}{dt}\,p(\vec{\ell},t) = \sum_{\vec{j}\in\{\vec{\ell}\}} W(\vec{j},\vec{\ell})\{p(\vec{j},t) - p(\vec{\ell},t)\}\,. \tag{8.11}$$

If we take the Laplace transform of Eq. (8.11), writing

$$\hat{p}(\vec{\ell},u) = \int_0^\infty e^{-tu}\,p(\vec{\ell},t)dt\;, \tag{8.12}$$

we find that

$$u\hat{p}(\vec{\ell},u) - \delta_{\vec{\ell},\vec{0}} = \sum_{\vec{j}\in\{\vec{\ell}\}} W(\vec{j},\vec{\ell})\{\hat{p}(\vec{j},u) - \hat{p}(\vec{\ell},u)\}\,. \tag{8.13}$$

Here we have assumed that the initial condition is

$$p(\vec{\ell},t) = \delta_{\vec{\ell},\vec{0}}\,. \tag{8.14}$$

If we rewrite Eq. (8.13) in the form

$$\sum_{\vec{j}\in\{\vec{\ell}\}} W(\vec{j},\vec{\ell})\{\hat{p}(\vec{j},u) - \hat{p}(\vec{\ell},u)\} + u\{0 - \hat{p}(\vec{\ell},u)\} + \delta_{\vec{\ell},\vec{0}} = 0\;, \tag{8.15}$$

we can interpret it in terms of an electrical circuit. Taking $W(\vec{j},\vec{\ell})$ as the conductance of the bond joining sites $\vec{j}$ and $\vec{\ell}$, u as the conductance between site $\vec{\ell}$ and ground (zero potential) and $\hat{p}(\vec{\ell},u)$ as the potential at site $\vec{\ell}$, we recognize Eq. (8.15) as Kirchhoff's law of current conservation in an electrical network; the Kronecker delta represents a current source of unit strength at $\vec{\ell} = \vec{0}$. The long-time limit $t \to \infty$ corresponds to letting $u \to 0$, and in this case the problem reduces to the random resistor network discussed in section 7.

We noted in section 7 the trivial exact solution of the one-dimensional random resistor problem: a random chain with individual bond conductances g_j (independently and identically distributed) is equivalent to a uniform chain of conductivity $\langle g^{-1}\rangle^{-1}$. One might therefore reasonably anticipate that in the long time limit, Eq. (8.11) is equivalent to a uniform master equation with all rates equal to

$$W_\infty \equiv \langle W^{-1}\rangle^{-1} = \{\int_0^\infty dw\, w^{-1} f(w)\}^{-1} . \tag{8.16}$$

Consequently (cf. Eq. (4.9)), the ensemble-averaged mean-squared displacement should be given as $t \to \infty$ by

$$\langle R^2(t)\rangle_\omega = 2W_\infty t . \tag{8.17}$$

This hand-waving argument tells us nothing about the behaviour of the transport when $W_\infty = 0$, as will happen if slow bonds are relatively abundant, e.g. if

$$f(w) \sim cw^{-\alpha} \quad , \quad 0 < \alpha < 1 \tag{8.18}$$

or even

$$f(w) \sim c > 0 \tag{8.19}$$

as $w \to 0$. However a very elegant exact solution of the one-dimensional problem is available, which confirms the 'handwaving' results when $W_\infty > 0$, and extends to the case $W_\infty = 0$.

The classic account of the symmetric $[W(j,\ell) = W(\ell,j)]$ one-dimensional random master equation (8.9) is a paper of Alexander, Bernasconi, Schneider and Orbach [188], which describes many physical systems for which randomized master equations provide natural models, and which summarizes and extends work on the problem in a series of papers of Alexander, Bernasconi, Beyeler, Schneider, Strässler, Orbach and Wyss [189]. More recent studies of the problem, which fill several gaps in the earlier analysis, have been made by Schneider and Bernasconi [190,191]; see also Anshelevich and Vologodskii [192].

The approach described in Alexander _et al_. [188] is based on obtaining a recursion relation for the admittances

$$G_\ell \equiv W(\ell,\ell+1)[\hat{p}(\ell,u) - \hat{p}(\ell+1,u)]/\hat{p}(\ell,u) \tag{8.20}$$

of semi-infinite chains in the electrical circuit analogue:

$$G_\ell = \{1/W(\ell,\ell+1) + 1/[u + G_{\ell+1}]\}^{-1} \ , \quad \ell \geq 0 \ . \tag{8.21}$$

A formal solution of the problem can therefore be constructed in terms of continued fractions, but what is needed is information upon the statistical distribution of this continued fraction, rather than its numerical value in a particular realization. It is easy to show that the probability density function $\phi_u(g)$ for G_ℓ satisfies the integral equation

$$\phi_u(g) = \int_0^\infty d\gamma\, \phi_u(\gamma) \int_0^\infty dw\, f(w)\, \delta(g - [\frac{1}{w} + \frac{1}{\gamma + u}]^{-1}) \ . \tag{8.22}$$

A closed form solution of (8.22) is not available. However, since it can be shown [188] that

$$\langle\hat{p}(0,u)\rangle_\omega = \int_0^\infty dg \int_0^\infty dg'\, \frac{\phi_u(g)\phi_u(g')}{u + g + g'} \ , \tag{8.23}$$

a knowledge of the asymptotic properties of $\phi_u(g)$ near $g = 0$ suffices to establish the small u behaviour of $\langle\hat{p}(0,u)\rangle_\omega$, and hence (via Tauberian theorems) to determine exactly the asymptotic behaviour of the ensemble averaged probability $\langle p(0,t)\rangle_\omega$ of occupying the starting site at time t.

For an ordered one-dimensional system we know from Eq. (4.10) that $p(0,t) \propto t^{-1/2}$. Alexander _et al_. [188,189] have shown that so long as $W_\infty = \langle W^{-1}\rangle^{-1}$ (defined by Eq. (8.16)) is finite,

$$\langle p(0,t)\rangle_\omega \sim (4\pi W_\infty t)^{-1/2} \quad \text{as} \quad t \to \infty \ , \tag{8.24}$$

so that the ensemble-averaged system behaves exactly like a uniform system, provided that the rate disorder is sufficiently mild. However, in the case (8.18), $\langle p(0,t)\rangle_\omega$ decays more slowly:

$$\langle p(0,t)\rangle_\omega \propto t^{-(1-\alpha)/(2-\alpha)} . \tag{8.25}$$

The ensemble averaged mean-squared displacement $\langle R^2(t)\rangle_\omega$ can also be determined, and it is found that when $W_\infty > 0$,

$$\langle R^2(t)\rangle_\omega \sim 2W_\infty t ; \tag{8.26}$$

however if the transition rate distribution $f(w)$ has the asymptotic behaviour (8.17) or (8.18) near $w = 0$ one finds instead that as $t \to \infty$,

$$\langle R^2(t)\rangle_\omega \propto t^{2(1-\alpha)/(2-\alpha)} \tag{8.27}$$

and

$$\langle R^2(t)\rangle_\omega \propto t/\ell n\, t \tag{8.28}$$

respectively. Motion is thus <u>subdiffusive</u> in the sense defined in section 4 if the rate disorder is "sufficiently bad". The occurrence of subdiffusive behaviour suggests that the ensemble-averaged properties of the system might be described by a continuous-time random walk process having a waiting-time density $\psi(t)$ with infinite mean, or equivalently a generalized master equation with an appropriately chosen memory kernel (cf. section 4).

Several algorithms exist for determining a "best-matched" memory kernel. Alexander, Bernasconi <u>et al</u>. [188,189g] determine an approximate Laplace transformed memory kernel $\hat{W}_0(u)$ by writing as a trial solution of the exact integral equation (8.22)

$$\phi_u(g) = \delta(g - g_{eff}(u)) , \tag{8.29}$$

where $\hat{W}_0(u) = g_{eff}(u)[g_{eff}(w + u)]/u$, and determining $g_{eff}(u)$ in a self-consistent fashion. They find using this approach the approximations

$$p_{eff}(0,t) \sim (4\pi W_\infty t)^{-1/2} \tag{8.30}$$

when $W_\infty > 0$, and

$$p_{eff}(0,t) \propto t^{-(1-\alpha)/(2-\alpha)} \tag{8.31}$$

if $f(w)$ has the behaviour (8.18) near $w = 0$. For well-behaved distributions, i.e. $W_\infty > 0$, both the power of t and the prefactor in the asymptotic form of the probability of occupancy of the origin are reproduced correctly, while for the ill-behaved case (8.18), the power of t is correct, but not the prefactor.

An alternative approach to the estimation of an appropriate memory kernel is the effective medium approximation of Kirkpatrick [147] discussed briefly in section 7, suitably modified to allow for the connections of all sites to ground. An advantage of this approach is that it is not restricted to one-dimensional lattices and is therefore applicable to the problem of the ant in the labyrinth defined earlier. We shall sketch the approach here. It was proposed almost simultaneously by Bryksin [193], Odagaki and Lax [194], Summerfield [195] and Webman [196], and has been further addressed by Haus et al. [197,198], Movaghar et al. [199], Webman and Klafter [200], and Sahimi et al. [201].

The simplest way to implement the lowest-order Kirkpatrick-type approximation is as follows. We compare the disordered system (8.11) with a uniform or 'ordered' system described by a generalized master equation with memory kernel $W_0(t)$:

$$\frac{\partial}{\partial t} p_0(\vec{\ell},t) = \int_0^t \sum_{\vec{j}\varepsilon\{\vec{\ell}\}} W_0(t-\tau)[p_0(\vec{j},\tau) - p_0(\vec{\ell},\tau)]d\tau . \tag{8.32}$$

Taking Laplace transforms, we find that

$$u\hat{p}_0(\vec{\ell},u) - \delta_{\vec{\ell},\vec{0}} = \hat{W}_0(u) \sum_{\vec{j}\varepsilon\{\vec{\ell}\}} [\hat{p}_0(\vec{j},u) - \hat{p}_0(\vec{\ell},u)] . \tag{8.33}$$

If we subtract Eq. (8.33) from Eq. (8.13) we eliminate the initial condition and obtain the equation

$$\begin{aligned} u\{\hat{p}(\vec{\ell},u) - \hat{p}_0(\vec{\ell},u)\} = & \sum_{\vec{j}\varepsilon\{\vec{\ell}\}} W(\vec{j},\vec{\ell})\{\hat{p}(\vec{j},u) - \hat{p}(\vec{\ell},u)\} \\ & - \sum_{\vec{j}\varepsilon\{\vec{\ell}\}} \hat{W}_0(u)\{\hat{p}_0(\vec{j},u) - \hat{p}_0(\vec{\ell},u)\} . \end{aligned} \tag{8.34}$$

A little rearrangement enables us to exhibit explicitly the effects of the fluctuation of $W(\vec{j},\vec{\ell})$ about the uniform system:

$$(z+\varepsilon)\{\hat{p}(\vec{\ell},u)-\hat{p}_0(\vec{\ell},u)\}-\sum_{\vec{j}\varepsilon\{\vec{\ell}\}}[\hat{p}(\vec{j},u)-\hat{p}_0(\vec{j},u)]$$

$$=\sum_{\vec{j}\varepsilon\{\vec{\ell}\}}\Delta(j,\ell;u)[\hat{p}(j,u)-\hat{p}(\ell,u)]\ , \qquad (8.35)$$

where we have written

$$\varepsilon = u/\hat{W}_0(u) \quad \text{and} \quad \Delta(\vec{j},\vec{\ell};u)=[W(\vec{j},\vec{\ell})-\hat{W}_0(u)]/\hat{W}_0(u)\ , \qquad (8.36)$$

and assumed that the lattice is periodic, with all sites having the same coordination number z.

Effective medium approximations are now constructed by retaining only a finite number of terms on the right hand side of Eq. (8.35), in the hope that the solution at site $\vec{\ell}$, while influenced strongly by fluctuations in transition rates of nearby bonds, is less sensitive to the rates associated with more distant bonds. The general analysis may be formalized in terms of 'bond-bond Green functions' [202], but we shall discuss here only the simplest approximation, called the single-bond effective medium approximation. It is assumed that only a single bond (joining sites $\vec{a}$ and $\vec{b}$ say) has transition rate or conductivity W differing from that of the uniform system $(\hat{W}_0(u))$, so that

$$(z+\varepsilon)\,[\hat{p}(\vec{\ell},u)-\hat{p}_0(\vec{\ell},u)]-\sum_{\vec{j}\varepsilon\{\vec{\ell}\}}[\hat{p}(\vec{j},u)-\hat{p}_0(\vec{j},u)]$$

$$=\Delta\delta_{\vec{\ell},\vec{a}}\,[\hat{p}(\vec{b},u)-\hat{p}(\vec{a},u)]+\Delta\delta_{\vec{\ell},\vec{b}}[\hat{p}(a,u)-\hat{p}(b,u)]\ , \qquad (8.37)$$

where we have suppressed the u-dependence of $\Delta \equiv [W-\hat{W}_0(u)]/\hat{W}_0(u)$. If we now introduce a Green function $G_{\vec{\ell}\vec{m}}(\varepsilon)$, the unique solution of the equation

$$(z+\varepsilon)\;G_{\vec{\ell}\vec{m}}(\varepsilon)-\sum_{\vec{j}\varepsilon\{\vec{\ell}\}}G_{\vec{j}\vec{m}}(\varepsilon)=-\delta_{\vec{\ell},\vec{m}} \qquad (8.38)$$

which decays as the distance between ℓ and m grows, we readily extract the formal 'solution'

$$\hat{p}(\vec{\ell},u) - \hat{p}_0(\vec{\ell},u) = \Delta[G_{\vec{\ell}\vec{a}} - G_{\vec{\ell}\vec{b}}][\hat{p}(\vec{a},u) - \hat{p}(\vec{b},u)] . \tag{8.39}$$

By setting $\vec{\ell} = \vec{a}$ and $\vec{\ell} = \vec{b}$ in turn in Eq. (8.39), we deduce that

$$\hat{p}(\vec{a},u) - \hat{p}(\vec{b},u) = [1 - \Delta\gamma(\varepsilon)]^{-1}[\hat{p}_0(\vec{a},u) - \hat{p}_0(\vec{b},u)] , \tag{8.40}$$

where

$$\gamma(\varepsilon) = \{G_{\vec{a}\vec{a}} + G_{\vec{b}\vec{b}} - G_{\vec{a}\vec{b}} - G_{\vec{b}\vec{a}}\} . \tag{8.41}$$

The object of the exercise all along is of course the determination of $\hat{W}_0(u)$ and Eq. (8.40) enables us to make a 'self-consistent' determination of it. We have one random variable, the rate W associated with the bond joining sites $\vec{a}$ and $\vec{b}$, and we demand that the average of $\hat{p}(\vec{a},u) - \hat{p}(\vec{b},u)$ with respect to W should be $\hat{p}_0(\vec{a},u) - \hat{p}_0(\vec{b},u)$.

This gives an equation for $\hat{W}_0(u)$:

$$\int_0^\infty \frac{dw\ f(w)}{1 - \gamma(u/\hat{W}_0(u))[w - \hat{W}_0(u)]/\hat{W}_0(u)} = 1 . \tag{8.42}$$

In electrical circuit terminology, the self-consistency condition used here is the requirement that the average fluctuation in voltage induced by W differing from $\hat{W}_0(u)$ should be zero. For the present single-bond level of approximation, exactly the same self-consistency equation is obtained if we insist instead that the average fluctuation in current be zero. (In higher order approximations, this desirable coincidence is lost.)

Although Eq. (8.42) is not amenable to exact solution, it is possible to extract the small u behaviour of $\hat{W}_0(u)$ (see e.g. [201]). The Green function defined by Eq. (8.38) is of course closely related to fundamental solutions of the Pólya random walk problem of Section 2, and it can be shown [201] that for periodic lattices,

$$\gamma(\varepsilon) = -2/z + (2\varepsilon/z)\ G(\varepsilon) \tag{8.43}$$

where

$$G(\varepsilon) = \varepsilon^{-1/2}(4+\varepsilon)^{-1/2} \quad \text{in one dimension} \tag{8.44a}$$

and, as $\varepsilon \to 0$,

$$G(\varepsilon) = \begin{cases} A \log (1/\varepsilon) + O(1) & \text{in two dimensions} \\ B + O(\varepsilon^{1/2}) & \text{in three dimensions} \end{cases} \tag{8.44b}$$

with A and B lattice-dependent constants. In the limit $u \to 0$, Eq. (8.42) reduces to Kirkpatrick's effective medium approximation for a random resistor network. (An advantage of the time-dependent problem is that so long as $\varepsilon > 0$ the convergence of infinite series arising in general Green function manipulations is much better than in the case $\varepsilon = 0$, and subtle problems of conditional convergence and the need to prescribe boundary conditions are largely circumvented.)

In two or higher dimensions, the effective medium approximation does not in general yield exact results, but is numerically accurate except near the percolation threshold, and it does have the advantage that it predicts the existence of a percolation threshold for the transition rate distribution

$$f(w) = (1-p)\delta_+(w) + p\delta(w-\Omega) \tag{8.45}$$

characterizing the problem of the ant in the labyrinth:

$$\hat{W}_0(0) = \begin{cases} (p-p^*)\Omega/(1-p^*) & p > p^* \\ 0 & p < p^* \end{cases} \tag{8.46}$$

with $p^* = 2/z$. If a random walker ('the ant') is introduced at ('parachuted onto') a randomly chosen site, his diffusion constant, defined up to a geometrical constant by

$$D = \lim_{t\to\infty} t^{-1} \langle R^2(t)\rangle_\omega \propto \hat{W}_0(0), \tag{8.47}$$

is predicted by the effective medium approximation to (i) vanish linearly with the deviation from the percolation threshold, above the percolation-threshold and (ii) be identically zero below the percolation threshold. Right at the percolation threshold, it is predicted [201] that

$$\langle R^2(t)\rangle_\omega \sim \begin{cases} t^{1/2} & \text{in two dimensions} \\ t^{1/2}(\ell n\, t)^{-1/2} & \text{in three dimensions}\,. \end{cases} \tag{8.48}$$

This has the appearance of subdiffusive motion (cf. Section 4), but of course since the ant is certain to arrive on a finite connected cluster, in any particular realization $R^2(t)$ approaches a constant. The growth with t of $\langle R^2(t)\rangle_\omega$ reflects the divergence of the mean cluster size at the percolation threshold. Given any time t, in a sufficiently large number of realizations at least one ant will have failed to probe the boundary of his prison! (Alternative analyses of the problem at $p = p_c$ have been given [186]; see also Redner's first article in this volume.)

We comment briefly here on the problem of the operational definition of the diffusion coefficient for ant-in-labryinth problems. For rate distributions of the form (8.45), the probability that the ant is parachuted onto a cluster of infinite size (i.e. not imprisoned) is $P(p)$, the percolation probability defined in section 6, vanishing as $(p - p_c)^\beta$ as the percolation threshold p_c is approached from above. Thus the diffusion coefficient for free labyrinthine ants is $D^* = D/P(p)$. If as $p \to p_c^+$, D vanishes as $(p - p_c)^t$, with t the conductivity exponent (see section 7: the effective medium approximation predicts that $t = 1$), we see that $D^* \propto (p - p_c)^{t-\beta}$. The quantity D^* is the diffusion constant appropriate to the case in which the ants enter a large piece of labyrinth from one of its boundaries. The distinction between the quantities D and D^*, both of which are experimentally relevant, has been clearly drawn by Straley [184].

In one dimension, the Kirkpatrick style EMA reproduces a number of exact results, as does the somewhat different effective-medium type approximation of Alexander _et al_. [188]. The reasons for this success have been examined by Webman and Klafter [200]. A number of other approaches to the one-dimensional problem have been given including a 'replica trick' calculation of Stephen and Kariotis [203] and a renormalization calculation of Machta [204]; see also Igarashi [205] and Zwanzig [206]. Haus _et al_. [197] have considered higher order effective medium approximations which enable the properties of the one-dimensional problem to be predicted throughout the time domain.

We have insufficient space here to discuss the variety of other models for random motion in random lattices which have been considered recently, several (but not all) of which have been successfully analysed using effective medium techniques. For randomized master equation systems in which bonds are directed, or rates asymmetric, quite subtle effects can arise [178,191,207-209], including resonant or oscillatory motion [208]. Systems with absorption or loss of particles can also be analysed [199,210]; a term $-\nu(\ell)p(\ell,t)$, with $\nu(\ell)$ either random of fixed, is added to the right hand side of Eq. (8.10). Many of these problems are tractable in one dimension, or on a Bethe lattice [184,211].

In Part A of this introductory article we have passed from discrete models of Markovian (memoryless) random processes, where the techniques of classical analysis render many problems tractable in principle (though nontrivial in practice), to models of correlated random conformation (section 5) or random topology (section 6) which are of extraordinary difficulty. It might be anticipated that the marriage of these models, as discussed in this concluding section of part A, might compound the difficulties of each, and ensure very slow progress. While this is largely the case, some real achievements have been made, and some simple concepts have emerged. One, which arises naturally from effective medium treatments, is the replacement of microscopic disorder and absence of memory with macroscopic spatial uniformity, but temporal memory. This is not necessarily always possible [188], but is in accord with a programme instituted a decade ago by Scher and Lax [68], who proposed that transport in local disordered amorphous semiconductors be modelled by continuous-time random walks with appropriately chosen waiting time distributions. Klafter and Silbey [212] have shown by a formal projection operator technique that for a class of randomized master equations, the generalized master equation (equivalent to a continuous-time random walk - see section 4) <u>is</u> the correct description of the ensemble-averaged system (although the memory kernel cannot be computed in practice, and must consequently be modelled with finesse). There is a grandeur in this view of nature, with its concurrent disorder in process and structure, being described in terms of such simple concepts as random walk and percolation ; and that through the subtle interplay of random process and random system, with such simple models phenomena both beautiful and important have been, and are being understood [213].

References for Part A

1. K. Pearson, "The problem of the random walk", Nature 72 (1905) 294. In the less abstract half of the random walk literature it has become something of a tradition to commence articles with a direct quote of all or part of this letter, or at least to reproduce it in a footnote. Physicists may be amused to note that on the same page as Pearson's letter, J.H. Jeans argues at some length that Planck's constant h cannot possibly have a non-zero value!

2. Lord Rayleigh, "On the resultant of a large number of vibrations of the same pitch and of arbitrary phase", Phil. Mag. 10 (1880) 73-78, reprinted in Scientific Papers Vol 1, pp. 491-496 (Dover, New York, 1964). Rayleigh's derivation is elementary, but insightful: he analyses first vibrations with random phases 0 or π, then vibrations with phases 0, $\pi/2$, π, $3\pi/2$, and then establishes that his conclusions hold more generally. Later ["On James Bernoulli's theorem in probabilities" Phil. Mag. 47 (1889) 246-251, reprinted in Scientific Papers Vol 4, pp. 370-375 (Dover, New York, 1964), and The Theory of Sound, Volume 1, 2nd edition, pp. 35-42 (New York, Dover, 1945)] he obtained the same results by approximating a difference equation describing the process by the diffusion equation.

3. Lord Rayleigh, "The problem of the random walk", Nature 72 (1905) 318, reprinted in Scientific Papers, Vol 5, p. 256.

4. K. Pearson, "The problem of the random walk", Nature 72 (1905) 342. Pearson concludes that: "the lesson of Lord Rayleigh's solution is that in open country the most probable place to find a drunken man who is at all capable of keeping on his feet is somewhere near his starting point!"

5. J.C. Kluyver, "A local probability problem", Proceedings of the Section of Sciences, Koninklijke Akademie van Wetenschappen te Amsterdam 8 (1906) 341-350 (Amsterdam, Johannes Müller, 1906: cover to cover translation of the original Dutch).

6. A.A. Markoff, Wahrscheinlichkeitsrechnung pp. 67-76 and pp. 173-177 (Leipzig, B.G. Teubner, 1912: translation from the Russian). Chandrasekhar [17] credits Markoff for the two-dimensional transform solution of Pearson's problem, but the reader will not readily translate Markoff's analysis into the modern approach. Markoff addresses the abstract question of addition of two-dimensional random vectors, with no mention of random walk concepts (and no citation of Pearson, Rayleigh or Kluyver). The problem of addition of random vectors had apparently been considered earlier in papers of Ch.M. Schols cited by Markoff: "Théorie des erreurs dans le plan et dans l'espace", Ann. de l'École polyt. de Delft 2 (1886) and "Démonstration directe de la loi limite pour les erreurs dans le plan et dans l'espace", ibid. 3 (1887).

7. Lord Rayleigh, "On the problem of random vibrations, and of random flights in one, two, or three dimensions", Phil. Mag. 37 (1919) 321-347; reprinted in Scientific Papers, Vol. 6, p. 604.

8. M.N. Barber and B.W. Ninham, Random and Restricted Walks: Theory and Applications (New York, Gordon and Breach, 1970).

9. E.W. Montroll and B.J. West, "On an enriched collection of stochastic processes", in Fluctuation Phenomena (ed. E.W. Montroll and J.L. Lebowitz) pp. 61-175 (Amsterdam, North-Holland, 1979).

10. G.H. Weiss and R.J. Rubin, "Random walks: theory and selected applications", Adv. Chem. Phys. 52 (1983) 363-505.

11. M. von Smoluchowski, "Drei Vorträge über Diffusion, Brownsche Molekularbewegung und Koagulation von Kolloidteilchen, Phys. Zeit. 17 (1916) 557-585; see the footnote to p. 558.

12. G. Pólya, "Über eine Aufgabe der Wahrscheinlichkeitsrechnung betreffend die Irrfahrt im Straßennetz", Math. Ann. 83 (1921) 149-160.

13. G. Pólya, "Sur la promenade au hasard dans un réseau de rues", Colloque consacré à la Théorie des Probabilités (organized by R. Wavre and M. Fréchet) pp. 25-44 (Paris, Hermann et Cie, 1938).

14. E. Lukacs, Characteristic Functions, 2nd edition (London, Griffin, 1970).

15. G.N. Watson, A Treatise on the Theory of Bessel Functions, 2nd edition pp. 419-421 (Cambridge University Press, 1944).

16. S. Bochner and K. Chandrasekharan, Fourier Transforms, Chapter 2 (Princeton University Press, 1949).

17. S. Chandrasekhar, "Stochastic problems in physics and astronomy", Rev. Mod. Phys. 15 (1943) 1-89; reprinted in N. Wax, ed., Selected Papers on Noise and Stochastic Processes (New York, Dover, 1954).

18. S.A. Vincenz and J. McG. Bruckshaw, "Note on the probability distribution of a small number of vectors", Proc. Camb. Phil. Soc. 56 (1960) 21-26.

19. L.R.G. Treloar, "The statistical length of long-chain molecules", Trans. Faraday Soc. 42 (1946) 77-82.

20. W. Feller, An Introduction to Probability Theory and its Applications, Vol. 2, 2nd edition (New York, Wiley, 1971).

21. S. Dvořák, "Treloar's distribution and its numerical implementation", J. Phys. A 5 (1972) 78-84.

22. R. Barakat, "Isotropic random flights", J. Phys. A. 6 (1973) 796-804.

23. R. Barakat, "Isotropic random flights: random numbers of flights", J. Phys. A 15 (1982) 3073-3082.

24. R.J. Nossal and G.H. Weiss, "A generalized Pearson random walk allowing for bias", J. Stat. Phys. 10 (1974) 245-253.

25. B.D. Hughes, E.W. Montroll and M.F. Shlesinger, "Fractal random walks", J. Stat. Phys. 28, (1982) 111-126.

26. F. Spitzer, Principles of Random Walk, 2nd edition (New York, Springer-Verlag, 1976).

27. E.W. Montroll, "Random walks in multidimensional spaces, especially on periodic lattices", J. Soc. Indust. Appl. Math. 4 (1956) 241-260.

28. E.W. Montroll, "Random walks on lattices", Proc. Symp. Appl. Math. 16 (1964) 193-220.

29. E.W. Montroll and G.H. Weiss, "Random walks on lattices. II", J. Math. Phys. 6 (1965) 167-181.

30. E.W. Montroll, "Random walks on lattices. III. Calculation of first-passage times with application to exciton trapping on photosynthetic units", J. Math. Phys. 10 (1969) 753-765.

31. S. Ishioka and M. Koiwa, "Random walks on diamond and hexagonal close packed lattices", Phil. Mag. A 37 (1978) 517-533.

32. M. Koiwa and S. Ishioka, "Random walks on three-dimensional lattices with the coordination number four", Phil. Mag. A. 40 (1979) 625-635.

33. F.S. Henyey and V. Seshadri, "On the number of distinct sites visited in 2D lattices", J. Chem. Phys. 76 (1982) 5530-5534.

34. B.D. Hughes and M. Sahimi, "Random walks on the Bethe lattice", J. Stat. Phys. 29 (1982) 787-794. See also P.W. Kasteleyn, "Some aspects of random walks on groups", to be published in Proceedings of the XI International Colloquium on Group Theoretical Methods in Physics (Istanbul, 1982), and S. Sawyer, "Isotropic random walks in a tree", Z. Wahrscheinlichkeitstheorie verw. Gebiete 42 (1978) 279-292.

35. B.D. Hughes, M. Sahimi and H.T. Davis, "Random walks on pseudolattices", Physica A, in press.

36. G.S. Joyce, "Exact results for a body-centered cubic lattice Green's function with applications in lattice statistics. I", J. Math Phys. 12 (1971) 1390-1414.

37. G.S. Joyce, "On the simple cubic lattice Green function", Phil Trans. R. Soc. Lond. A 273 (1973) 583-610.

38. B.D. Hughes and M.F. Shlesinger, "Lattice dynamics, random walks and non-integral effective dimensionality", J. Math. Phys. 23 (1982) 1688-1692.

39. B.D. Hughes, M.F. Shlesinger and E.W. Montroll, "Random walks with self-similar clusters", Proc. Natl. Acad. Sci. USA 78 (1981) 3287-3291.

40. G.N. Watson, "Three triple integrals", Quart. J. Math (Oxford) 10 (1939) 266-276; see also [8]. Actually Watson states that the value of IBCC I_{BCC} "appears to be fairly well known"; the reduction of this integral to known functions is very much simpler than that for I_{FCC} or I_{SC}.

41. M.L. Glasser and I.J. Zucker, "Extended Watson integrals for the cubic lattices", Proc. Natl. Acad. Sci. USA 74 (1977) 1800-1801, and "Lattice sums", Theoretical Chemistry: Advances and Perspectives Vol. 5 pp. 67-139 (New York, Academic Press 1980). In an unpublished erratum, Glasser and Zucker correct their published formula to that stated in our Eq. (2.23).

42. K. Lindenberg, V. Seshadri, K.E. Shuler and G.H. Weiss, "Lattice random walks for sets of random walkers. First passage times". J. Stat. Phys. 23 (1980) 11-25.

43. J.E. Gillis and G.H. Weiss, "Expected number of distinct sites visited by a random walk with an infinite variance", J. Math. Phys. 11 (1970) 1307-1312.

44. W. Th. F. den Hollander and P.W. Kasteleyn, "Random walks with 'spontaneous emission' on lattices with periodically distributed imperfect traps", Physica 112A (1982) 523-543, and "Random walks on lattices with points of two colours. I". Physica 117A (1983) 179-188.

45. M.D. Hatlee and J.J. Kozak, "Random walks on finite lattices with traps", Phys. Rev. B 21 (1980) 1400-1407, and "Random walks on finite lattices with traps. II. The case of a partially absorbing trap", ibid. 23 (1981) 1713-1718; C.A. Walsh and J.J. Kozak, "Exact algorithm for d-dimensional walks on finite and infinite lattices with traps", Phys. Rev. Lett. 47 (1981) 1500-1502.

46. M.F. Shlesinger and B.D. Hughes, "Analogs of renormalization group transformations in random processes", Physica 109A (1981) 597-608.

47. R. Rammal and G. Toulouse, "Random walks on fractal structures and percolation clusters", J. Physique Letts. 44 (1983) L13-L22.

48. U. Landman, E.W. Montroll and M.F. Shlesinger, "Random walks and master equations with internal degrees of freedom", Proc. Natl. Acad. Sci. USA 74 (1977) 430-433; U. Landman and M.F. Shlesinger, "Stochastic theory of multistate diffusion in perfect and defective lattice systems: I. Mathematical formalism and II. Case Studies", Phys. Rev. B. 19 6207-6219, 6220-6237.

49. J.M. Hill, "A discrete random walk model for diffusion in media with double diffusivity", J. Austral. Math. Soc., Series B 22 (1980) 58-74.

50. P.T. Cummings and G. Stell, "Random flights in Euclidean space. I. Generalized analysis and results for flights with prescribed hit expectance density about the origin", preprint, State University of New York at Stony Brook (1983).

51. G.S. Joyce, "Critical properties of the spherical model", in Phase Transitions and Critical Phenomena, Vol. 2, ed. C. Domb and M.S. Green, pp. 375-442 (London, Academic Press, 1972); for some other connections between random walks and critical phenomena, see H. Silver, N.E. Frankel and B.W. Ninham, "A class of mean field models", J. Math. Phys. 13 (1972) 468-474.

52. E.W. Montroll, "On the dynamics of the Ising model of cooperative phenomena", Proc. Natl. Acad. Sci. USA 78 (1981) 36-40; E.W. Montroll and H. Reiss, "Phase transition versus disorder: A criterion derived from a two-dimensional dynamic ferromagnetic model", ibid 2659-2663.

53. J. Gillis, "Centrally biased discrete random walk", Quart. J. Math. (Oxford) 7 (1956) 144-152. See also [34].

54. V. Seshadri and B.J. West, "Fractal dimensionality of Lévy processes", Proc. Natl. Acad. Sci. USA 79 (1982) 4501-4505.

55. P. Lévy, Théorie de l'addition des variables aléatoires (Paris, Gauthier-Villars, 1937).

56. B.V. Gnedenko and A.N. Kolmogorov, Limit distributions for sums of independent random variables, revised edition (Reading, Massachusetts, Addison-Wesley, 1968).

57. B.B. Mandelbrot, Fractals: Form, Chance and Dimension (San Francisco, Freeman, 1977).

58. B.B. Mandelbrot, The Fractal Geometry of Nature (San Francisco, Freeman, 1982).

59. R.G. Laha and V.K. Rohatgi, Probability Theory (New York, Wiley, 1979).

60. I. Oppenheim, K.E. Shuler and G.H. Weiss, Stochastic processes in chemical physics: The master equation (Cambridge, Massachusetts, M.I.T. Press, 1977). A number of classic articles on master equations are reprinted in this volume.

61. R.W. Zwanzig, "Statistical mechanics of irreversibility", Lectures in Theoretical Physics (Boulder) Vol. 3, pp. 106-141 (New York, Interscience, 1961). This article, and some others on the same topic are included in [60].

62. V.M. Kenkre, "The master equation approach: coherence, energy, energy transfer, annihilation and relaxation", in V.M. Kenkre and P. Reineker, Exciton dynamics in molecular crystals and aggregates, Springer Tracts in Modern Physics No. 94, pp. 1-109 (Berlin, Springer, 1982).

63. P. Lévy, "Processus Semi-Markoviens", Proceedings of the International Congress of Mathematicians, Amsterdam 1954, Vol. 3, pp. 416-426 (Amsterdam, North-Holland, 1956).

64. W. Feller, "On Semi-Markov processes", Proc. Natl. Acad. Sci. USA 51 (1964) 653-659.

65. W.L. Smith, "Regenerative stochastic processes", Proc. Roy. Soc. Lond A 232 (1955) 6-31.

66. R. Pyke, "Markov renewal processes: definitions and preliminary properties", Ann. Math. Stat. 32 (1961) 1231-1242.

67. B.W. Connolly, "On randomized random walks", SIAM Rev. 13 (1971) 81-99; in this paper Connolly allows steps to the left and right to have different time constants α in Eq. (4.14).

68. H. Scher and M. Lax, "Stochastic transport in a disordered solid. I. Theory", Phys. Rev. B 7 (1973) 4491-4502.

69. M.F. Shlesinger, J. Klafter and Y.M Wong, "Random walks with infinite spatial and temporal moments", J. Stat. Phys. 27 (1982) 499-512.

70. V.M. Kenkre, E.W. Montroll and M.F. Shlesinger, "Generalized master equations for continuous-time random walks", J. Stat. Phys. 9 (1973) 45-50.

71. D. Bedeaux, K. Lakatos-Lindenberg and K.E. Shuler, "On the relation between master equations and random walks and their solutions", J. Math. Phys. 12 (1971) 2116-2123.

72. S-K. Ma, Modern Theory of Critical Phenomena, pp. 410-414 (Reading, Massachusetts, W.A. Benjamin, 1976).

73. D.J. Amit, G. Parisi and L. Peliti, "Asymptotic behavior of the 'true' self-avoiding walk", Phys. Rev. B. 27 (1983) 1635-1645. They consider the case in which the probability of stepping to a nearest-neighbour site j of the current site i is $p(j,i) = \exp(-gn_j)/\sum_\ell \exp(-gn_\ell)$, the sum being taken over all nearest-neighbours of site i, $g > 0$ and n_ℓ being the number of previous visits to site ℓ . A renormalization treatment of this problem has been given by S.P. Obukhov and L. Peliti, "Renormalization of the 'true' self-avoiding walk", J. Phys. A. 16 (1983) L147-L151; see also L. Pietronero, "Critical dimensionality and exponent of the 'true' self-avoiding walk", Phys. Rev. B. 27 (1983) 5887-5889. It appears that the analogue of the exponent ν defined by (5.9) is independent of the dimensionality E for $E > 2$.

74. H. Yamakawa, Modern Theory of Polymer Solutions (New York, Harper and Row, 1971). See also P.G. de Gennes, "Some conformation problems for long macromolecules", Rep. Prog. Phys. 32 (1969) 187-205; S.G. Whittington, "The excluded volume effect in polymers", J. Phys. A. 3 (1970) 28-32; K.F. Freed, "Functional integrals and polymer statistics", Advances in Chemical Physics 22 (1972) 1-128.

75. P.G. de Gennes, Scaling Concepts in Polymer Physics (Ithaca, New York, Cornell University Press, 1979).

76. C. Domb, "Self-avoiding walks on lattices", Advances in Chemical Physics 15 (1969) 229-259. This review emphasises series expansion analyses of the large n forms of c_n, u_n and $\langle R_n^2 \rangle$.

77. C. Domb, "From random to self-avoiding walks", J. Stat. Phys. 30 (1983) 425-436. This article covers some developments since the appearance of [76]. (Domb also discusses a model which interpolates between ordinary Pólya and self-avoiding walks due to C. Domb and G.S. Joyce, "Cluster expansion for a polymer chain", J. Phys. C. 5 (1972) 956-976. See also F. Family and H. Gould, "Polymer chain statistics and universality; crossover from random to self-avoiding walks", J. Phys. A. (1983) in press.)

78. D.S. McKenzie, "Polymers and scaling", Physics Reports 27 (1976) 35-88.

79. S.G. Whittingon, "Statistical mechanics of polymer solution and polymer adsorption", Advances in Chemical Physics 51 (1982) 1-48.

80. J.M. Hammersley, G.M. Torrie and S.G. Whittington, "Self-avoiding walks interacting with a surface", J. Phys. A 15 (1982) 539-571.

81. B. Vincent and S.G. Whittington, "Polymers at interfaces and in disperse systems", in Surface and Colloid Science, ed. E. Matijevic, Vol. 12, pp. 1-117 (New York, Plenum, 1982).

82. M.F. Shlesinger, "Weierstrassian Lévy flights and self-avoiding random walks", J. Chem. Phys. 78 (1983) 416-419; S. Havlin and D. Ben-Avraham, "Fractal dimensionality of polymer chains", J. Phys. A 15 (1982) L311-L316; "New method of analysing self-avoiding walks in four dimensions", ibid. L317-L320; "New approach to self-avoiding walks as a critical phenomenon", ibid. L321-L328; "Theoretical and numerical study of fractal dimensionality in self-avoiding walks", Phys. Rev. A 26 (1982), 1728-1734.

83. J.M. Hammersley and K.W. Morton, "Poor Man's Monte Carlo", J. Roy. Stat. Soc. B 16 (1954), 23-38.

84. S.R. Broadbent and J.M. Hammersley, "Percolation processes. I. Crystals and mazes", Proc. Camb. Phil. Soc. 53 (1954) 629-641.

85. J.M. Hammersley, "Percolation processes. II. The connective constant", Proc. Camb. Phil. Soc. 53 (1957) 642-645.

86. J.M. Hammersley, "The number of polygons on a lattice", Proc. Camb. Phil. Soc. 57 (1961) 516-523.

87. J.M. Hammersley and D.J.A. Welsh, "Further results on the rate of convergence to the connective constant of the hypercubical lattice", Quart. J. Math. Oxford 13 (1962) 108-110. These authors establish that for hypercubic lattices of dimension E, $\kappa n < \log c_n < \kappa n + \gamma\sqrt{n} + \log E$, with γ independent of E.

88. H. Kesten, "On the number of self-avoiding walks", J. Math. Phys. 4 (1963) 960-969; "On the number of self-avoiding walks. II", ibid. 5 (1964) 1128-1137.

89. M.E. Fisher, "The shape of a self-avoiding walk or polymer chain", J. Chem. Phys. 44 (1966) 616-622.

90. J.M. Hammersley, "Long-chain polymers and self-avoiding random walks. I and II", Sankhya, Series A 25 (1963) 29-38, 269-272. It may be noted that $z - 1$ is (trivially) the connective constant of a Bethe lattice with coordination number z. Since the coordination number of the hypercubic lattice is $2E$, Eq. (5.11) shows that as $E \to \infty$, μ approaches the Bethe lattice result.

91. M.G. Watts, "Application of the method of Padé approximants to the excluded volume problem", J. Phys. A 8 (1975) 61-66.

92. P.G. de Gennes, "Exponents for the excluded volume problem as derived by the Wilson method", Phys. Lett. 38A (1972) 339-340.

93. R.G. Bowers and A. McKerrell, "An exact relation between the classical n-vector model ferromagnet and the self-avoiding walk problem", J. Phys. C. 6 (1973) 2721-2732.

94. J. des Cloizeaux, "Lagrangian theory for a self-avoiding random chain", Phys. Rev. A. 10 (1974) 1665-1669; also "The Lagrangian theory of polymer solutions at intermediate concentrations", J. Physique 36 (1975) 281-291, and "Théories Lagrangiennes à zéro composante. Application à l'étude des polymères en solution (chaînes avec volume exclu) et des propriétés d'électrons soumis à un potentiel aléatoire" J. Physique 37, suppl. Cl (1976) 255-269.

95. G. Sarma, Appendix to M. Daoud, J.P. Cotton, B. Farnoux, G. Jannink, G. Sarma, H. Benoit, R. Duplessix, C. Picot and P.G. de Gennes, "Solutions of flexible polymers. Neutron experiments and interpretation", Macromolecules 8 (1975) 804-818; see also G. Sarma, "Conformation des polymères en solution", in Ill-Condensed Matter, ed. R. Balian, R. Maynard and G. Toulouse, pp. 537-552 (Amsterdam, North-Holland, 1979).

96. P.D. Gujrati, "A new mapping between self-avoiding walks and the $n \to 0$ limit", J. Phys. A 14 (1981) L345-L348; "Correct correspondence between self-avoiding random walks and a magnetic system as $n \to 0$ on a lattice", Phys. Rev. B 24 (1981) 2854-2856; "Magnetic analog of self-avoiding walks (polymer chains) on a lattice", Phys. Rev. A 24 (1981) 2096-2108.

97. J. Wheeler and P. Pfeuty, "Polymer statistics, the n-vector model, and thermodynamic stability", Phys. Rev. A 23 (1981) 1531-1534 (Erratum, ibid. 24 (1981) 646); also "The $n \to 0$ vector model and equilibrium polymerization", Phys. Rev. A 24 (1981) 1050-1062.

98. B. Widom, "Equation of state in the neighborhood of the critical point", J. Chem. Phys. 43 (1965) 3898-3905; L.P. Kadanoff, W. Götze, D. Hamblen, R. Hecht, E.A.S. Lewis, V.V. Palcizuskas, M Rayl, J. Swift, D. Aspnes and J. Kane, "Static phenomena near critical points: Theory and experiment", Rev. Mod. Phys. 39 (1967) 395-431.

99. S. Havlin and D. Ben-Avraham, to "Corrections to scaling in self-avoiding walks", Phys. Rev. A 27 (1983) 2759-2762 give the estimate $\nu = 0.588 \pm 0.003$ in three dimensions, using a method based on fractal dimensionality, while J.C. le Guillou and Z. Zinn-Justin, "Critical exponents from field theory", Phys. Rev. B. 21 (1980) 3976-3998 give $\nu = 0.588 \pm 0.001$. Evidence is thus accumlating that in three dimensions the 'accepted' values of the exponents are not precisely correct.

100. H.J. Hilhorst, "Renormalization of the self-avoiding walk on a lattice", Phys. Lett. 56A (1976) 153-154, and "Real-space renormalization of the self-avoiding walk by a linear transformation", Phys. Rev. B. 16 (1977) 1253-1265; J.C. le Guillou and J. Zinn-Justin, "Critical exponents for the n-vector model in three dimensions from field theory", Phys. Rev. Lett. 39 (1977) 95-98, and "Critical exponents from field theory", Phys. Rev. B 21 (1980) 3976-3998.

101. B. Derrida, "Phenomenological renormalization of the self avoiding walk in two dimensions", J. Phys. A 14 (1981) L5-L9.

102. B. Nienhuis, "Exact critical point and critical exponents of O(n) models in two dimensions", Phys. Rev. Lett. 49 (1982) 1062-1065. For a reconciliation of Nienhuis' results with numerically determined values of critical exponents see I. Majid, Z.V. Djordjevic and H.E. Stanley, "Correlation length exponent for the O(n) model in two dimensions for $n = 0$", Phys. Rev. Lett. 51 (1983) 143, and A.J. Guttmann, "On two dimensional self-avoiding walks", preprint, University of Newcastle (1983).

103. P.W. Kasteleyn, "A soluble self-avoiding walk problem", Physica 29 (1963) 1329-1337; see also M.N. Barber, "Asymptotic results for self-avoiding walks on a Manhattan lattice", Physica 48 (1970) 237-241.

104. B.K. Chakrabarti and S.S. Manna, "Critical behaviour of directed self-avoiding walks", J. Phys. A 16 (1983) L113-L116. The numerical estimate $\nu \approx 0.86$ for a two dimensional directed self-avoiding walk given in this paper has been shown to be inaccurate by S. Redner and I. Majid ("Critical properties of directed self-avoiding walks", J. Phys. A 16 (1983) L307-L310), who derive the exact result that $\nu = 1$. Directed self-avoiding walks can be used to construct rigorous bounds on the connective constant of a lattice: see M.E. Fisher and M.F. Sykes, "Excluded volume problem and the Ising model of ferromagnetism", Phys. Rev. 114 (1959) 45-58.

105. F.W. Wiegel, "Markovian nature of the two-dimensional self-avoiding walk problem", Physica 98A (1979) 345-351; see also "On a remarkable class of two-dimensional random walks", J. Math. Phys. 21 (1980) 2111-2113.

106. M. Eden, "A two-dimensional growth process", Proceedings of the 4th Berkeley Symposium on Mathematical Statistics and Probability, ed. J. Neyman, Vol. 4, pp. 223-239 (Berkeley, University of California Press, 1961).

107. T.A. Witten and L.M. Sander, Diffusion-limited aggregation, a kinetic critical phenomenon", Phys. Rev. Lett. 47 (1981) 1400-1403; P. Meakin, "Diffusion controlled cluster growth in two, three and four dimensions", Phys. Rev. A 27 (1983) 604-607; P. Meakin, "Diffusion-controlled deposition on fibers and surfaces", Phys. Rev. A 27 (1983) 2616-2623; T.A. Witten and L.M. Sander, "Diffusion-limited aggregation", Phys. Rev. B 27 (1983) 5686-5697.

108. S.R. Broadbent, in "Discussion on the Symposium on Monte Carl Methods", J. Roy. Stat. Soc. Series B 16 (1954) 68.

109. J.M. Hammersley, "Percolation processes: lower bounds for the critical probability", Ann. Math. Stat. 28 (1957) 790-795.

110. J.M. Hammersley, "Comparison of atom and bond percolation processes", J. Math. Phys. 2 (1961) 728-733; for some two-dimensional lattices the inequality $p_c^B < p_c^S$ was given earlier by M.E. Fisher, "Critical probabilities for cluster size and percolation problems", J. Math Phys. 2 (1961) 620-627.

111. J.M. Hammersley, "Bornes supérieures de la probabilité critique dans un processus de filtration", Le Calcul des Probabilités et ses Applications, pp. 17-34 (Centre Nationale de la Recherche Scientifique, Paris, 1959); discussion ibid. pp. 35-37.

112. M.E. Fisher and J.W. Essam, "Some cluster size and percolation problems", J. Math. Phys. 2 (1961) 609-619.

113. J.W. Essam, "Percolation and cluster size", in Phase Transitions and Critical Phenomena, Vol. 2 (ed. C. Domb and M.S. Green) pp. 197-270 (London, Academic Press, 1972).

114. M.F. Sykes and J.W. Essam, "Exact critical probabilities for site and bond problems in two dimensions", J. Math. Phys. 5 (1964) 1117-1127.

115. H. Kesten, "The critical probability of bond percolation on the square lattice equals ½", Comm. Math. Phys. 74 (1980) 41-59.

116. H. Kesten, Percolation Theory for Mathematicians (Boston, Birkhaüser, 1982).

117. J.C. Wierman, "Bond percolation on honeycomb and triangular lattices", Adv. Appl. Prob. 13 (1981) 298-313.

118. J.C. Wierman, "Percolation theory", Ann. Prob. 10 (1982) 509-524.

119. J.W. Essam, "Percolation theory", Rep. Prog. Phys. 43 (1980) 833-912.

120. D. Stauffer, "Scaling theory of percolation clusters", Phys. Rep. 54 (1979), 1-74.

121. D. Stauffer, "Scaling properties of percolation clusters", Lecture Notes in Physics 149 (1981) 9-25.

122. H. Kesten, "Analyticity properties and power law estimates of functions in percolation theory", J. Stat. Phys. 25 (198) 717-756.

123. P.W. Kasteleyn and C.M. Fortuin, "Phase transitions in lattice systems with random local properties", J. Phys. Soc. Japan Suppl. 26 (1969) 11-14. See also C.M. Fortuin and P.W. Kasteleyn, "On the random cluster model. I. Introduction and relation to other models", Physica 57 (1972) 536-564; F.Y. Wu, "Percolation and the Potts model", J. Stat. Phys. 18 (1978) 115-123; and S. Alexander, "Hamiltonian formulation of bond percolation: an alternative derivation", J. Phys. A 11 (1978) 1803-1806.

124. A.G. Dunn, J.W. Essam and J.M. Loveluck, "Scaling theory for the pair connectedness in percolation models", J. Phys. C 8 (1975) 743-750.

125. P. Pfeuty and G. Toulouse, Introduction to the Renormalization Group and to Critical Phenomena (New York, Wiley, 1977).

126. G.S. Rushbrooke, "On the thermodynamics of the critical region of the Ising problem", J. Chem. Phys. 39 (1963) 842-843.

127. R.B. Griffiths, "Thermodynamic inequality near the critical point for ferromagnets and fluids", Phys. Rev. Lett. 14 (1965) 623-624.

128. G. Toulouse, "Perspectives from the theory of phase transitions", Nuovo Cimento 23B (1974) 234-240. The existence of the upper critical dimension E_c, and the result that $E_c = 6$ for percolation theory, are generally believed by physicists, but like most "known results" of percolation theory, rigorous proofs are lacking. Monte Carlo simulations of S. Kirkpatrick, "Percolation processes in higher dimensions: approach to the mean-field limit", Phys. Rev. Lett. 36 (1976) 69-72 support Toulouse's results.

129. A.B. Harris, T.C. Lubensky, W.K. Holcomb and C. Dasgupta, "Renormalization group approach to percolation problems", Phys. Rev. Lett. 35 (1975) 327-330 (errata, ibid., 1397).

130. Mean field theories of ferromagnetic models neglect fluctuations, by assuming that partial ordering of the spins produces a uniform local magnetic field. See, for example, ref. [72], pp. 34-39. Mean field calculations predict that $\alpha = \alpha'$ and $\gamma = \gamma'$.

131. R. Bass and M.J. Stephen, "Voltage correlations in a random Bethe lattice", J. Phys. C. 8 (1975) L281-L284 propose a definition of $\xi(p)$ which gives a finite correlation length at $p = p_c$; alternative definitions are given by A. Coniglio, "Some cluster size and percolation problems for interacting spins", Phys. Rev. B 13 (1976) 2194-2207, and J.P. Straley, "The ant in the labyrinth: diffusion in random networks near the percolation threshold", J. Phys. C 13 (1980) 2991-3002.

132. M.J. Stephen, "Site-cluster distributors and equation of state for the bond percolation model", Phys. Rev B 15 (1977) 5674-5680; see also R.G. Priest and T.C. Lubensky, "Critical properties of two tensor models with application to the percolation problem", Phys. Rev. B 13 (1976) 4159-4171 (errata ibid. 14 (1976) 5125).

133. M.P.M. den Nijs, "A relation between the temperature exponents of the eight-vertex and q-state Potts model", J. Phys. A 12 (1979) 1857-1868; see also "Extended scaling relation for the magnetic critical exponents of the q-state Potts model", Phys. Rev. B 27 (1983) 1674-1679.

134. B. Nienhuis, E.K. Riedel and M. Schick, "Magnetic exponents of the two dimensional q-state Potts model", J. Phys. A. 13 (1980) L189-L192.

135. R.B. Pearson, "Conjecture for the extended Potts model magnetic eigenvalue", Phys. Rev. B 22 (1980) 2579-2580.

136. R.B. Pearson, "Number theory and critical exponents", Phys. Rev. B 22 (1980) 3465-3470. With a few modest assumptions, Pearson shows that if the critical exponents α and β are rational, and therefore able to be written as the ratio m/n of two relatively prime integers m and n, the value of n can be predicted.

137. Applications oriented reviews include:
(a) H.L. Frisch and J.M. Hammersley, "Percolation processes and related topics", J. Soc. Indust. Appl. Math. 4 (1963) 894-918;
(b) H.E. Stanley, "New directions in percolation theory including possible applications to the real world", Lecture Notes in Physics, 149 (1981) 59-83.
(c) G. Deutscher, "Experimental relevance of percolation", Lecture Notes in Physics 149 (1981) 26-40.

138. The article by V.K.S. Shante and S. Kirkpatrick, "An introduction to percolation theory", Adv. Phys. 20 (1971) 325-357 remains a good introduction to the subject, and addresses applications of interest in condensed matter physics. For more recent developments in this area see the following articles in Ill-Condensed Matter (ed. R. Balian, R. Maynard and G. Toulouse: Amsterdam, North-Holland, 1979): D.J. Thouless, "Percolation and localization" (pp. 1-62); S. Kirkpatrick, "Models of disordered materials" (pp. 321-403); T.C. Lubensky, "Thermal and geometrical critical phenomena in random systems" (pp. 405-475).

139. The structure of the connected component as $p \to p_c^+$ is examined by S. Kirkpatrick, "The geometry of the percolation threshold", in AIP Conference Proceedings Vol. 40, ed. J.C. Garland and D.B. Tanner, pp. 99-117 (New York, American Institute of Physics, 1978); see also Redner's first article in this volume. Illustrations from a motion picture recording the growth of connectivity as p increases are given by C. Domb, E. Stoll and T. Schneider, "Percolation clusters", Contemp. Phys. 21 (1980) 577-592.

140. The problem of polymer gelation, interpreted as a percolation process, is reviewed by D. Stauffer, A. Coniglio and M. Adam, "Gelation and critical phenomena", Adv. Polymer Sci. 44 (1982) 103-158; this article also contains a useful survey of variants of the basic percolation model.

141. The following articles are written with emphasis on fundamental mathematical problems in percolation theory, including the problem of "first passage percolation": D.J.A. Welsh, Percolation and related topics", Sci. Prog. Oxf. 64 (1977) 65-83; J.M. Hammersley and D.J.A. Welsh, "First passage percolation, subadditive processes, stochastic networks, and generalized renewal processes", in Bernoulli-Bayes-Laplace Anniversary Volume (ed. J. Neyman and L.M. le Cam), pp. 61-110 (New York, Springer-Verlag, 1965); J.M. Hammersley and D.J.A. Welsh, "Percolation theory and its ramifications", Contemp. Phys. 21 (1981) 593-605; see also the monograph by R.T. Smythe and J.C. Wierman, "First passage percolation on the square lattice", Lecture Notes in Mathematics 671 (1978).

142. P.H. Winterfeld, L.E. Scriven and H.T. Davis, "Percolation and conductivity of random two-dimensional composites", J. Phys. C. 14 (1981) 2361-2376.

143. S.W. Haan and R. Zwanzig, "Series expansions in a continuum percolation problem", J. Phys. A. 10 (1977) 1547-1555.

144. E.T. Gawlinski and H.E. Stanley, "Continuum percolation in two dimensions: Monte Carlo tests of scaling and universality for non-interacting discs", J. Phys. A 14 (1981) L291-L299.

145. R. Zallen, "Stochastic geometry", in Fluctuation phenomena (ed. E.W. Montroll and J.L. Lebowitz) pp. 177-228 (Amsterdam, North-Holland, 1979). See also a book by Y. Waseda, The structure of non-cyrstalline materials (New York, McGraw-Hill, 1980).

146. S.A. Roach, The theory of random clumping (London, Methuen, 1968).

147. S. Kirkpatrick, "Percolation and conduction", Rev. Mod. Phys. 45 (1973) 574-588.

148. J.P. Clerc, G. Giraud, J. Roussenq, R. Blanc, J.P. Carton, E. Guyon, H. Ottavi and D. Stauffer, "La percolation: modèles, simulations analogiques et numériques", Annales de Physique 8 (1983) 3-105.

149. In principle one should prove that the critical value of p defined via the conductivity coincides with the topologically defined bond percolation threshold of section 6. Kesten [116] has given a proof of this for the square lattice.

150. B. Derrida and J. Vannimenus, "A transfer matrix approach to random resistor networks", J. Phys. A. 15 (1982) L557-L564.

151. P.G. de Gennes, "On a relation between percolation theory and the elasticity of gels", J. Physique Lett. 37 (1976) L1-L2.

152. A.B. Harris and R. Fisch, "Critical behavior of random resistor networks", Phys. Rev. Lett. 38 (1977) 796-799.

153. R.B. Stinchcombe, "The branching model for percolation theory and electrical conductivity", J. Phys. C. 6 (1973) L1-L5, and "Conductivity and spin-wave stiffness in disordered systems - an exactly soluble model", J. Phys. C. 7 (1974) 179-203; see also J. Heinrichs and N. Kumar, "Simple exact treatment of conductance in a random Bethe lattice", J. Phys. C 8 (1975) L510-L516.

154. J.P. Straley, "Random resistor tree in an applied field", J. Phys. C 10 (1977) 3009-3013.

155. S. Alexander and R. Orbach, "Density of states on fractals: 'fractons'", J. Physique Lett. 43 (1982) L625-L631; see especially the note added in proof.

156. See, e.g., D.A.G. Bruggeman, "Berechnung verschiedener physikalischer Konstanten von heterogenen Substanzen. I. Dielektrizitatskonstanten und Leitfahigkeiten der Mischkorper aus isotropen Substanzen", Ann. Phys. (Leipzig) 24 (1935) 636-697, and R. Landauer, "The electrical resistance of a binary mixture", J. Appl. Phys. 23 (1952) 779-784.

157. A substantial review of the coherent potential approximation has been given by F. Yonezawa and K. Morigaki, "Coherent potential approximation - Basic concepts and applications", Prog. Theor. Phys. Suppl. 53 (1973) 1-75. See also the short expository article by J.A. Krumhansl, "It's a random world", in Amorphous Magnetism (ed. H.O. Hooper and A.M. de Graaf) pp. 15-25 (New York, Plenum, 1973), and for more recent references, F. Yonezawa, "Transport properties of liquid non-simple metals", J. Physique 41 suppl. C8 (1980) 447-457. For a proof of the equivalence of the coherent potential approximation and the effective medium approximation see M. Hori and F. Yonezawa, "Statistical theory of effective electrical, thermal, and magnetic properties of random heterogeneous materials. IV. Effective medium theory and cumulant expansion method", J. Math. Phys. 16 (1975) 352-364. These authors propose a different approximation technique, employing cumulants, which predicts that $p_c = 1 - \exp(-2/z)$ for a lattice of coordination number z. For large z this reduces to the effective medium result (7.12). Although for two-dimensional lattices, their prediction is less accurate then (7.12), the situation is reversed in three dimensions.

158. S. Kirkpatrick, "Classical transport in disordered media: scaling and effective-medium theories", Phys. Rev. Lett. 27 (1971) 1722-1725.

159. M. Sahimi, B.D. Hughes, L.E. Scriven and H.T. Davis, "Real-space renormalization and effective medium approximation to the percolation conduction problem", Phys. Rev. B. 28 (1983), 307-311. For other improvements of the basic effective medium approximation based on finite clusters of bonds see, for example: L. Turban, "On the effective-medium approximation for bond-percolation conductivity", J. Phys. C 11 (1978) 449-459, T. Nagatani, "A two-bond theory of conductivity in bond disordered resistor networks", J. Phys. C 14 (1981) 3383-3391 and references cited therein. An ad hoc, but effective alternative approach has been given by M. Nakamura, "A method to improve the effective medium theory towards percolation problem", J. Phys. C 15 (1982) L749-L752.

160. J.W. Essam, C.M. Place and E.H. Sondheimer, "Self consistent calculation of the conductivity in a disordered branching network", J. Phys. C 8 (1974) L258-L260.

161. P.M. Kogut and J.P. Straley, "Distribution-induced non-universality of the percolation conductivity exponents", J. Phys. C 12 (1979) 2151-2159; see also A. Ben-Mizrahi and D.J. Bergman, "Non-universal critical behaviour of random resistor networks with a singular distribution of conductances", J. Phys. C. 14 (1981) 909-922, and J.P. Straley, "Non-universal threshold behaviour of random resistor networks with anomalous distributions of conductances", J. Phys. C. 15 (1982) 2343-2345, where renormalization group arguments are used.

162. J. Bernasconi and H.J. Weisman, "Effective-medium theories for site-disordered resistance networks", Phys. Rev. B 13 (1976) 1131-1139; T. Joy and W. Strieder, "Effective medium theory of site percolation in a random simple triangular conductance network", J. Phys. C 11 (1978) L867-L870 (errata ibid. 12 (1979) L53).

163. J.P. Straley, "Critical phenomena in resistor networks", J. Phys. C. 9 (1976) 783-795; see Ref. [154] for a reinterpretation of some of these results with the now preferred definition of σ for the Bethe lattice. Also see Ref. 151 and R. Fisch and A.B. Harris, "Critical behavior of random resistor networks near the percolation threshold", Phys. Rev. B. 18 (1978) 416-420, where the formal relation of Kasteleyn and Fortuin [123] between the q-state Potts model in the limit $q = 0$ and the resistance between sites of a homogeneous lattice is exploited.

164. J.P. Straley, "Critical exponents for the conductivity of random resistor networks", Phys. Rev. B. 15 (1977) 5733-5737.

165. J.P. Straley, "Threshold behaviour of random resistor networks: a synthesis of theoretical approaches", J. Phys. C. 15 (1982) 2333-2341.

166. See, for example, J. Koplik, "Creeping flow in two-dimensional networks", J. Fluid Mech. 119 (1982) 219-247. Koplik's analysis (see also his paper, "On the effective medium theory of random linear networks", J. Phys. C 14 (1981) 4821-4837) shows that for nonpercolative distributions, i.e. when $f(g)$ has no delta function at $g = 0$, the effective medium approximation is remarkably accurate. Indeed for a special class of such distributions, the effective medium approximation is exact for the square lattice: see J. Marchant and R. Gabillard, "Sue le calcul d'un réseau résistif aléatoire, C.R. Acad. Sci. Paris B281 (1975) 261-264.

167. G.R. Jerauld, J.C. Hatfield, L.E. Scriven and H.T. Davis, "Percolation and conduction on Voronoi and triangular networks: a case study in topological disorder", to appear in J. Phys. C.

168. G.C. Koerber, Properties of Solids (Englewood Cliffs, N.J., Prentice-Hall, 1962) pp. 69-71. See also P.G. Sherman, Diffusion in solids (New York, McGraw-Hill, 1963) and Y. Adda and J. Philibert, La diffusion dans les solides, 2 volumes (Paris, Presses Universitaires de France, 1966).

169. Since the electrical transport properties of semiconductors are of immense technical importance (e.g. in the Xerox process) many experiments have been made. See, for example, references cited by H. Scher and M. Lax, "Stochastic transport in a disordered solid. II. Impurity conduction", Phys. Rev. B 7 (1973) 4502-4519.

170. Dispersion in disordered porous media is reviewed in J.J. Fried and M.A. Combernous, "Dispersion in porous media", Advances in Hydroscience 7 (1971) 169-282; for more recent references see e.g. M. Sahimi, L.E. Scriven and H.T. Davis, "Dispersion in disordered porous media", Chem. Eng. Comm., in press. The simpler case of dispersion in spatially periodic porous media has been analysed exhaustively by H. Brenner, "Dispersion resulting from flow through spatially periodic porous media", Phil. Trans. R. Soc. Lond. A 297 (1980) 81-133, and H. Brenner and P.M. Adler, "Dispersion resulting from flow through spatially periodic porous media. II. Surface and intraparticle transport", ibid. 307 (1982) 169-200.

171. M.V. Kozlov, "Random walk in a one-dimensional random medium", Theory Prob. Appl. 18 (1973) 387-388.

172. F. Solomon, "Random walks in a random environment", Ann. Prob. 3 (1975) 1-31.

173. D.E. Temkin, "One-dimensional random walks in a two-component chain", Soviet Math. Dokl. 13 (1972) 1172-1176.

174. Ya. G. Sinai, "The limiting behavior of a one-dimensional random walk in a random environment", Theory Prob. Appl. 27 (1982) 256-268. Stated precisely Sinai's key result is as follows. Assume that $\langle \log \sigma \rangle = 0$. Let $\varepsilon > 0$ and $\delta > 0$ be given. For all sufficiently large n there exist a set C_n in the space of all realizations ω and a point $m(n) = m(n;\omega)$ for each $\omega \in C_n$ such that (i) the probability that a given realization belongs to C_n exceeds $1 - \varepsilon$; and (ii) with Pr denoting probability measure over all walks in a given environment ω, if $\omega \in C_n$, then

$$\Pr(|X_n/\log^2 n - m(n)| < \delta) \to 1 \quad \text{as} \quad n \to \infty .$$

The convergence is uniform in $\omega \in C_n$, and as $n \to \infty$ the probability distributions for $m(n)$ converge weakly to some limit distribution.

175. H. Kesten, M.V. Kozlov and F. Spitzer, "A limit law for random walk in a random environment", Compositio Math. 30 (1975) 145-168.

176. A distribution is called 'arithmetic' (see Ref. 20, p. 138) if it is concentrated on the set of points $0, \pm\lambda, \pm 2\lambda, \ldots$ for some positive λ called the 'span'. In the example (8.7), $\log \sigma$ has probability density function $\psi(\eta) = (1-p)\delta(\eta - \log\{a/[1-a]\}) + p\delta(\eta + \log\{a/[1-a]\})$, i.e. an arithmetic distribution of span $\log\{a/[1-a]\}$, and so is not covered by the results of Kesten et al.

177. B. Derrida and Y. Pomeau, "Classical diffusion in a random chain", Phys. Rev. Lett. 48 (1982) 627-630.

178. B. Derrida, "Velocity and diffusion coefficient of a periodic one-dimensional hopping model", J. Stat. Phys. 31 (1983) 433-450.

179. S.A. Kalikow, "Generalized random walk in a random environment", Ann. Prob. 9 (1981) 753-768.

180. V.V. Anshelevich, K.M. Khanin and Ya G. Sinai, "Symmetric random walks in random environments", Comm. Math. Phys. 85 (1982) 449-470. The analysis of this paper requires symmetry in local transition probabilities, but not isotropy. All of the results are established for anisotropic systems, and transitions are not restricted to nearest-neighbour sites.

181. P.G. de Gennes, "La percolation: un concept unificateur", La Recherche 7 (1976) 919-927.

182. C.D. Mitescu and J. Roussenq, "Une fourmi dans un labyrinthe: diffusion dans un système de percolation", C.R. Acad. Sci. Paris 283A (1976) 999-1001.

183. C.D. Mitescu, H. Ottavi and J. Roussenq, "Diffusion on percolation lattices: the layrinthine ant", in AIP Conference Proceedings Vol. 40 (ed. J. Garland and D.B. Tanner) pp. 377-381 (New York, American Institute of Physics, 1978).

184. J.P. Straley, "The any in the labyrinth: diffusion in random metworks near the percolation threshold", J. Phys. C 13 (1980) 2991-3002.

185. Y. Gefen, A. Aharony and S. Alexander, "Anomalous diffusion on percolating clusters", Phys. Rev. Lett. 50 (1983) 77-80.

186. D. Ben-Avraham and S. Havlin, "Diffusion on percolation clusters at criticality" J. Phys. A 15 (1982) L691-L697, and M. Sahimi and G.R. Jerauld, "Random walks on percolation clusters at the percolation threshold", submitted to J. Phys. C.

187. Alternative models with continuous time-dependence are easily defined but not considered here. See, for example, G. Ritter, "A continuous-time analogue of random walk in a random environment", J. Appl. Prob. 17 (1980) 259-264, and B.D. Hughes, M. Sahimi, L.E. Scriven and H.T. Davis, "Transport and conduction in random systems", to appear in Int. J. Engng. Sci.

188. S. Alexander, J. Bernasconi, W.R. Schneider and R. Orbach, "Excitation dynamics in random one-dimensional systems", Rev. Mod. Phys. 53 (1981) 175-198. (These authors also consider the randomized rate equation $c_\ell dP_\ell/dt = W\{P_{\ell+1} + P_{\ell-1} - 2P_\ell\}$ with W fixed, but $\{C_\ell\}$ a set of independently and identically distributed random variables.)

189. (a) J. Bernasconi, S. Alexander and R. Orbach, "Classical diffusion in a one-dimensional disordered lattice", Phys. Rev. Lett 41 (1978) 185-187;
(b) S. Alexander, J. Bernasconi and R. Orbach, "Spectral diffusion in a one-dimensional percolation model", Phys. Rev. B 17 (1978) 4311-4314;
(c) S. Alexander, J. Bernasconi and R. Orbach, "Low energy density of states for disordered chains", J. Physique 39 Suppl C6 (1978) 706-707;
(d) S. Alexander, J. Bernasconi, W.R. Schneider and R. Orbach, "Excitation dynamics in random one-dimensional systems", in Physics in One Dimension, ed. J. Bernasconi and T. Schneider, pp. 277-288 (Berlin, Springer-Verlag, 1981);
(e) J. Bernasconi and H.U. Beyeler, "Some comments on hopping in random one-dimensional systems", Phys. Rev. B 21 (1980) 3745-3747;
(f) J. Bernasconi, H.U. Beyeler, S. Strässler and S. Alexander, "Anomalous frequency-dependent conductivity in disordered one-dimensional systems", Phys. Rev. Lett. 42 (1979) 819-822;
(g) J. Bernasconi, W.R. Schneider and W. Wyss, "Diffusion and hopping conductivity in disordered one-dimensional lattice systems", Z. Phys. B 37 (1980) 175-184.
(h) J. Bernasconi and W.R. Schneider, "Classical hopping conduction in random one-dimensional systems: non universal limit theorems and quasilocalization effects", Phys. Rev. Lett. 47 (1981) 1643-1647.

190. (a) W.R. Schneider and J. Bernasconi, "Diffusion in one-dimensional lattice systems with random transfer rates", Lecture Notes in Physics 153 (1982) 389-393.
(b) W.R. Schneider, "Hopping transport in disordered one-dimensional lattice systems: random walk in a random medium", Lecture Notes in Physics 173 (1982) 289-303.

191. J. Bernasconi and W.R. Schneider, "Diffusion in random one-dimensional systems", J. Stat. Phys. 30 (1983) 355-362.

192. V.V. Anshelevich and A.V. Vologodskii, "Laplace operator and random walk on a one-dimensional nonhomogeneous lattice", J. Stat. Phys. 25 (1981) 419-430. These authors consider the master equation

$$(\partial/\partial t)P_\ell(t) = [1-\delta_{\ell,1}]a_{\ell-1}\,P_{\ell-1}(t) - (a_{\ell-1} + a_\ell)P_\ell(t) + [1-\delta_{\ell,N-1}]a_\ell P_{\ell+1}(t)$$

which describes motion on a finite linear chain, with the zeroth and Nth sites absorbing boundaries. The ℓth site is assigned the coordinate ℓ/N. When the limit $N \to \infty$ is taken in an appropriate manner, the solution of the master equation is shown to approach the solution of the diffusion equation with absorbing boundaries,

$$(\partial/\partial t)\,p(x,t) = a(\partial^2/\partial x^2)p(x,t)\ ,\ p(0,t)= p(1,t) = 1\ ,$$

so long as $a = \lim_{N\to\infty} N(\sum_{\ell=0}^{N-1} a_\ell^{-1})$ exists and is non-zero. For independent, randomly distributed coefficients a_ℓ, this implies that if a_ℓ^{-1} has finite mean $\langle a_\ell^{-1}\rangle$, then the effective diffusion coefficient is $a = \langle a_\ell^{-1}\rangle^{-1}$ with probability 1.

193. V.V. Bryksin, "Frequency dependence of the hopping conductivity of a one-dimensional system calculated by the effective-medium method", Sov. Phys. Solid St. 22 (1980) 1194-1199.

194. T. Odagaki and M. Lax, "Coherent-medium approximation in the stochastic transport theory of random media", Phys. Rev. B 24 (1981) 5284-5294. See also M. Lax and T. Odagaki, "Coherent medium approach to hopping conduction", Lecture Notes in Physics 154 (1982) 148-176.

195. S. Summerfield, "Effective medium theory of A.C. hopping conductivity for random bond lattice models, Solid St. Comm. 39 (1981) 401-402.

196. I. Webman, "Effective medium approximation for diffusion on a random lattice", Phys. Rev. Lett. 47 (1981) 1496-1499. See also I. Webman, "Effective medium approximation for diffusion or random networks", Lecture Notes in Physics 154 (1982) 297-303.

197. J.W. Haus, K.W. Kehr and K. Kitahara, "Long-time tail effects on particle diffusion in a disordered system", Phys. Rev. B 25 (1982) 4918-4921.

198. J.W. Haus, K.W. Kehr and K. Kitahara, "Transport in a disordered medium: analysis and Monte-Carlo simulation", Z. Phys. B 50 (1983) 161-169.

199. B. Movaghar, M. Grunewald, B. Pohlmann, D. Wurtz and W. Schirmacher, "Theory of hopping and multiple-trapping transport in disordered systems", J. Stat. Phys. 30 (1983) 315-334, and references therein to earlier work of these authors.

200. I. Webman and J. Klafter, "Diffusion in one-dimensional disordered systems: an effective-medium approximation", Phys. Rev. B 26 (1982) 5950-5952.

201. M. Sahimi, B.D. Hughes, L.E. Scriven and H.T. Davis, "Stochastic transport in disordered systems", J. Chem. Phys. 78 (1983) 6849-6864.

202. For the Kirkpatrick random resister problem, the Green function formalism has been developed by J. A. Blackman, "A theory of conductivity in disordered resistor networks", J. Phys. C 9 (1976) 2049-2071, and G. Ahmed and J. A. Blackman, "On theories of transport in disordered media" ibid. 12 (1976) 837-853. Its extension to the present problem may be found in Ref. 201.

203. M.J. Stephen and R. Kariotis, "Diffusion in a one-dimensional disordered system", Phys. Rev. B 26 (1982) 1917-2925.

204. J. Machta, Generalized diffusion coefficient in one-dimensional random walks with static disorder", Phys. Rev. B 26 (1982) 2917-2925; "Renormalization group approach to random walks on disordered lattices", J. Stat. Phys. 30 (1983) 305-314.

205. A. Igarashi," Hopping diffusion in a one-dimensional random system", Prog. Thear. Phys. 69 (1983) 1031-1034.

206. R. Zwanzig, "Non-Markoffian diffusion in a one-dimensional disordered lattice", J. Stat. Phys. 28 (1982) 127-133.

207. M.J. Stephen, "Diffusion on a directed percolating network", J. Phys. C 14 (1981) L1077-L1080.

208. J. Bernasconi and W.R. Schneider, "Diffusion on a one-dimensional lattice with random asymmetric transition rates", J. Phys. A 15 (1983) L729-L734.

209. M. Barma and D. Dhar, "Directed diffusion in a percolation network", J. Phys. C 16 (1983) 1451-1458.

210. See also B.D. Hughes, M. Sahimi, L.E. Scriven and H.T. Davis, "Transport and conduction in random systems", Int. J. Eng. Sci, in press. Such models are lattice versions of a classic colloid problem of M. Smoluchowski, ref 11; important recent papers with a physical chemistry orientation include B.U. Felderhof and J.M. Deutch, "Concentration dependence of the rate of diffusion - controlled reactions, J. Chem. Phys. 64 (1976) 4551-4558, P. Grassberger and I. Procaccia, "The long-time properties of diffusion in a medium with

static traps", ibid. 77 (1982) 6281-6284, M. Muthukumar, "Concentration dependence of diffusion controlled processes among static traps", ibid. 76 (1982) 2667-2671, S. Prager and H. L. Frisch, "Diffusion-controlled reactions on a two-dimensional lattice", ibid. 72 (1980) 2941- , and R.F. Keyser and J.B. Hubbard, "Diffusion in a medium with a random distribution of static traps", Phys. Rev. Lett. 51 (1983) 79-82.

211. B. Movaghar, B. Pohlmann and W. Schirmacher, "Random walk in disordered hopping systems", Solid State Comm. 34 (1980) 451-454.

212. J. Klafter and R. Silbey, "Derivation of the continuous-time random walk equation", Phys. Rev. Lett. 44 (1980) 55- .

213. We apologize to Charles Darwin for borrowing the structure, and the tone, of his concluding sentence in The Origin of Species, as our closing remark.

PART B: DIFFUSION IN DISORDERED MEDIA: VARIATIONAL BOUNDS

1. Statement of the Problem and Variational Formulation

A large class of diffusion problems of great practical importance arises in connection with diffusive transport in multi-phase media, such as porous solids or eutectic mixtures. Although the detailed phase geometry is sometimes represented as being regularly periodic [1], such simplification produces systems not essentially different from the lattices discussed in part A. The main interest here has accordingly focussed on diffusion in media with random phase geometries, characterized by the following general features:

i) Each phase α is a uniform continuum in which the diffusing solute has a diffusion coefficient D_α. We ignore complications caused by unequal partition of the solute between phases, i.e. the solute concentration is assumed to be continuous at all points on the phase boundaries.*

ii) The phase geometry is not known in complete detail, but limited information of a statistical nature is available, most often in the form of various averages and correlations (see for example the paper by Stell in this volume).

iii) The phase geometry is time-independent** and spatially homogeneous, that is all averages and correlations generated by it are independent of <u>absolute</u> position.

---

* Physically this is almost always a poor assumption, but it is readily avoided by replacement of the solute concentration c and diffusion coefficient D_α by the solute activity $\underline{a} \equiv c/\kappa_\alpha$ and adjusted diffusion coefficient $D_\alpha^{(a)} \equiv D_\alpha \kappa_\alpha$, where κ_α, the solute partition coefficient in phase α, is defined so that $\underline{a}$ rather than c varies continuously across phase boundaries. Stated in terms of $\underline{a}$ and $D_\alpha^{(a)}$, the problem remains unchanged.

** Temporal fluctuations are discussed in Section 6.

iv) There is a characteristic length ℓ that determines the scale of the phase geometry. We shall consider cubical samples of exterior dimensions $L \times L \times L$, where $L \gg \ell$, cut from a much larger body of the material in question.

v) The basic diffusion experiment on a sample is performed as follows: We maintain the front face of the cube at a uniform value c_1 of the solute concentration, and the back face at a different value c_2; all other faces are sealed so that no solute can pass through them. Under these conditions we allow the system to come to a steady state, and then measure the total amount $\dot{M}$ of solute passing per unit time through either the front or back face of the cube.

vi) The effective diffusion coefficient for the sample is now defined as

$$D_{es} \equiv |\dot{M}/(c_2 - c_1)L|$$
$$= |\langle J_z \rangle / \langle \partial c/\partial z \rangle| , \tag{1.1}$$

where we have chosen the z-axis of our coordinate frame normal to the front and back faces, $J_z(\vec{r})$ is the z-component of the local solute flux $\vec{J}(\vec{r})$ at $\vec{r} \equiv (x,y,z)$, $c(\vec{r})$ is the local solute concentration, and the angular bracket notation indicates an average over the sample volume V:

$$\langle \cdot \rangle = \frac{1}{V} \int_V (\cdot)\, d^3\vec{r}$$

vii) If we now repeat (v) and (vi) for a second sample of the material, again a cube, of the same size and orientation as the first, we will in general obtain a slightly different value of D_{es}, since the interior phase geometry will not be the same in detail. However, we expect that in the limit $L/\ell \to \infty$ this sample dependence will disappear:

$$D_{es} \xrightarrow[L \gg \ell]{} D_e^{(zz)} \left(1 + O\left(\frac{\ell}{L}\right)\right) , \tag{1.2}$$

where $D_e^{(zz)}$ is a true material property, the zz-component of the effective diffusion tensor $\mathbf{D_e}$ (we shall for the most part be concerned with isotropic materials, for which $\mathbf{D_e}$ can be replaced by the scalar D_e).

Although the existence of a sample-independent diffusion tensor is intuitively self-evident, proving the mathematical validity of the limit (1.2) for statistically homogeneous materials is no trivial matter (see the paper by Papanicolaou in this volume, and ref. 1 therein). Intuition also suggests that the boundary conditions imposed in (v) above are much stricter than necessary for our purpose: we expect for instance that if, instead of maintaining a strictly constant concentration over the front face of the sample, we allowed the concentration there to fluctuate around a mean value c_1, requiring only that the characteristic length of the fluctuation distribution be of order ℓ, we would still arrive at the same value of $D_e^{(zz)}$ in the limit $L/\ell \to \infty$. Similarly, it should not really be necessary to have the normal flux of solute vanish at every point on the lateral faces of the cube, so long as it vanishes in the mean, and correlations between fluctuations in the fluxes at different points vanish at separations large compared to ℓ. Even requiring uniform normal flux in place of uniform concentrations at the front and back faces of the sample should not change the limiting value of $D_e^{(zz)}$. All of these modifications in the boundary condition (v) affect only the concentration distribution in a "skin" of thickness comparable to ℓ at the exterior surface of the sample; the distributions in the sample core remain unchanged. Equation (1.2) merely states that $(D_{es} - D_e^{(zz)})/D_e^{(zz)} \to 0$ as the volume ratio of the skin and core regions.

In the absence of mathematical proof, what justifies this intuitive faith in the insensitivity, not only of $D_e^{(zz)}$, but also of $c(\vec{r})$ itself in the core region, to details of the surface distributions? Physicists and chemists are accustomed of course to highly disordered systems (such as liquids and gases, when viewed at the molecular level) showing astonishingly simple bulk behavior. However a more immediate basis for our belief is furnished by the example of materials with periodic phase geometries. The core $c(\vec{r})$ of a periodic material is clearly going to show the periodicity of the phase structure. Within each unit cell in the core, $c(\vec{r}) - \langle \partial c/\partial z \rangle z$ is the solution of a steady state-diffusion problem with periodic boundary conditions. Apart from a multiplicative constant determined by specification of a mean flux or concentration gradient, the concentration distribution in the interior of a large sample is therefore unaffected by sample size, or by surface fluctuations in $c(\vec{r})$ and $\vec{J}(\vec{r})$ whose spatial correlations do not extend too far beyond the dimensions of a unit cell. Only the first few layers of cells adjacent to the sample surface will show significant deviations from the core $c(\vec{r})$, and only in these surface layers will details of the boundary conditions exert any influence.

If a periodic structure generates periodic steady state concentration fluctuations in the interior core of a sufficiently large sample, it appears reasonable to expect a random but spatially homogeneous structure to generate spatially homogeneous fluctuations under the same conditions. Moreover the statistical properties of the concentration fluctuations in the sample core should not only remain the same from one sample to the next, but should for each sample be determined only by the mean solute flux or concentration gradient, just as in the periodic case. Throughout the remainder of part B we shall assume the validity of these statements, even though a formal proof is lacking.

Once we accept the existence of a sample-independent effective diffusion tensor, the problem becomes how to relate it to the detailed variations of $D(\vec{r})$. Before pursuing this question, we should perhaps remark that mathematically equivalent problems arise in many physical situations that involve different modes of transport (and sometimes no transport at all), for example:

- Electrical or thermal conduction [2]
- Dielectric behavior [3,4]
- Magnetic susceptibility [5]

The main result of this diversity of application has been to make the literature of the field harder to follow, since contributions may be scattered over a wide variety of headings. Here we will stay with the language of diffusive transport; moreover, from now on we restrict ourselves to the case of two-phase materials, so that the local diffusion coefficient $D(\vec{r})$ can take on only two values, D_α and D_β; it is then convenient to express $D(\vec{r})$ in terms of a step function $g(\vec{r})$, defined to be 1 in the α and 0 in the β phase:

$$D(\vec{r}) = D_\beta - (D_\beta - D_\alpha)g(\vec{r}) \tag{1.3}$$

The steady state concentration and flux distributions satisfy the diffusion equations

$$\vec{\nabla} \cdot \vec{J}(\vec{r}) = 0 \tag{1.3a}$$

$$\vec{J}(\vec{r}) = - D(\vec{r})\, \vec{\nabla} c(\vec{r}) \tag{1.3b}$$

for all $\vec{r}$ in $V \equiv (-\frac{L}{2} < x,y,z < \frac{L}{2})$; on the boundary between the α and β phases, these equations imply the continuity of c and of the normal component of $\vec{J}$. The diffusion experiment as specified in (v) above then requires

$$c(x,y,-\frac{L}{2}) = c_1 \ , \ c(x,y,\frac{L}{2}) = c_2 \tag{1.4a}$$

$$J_x(\pm \frac{L}{2},y,z) = J_y(x,\pm \frac{L}{2},z) = 0 \tag{1.4b}$$

on the sample surface. Unfortunately, $g(\vec{r})$ is not known to us in detail, and even if it were somehow supplied for the particular sample at hand, its complexity would be such as to prohibit an actual solution of (1.3) - (1.4). Obviously calculating $D_e^{(zz)} = - \langle J_z \rangle L/(c_2 - c_1)$ from an explicit evaluation of $\vec{J}(\vec{r})$ is not a viable option.

Beginning with Brown [3] in 1955, this led a number of authors [5,6] to ask instead the following question: given that only certain statistiscal information about $g(\vec{r})$ is available, what rigorous statements can be made about $\mathbf{D}_e$? Brown himself used a perturbation treatment to show that, for an isotropic material with $|D_\alpha - D_\beta| \ll \langle D \rangle$,

$$D_e = \langle D \rangle \left[1 - \frac{1}{3}\phi(1-\phi) \frac{(D_\alpha - D_\beta)^2}{\langle D \rangle^2} + O\left(\frac{(D_\alpha - D_\beta)^3}{\langle D \rangle^3}\right)\right], \tag{1.5}$$

where $\phi \equiv \langle g \rangle$ is the volume fraction of phase α *. The next term in the

---

* This series must not be confused with the expansion in powers of ϕ commonly used when the material has been formed by imbedding particles of one phase in a matrix of the other [7].

series involves already the three-point correlation

$$G(\vec{\rho},\vec{\rho}\,') \equiv \langle g(\vec{r})g(\vec{r}+\vec{\rho})g(\vec{r}+\vec{\rho}\,')\rangle \ , \tag{1.6}$$

and subsequent terms require correlations of ever increasing order.

One possible means of handling larger fluctuations is to follow the path taken by statistical mechanicists in the theory of dense fluids [8], that is to seek an approximate closure of Brown's series [6]. However a better alternative lies in the possibility [5,9] of making rigorous statements about diffusion rates in random media in terms of bounds on D_e derived from the Dirichlet and Thomson variational principles familiar in electrostatics.

The Dirichlet formulation of the diffusion problem (1.3)-(1.4) calls for minimization, with respect to all $c(\vec{r})$ satisfying (1.4a), of

$$S_D\{c\} \equiv \langle D(\vec{r})(\vec{\nabla}c(\vec{r}))^2\rangle/\langle \partial c/\partial z\rangle^2 \ , \tag{1.7}$$

and we have also a complementary formulation in the Thomson principle, which requires maximization, with respect to all $\vec{J}(\vec{r})$ satisfying (1.3a) and (1.4b), of the quantity

$$S_T(\vec{J}) \equiv \langle J_z\rangle^2/\langle \vec{J}^2(\vec{r})/D(\vec{r})\rangle \ . \tag{1.8}$$

Either of these variational problems generates Euler-Lagrange equations which, when combined with the subsidiary conditions on admissible trial functions, return us to the original diffusion problem (1.3)-(1.4) [10]. From these equations it follows furthermore that their solution $\bar{c}(\vec{r})$ or $\vec{\bar{J}}(\vec{r}) \equiv -D(\vec{r})\vec{\nabla}\bar{c}(\vec{r})$ gives equal values of S_D and S_T

$$S_D\{\bar{c}\} = S_T\{\vec{\bar{J}}\} = \left| \frac{\langle \bar{J}_z\rangle}{\langle \partial\bar{c}/\partial z\rangle} \right| = D_{es} \tag{1.9}$$

If, in place of $\bar{c}(r)$ and $\vec{\bar{J}}(\vec{r})$, we use trial functions $c(\vec{r})$ and $\vec{J}^*(\vec{r})$ satisfying the admissibility conditions (1.4a) and ((1.3a), (1.4b)) respectively, the resulting values of S_D and S_T must therefore satisfy the inequalities

$$S_D\{c^*\} \geq D_{es} \geq S_T\{\vec{J}^*\} \ . \tag{1.10}$$

As it stands (1.10) is perfectly rigorous, and it offers us the possibility of using even incomplete information on $D(\vec{r})$ to at least limit D_{es} to a finite range of values. For example, we can take $c^* = c_1 - (c_1 - c_2)(z + \frac{1}{2}L)/L$ and $\vec{J}^* = (0,0,\langle J_z\rangle)$, and so obtain the well-known bounds

$$\langle D\rangle > D_{es} > 1/\langle 1/D\rangle ; \qquad (1.11)$$

since $D(\vec{r})$ is spatially homogeneous, these bounds are independent of sample size, and therefore apply to $D_e^{(zz)}$ as well (we shall see in the next section that a much narrower squeeze can be had if the material is known to be isotropic).

In general however it will be inconvenient to work with trial functions that must, near the sample surface, be adjusted so as to satisfy either (1.4a) or (1.4b); we would really prefer to use for $c^*(\vec{r})$ and $\vec{J}^*(\vec{r})$ functions whose fluctuations are also spatially homogeneous, even within the surface layers. Why not simply drop (1.4) as a condition on our trial functions? If we do, then minimization of S_D and maximization, subject to (1.3a), of S_T lead to concentration distributions $c_D(\vec{r})$ and $c_T(\vec{r})$ which are no longer quite identical. Though both still satisfy (1.3) at all points in V, they do so under different boundary conditions:

$$\begin{aligned} \frac{\partial c_D}{\partial x} &= 0 \qquad \text{if} \quad x = \pm L/2 \\ \frac{\partial c_D}{\partial y} &= 0 \qquad \text{if} \quad y = \pm L/2 \\ D\frac{\partial c_D}{\partial z} &= \langle D\frac{\partial c}{\partial z}\rangle \qquad \text{if} \quad z = \pm L/2 \end{aligned} \qquad (1.12a)$$

and

$$\begin{aligned} c_T(x,y,-\tfrac{L}{2}) &= c_1 \\ c_T(x,y,\tfrac{L}{2}) &= c_2 \end{aligned} \qquad (1.12b)$$

$$c_T(\pm\tfrac{L}{2},y,z) = c_T(x,\pm\tfrac{L}{2},t) = c_1 - (c_1 - c_2)(z + \tfrac{1}{2}L)/L$$

Put into words, (1.12a) requires a uniform flux through each face of the sample, whereas (1.12b) requires the concentration on the sample surface to be a linear function of z alone. Since c_D and c_T have resulted from minimizing S_D and maximizing S_T under less stringent conditions than (1.4a) and (1.4b), we know that

$$S_T\{-D\,\vec{\nabla}\,c_T\} > D_{es} > S_D\{c_D\}. \tag{1.13}$$

It would appear therefore that relaxation of (1.4) as a requirement for admissibility of trial functions c^* or $\vec{J}^*$ produces S_D and S_T values which can lie on either side of D_{es}, and are therefore useless for bounding purposes.

Actually, however, our earlier discussion leads us to expect that c, c_D, and c_T differ appreciably only within a skin of thickness ℓ at the sample surface, and even there the differences should level off in magnitude as $L/\ell \to \infty$. It follows then that $S_D\{c_D\}$, $S_D\{c\}$, and $S_T\{-D\nabla c_T\}$ are all within a correction of order $D_e\ell/L$ of one another. Provided that we go to the limit of infinite sample size, we may therefore replace (1.10) by the more widely useful inequality

$$\lim_{L\to\infty} S_D\,\{c^*\} \geq D_e^{(zz)} \geq \lim_{L\to\infty} S_T\,\{\vec{J}^*\}\;, \tag{1.14}$$

where the only remaining admissibility condition is $\vec{\nabla}\cdot\vec{J}^* = 0$. Furthermore, no detriment to the quality of the bounds (1.14) should result if we limit ourselves to trial functions with spatially homogeneous fluctuations. (i.e. there exist spatially homogeneous $\vec{\nabla}c^*$ and $\vec{J}^*$ which, in the limit $L/\ell \to \infty$, lead to arbitrarily tight squeezes on $D_e^{(zz)}$).

Although these statements remain to be proven, we can at least show that spatially homogeneous trial gradients or fluxes will satisfy (1.14) if certain variances are bounded. For example, if

$$c_H(\vec{r}) = a + bz + c'(\vec{r})$$

is a trial concentration distribution with a spatially homogeneous $c'(\vec{r})$ fluctuation about a linear mean profile $a + bz$, we can convert it to a valid trial function c^* for the rigorous upper bound (1.10) through minor modification near the sample surface:

$$c^*(\vec{r}) = a + bz + (1 + h(z))c'(\vec{r}),$$

where

$$h(z) = \begin{cases} (\tfrac{1}{2}L - \ell + z)/\ell & -\tfrac{1}{2}L < z < -\tfrac{1}{2}L + \ell \\ 0 & -\tfrac{1}{2}L + \ell < z < \tfrac{1}{2}L - \ell \\ (\tfrac{1}{2}L - \ell - z)/\ell & \tfrac{1}{2}L - \ell < z < \tfrac{1}{2}L \end{cases}$$

We can now estimate, for any choice of ℓ, the difference

$$|S_D\{c^*\} - S_D\{c_H\}| = \frac{1}{b^2L^3}\int_V D(\vec{r})\,\vec{\nabla}(c^* + c_H)\cdot\vec{\nabla}(c^* - c_H)d^3\vec{r}$$

$$< \frac{2D_M}{b^2L}\left[\frac{\langle c'^2\rangle}{\ell} + 3\ell\langle(\vec{\nabla}c')^2\rangle\right] + O\left(\frac{1}{L^2}\right),$$

where D_M is the maximum value of $D(\vec{r})$. Thus if $\langle c'^2\rangle$ and $\langle(\nabla c')^2\rangle$ are bounded (which will always be the case for any reasonable trial function) it does not really matter that c_H fails to satisfy (1.4a) at the entrance and exit faces of a large sample, at least so far as $S_D\{c_H\}$ being an upper bound on D_e is concerned. Similar arguments can be given for the lower bound: if $\vec{J}_H(r)$ is a spatially homogeneous trial flux distribution and $\vec{B}_H$ is the vector potential associated with its fluctuations,

$$\vec{J}_H = \langle\vec{J}\rangle + \vec{\nabla}\times\vec{B}_H(\vec{r}),$$

we can construct from it a valid trial flux $\vec{J}^*$ for the lower bound (1.10) as follows:

$$\vec{J}^*(\vec{r}) = \langle\vec{J}\rangle + \vec{J}'_H(\vec{r}), \qquad \vec{J}'_H = \vec{\nabla}\times\vec{B}_H(\vec{r})$$

$$\vec{B}^*(r) = \begin{cases} \frac{1}{\ell}\,\zeta(\vec{r})\,\vec{B}_H(\vec{r}) & (\zeta < \ell) \\ \vec{B}_H(\vec{r}) & (\zeta > \ell) \end{cases}$$

where $\zeta(\vec{r})$ is the distance between $\vec{r}$ and the nearest point on the sample surface. Even rough estimation of $S(\vec{J}^*)$ shows that, if D_m is the minimum value of $D(\vec{r})$,

$$|S_T^{-1}\{\vec{J}_H\} - S_T^{-1}\{\vec{J}^*\}| < \frac{6}{D_m L}\left([\sqrt{\ell\langle J_H'^2\rangle} + \sqrt{\frac{1}{\ell}\langle B_H^2\rangle}]/|\langle\vec{J}\rangle|\right)^2$$

In the limit $L \to \infty$, therefore, boundedness of the variances in $\vec{J}_H'$ and $\vec{B}_H$ guarantees that homogeneous trial fluxes will produce lower bounds on D_e, even if (1.4b) is not satisfied at every point on the sample surface. What remains to be proven, of course, is that there exist homogeneous trial functions c_H and $\vec{J}_H$ that will narrow the gap between $S_D\{c_H\}$ and $S_T\{\vec{J}_H\}$ to zero for sufficiently large L. We should emphasize that there is no real doubt that this is so. Failure of upper and lower bounds to coincide in the following sections should be attributed solely to limitations on the choice of trial functions imposed by incomplete characterization of the material and considerations of computational tractability; no improvement, at least none that will survive the transition to infinite sample size, can be expected from strict adherence to the surface conditions (1.4).

2. Bounds Independent of Phase Geometry

Hashin and Shtrikman [5] set themselves the problem of how close the bounds (1.14) on D_e can be made if we know only D_α, D_β, the volume fraction ϕ of α-phase, and that the material is isotropic. To derive their results, we begin by rewriting the upper bound (1.14) in the form

$$D_e - \langle D\rangle \leq -2|\langle D\,\vec{\nabla}c'\rangle| + \langle D(\vec{\nabla}c')^2\rangle\,, \tag{2.1}$$

where $\vec{\nabla}c' \equiv (\vec{\nabla}c^* - \langle\vec{\nabla}c^*\rangle)/|\langle\vec{\nabla}c^*\rangle|$ is the relative fluctuation of the concentration gradient about its mean. Since $\langle\vec{\nabla}c'\rangle = 0$, we may multiply it by any constant λ and so generate another trial fluctuation $\lambda\vec{\nabla}c'$; for any choice of $\vec{\nabla}c'$, minimization with respect to λ gives

$$D_e - \langle D\rangle \leq - \langle D\vec{\nabla}c'\rangle^2/\langle D\vec{\nabla}c')^2\rangle$$

or (2.2)

$$D_e - \langle D\rangle \leq - \frac{(D_\alpha - D_\beta)^2\langle g\vec{\nabla}c'\rangle\rangle^2}{D_\alpha\langle(\vec{\nabla}c')^2\rangle - (D_\alpha - D_\beta)\langle(1 - g)(\vec{\nabla}c')^2\rangle}$$

So far only the isotropy of the material has been used; the set of trial functions $c'(\vec{r})$ still includes the true fluctuation distribution for which (2.2) becomes an equality. The next step virtually eliminates any likelihood of such an outcome by replacing the denominator on the right hand side with a larger quantity. If D_α is the greater of the two diffusion coefficients, then, since by Schwartz's inequality

$$\frac{\langle(1 - g)(\vec{\nabla}c')^2\rangle}{\langle 1 - g\rangle} \geq \frac{\langle(1 - g)\vec{\nabla}c')\rangle^2}{\langle 1 - g\rangle^2} = \frac{\langle g'\vec{\nabla}c'\rangle^2}{(1 - \phi)^2},$$

$$(g' \equiv g - \phi)$$

we can write the weaker inequality

$$D_e - \langle D\rangle \leq - (D_\alpha - D_\beta)^2\ (1 - \phi)/[D_\alpha(1 - \phi)\ \frac{\langle(\vec{\nabla}c')^2\rangle}{\langle g'\vec{\nabla}c'\rangle^2} - (D_\alpha - D_\beta)] \quad (2.3)$$

The weakening of the upper bound (2.3) relative to the original bound (2.2) is compensated by the disappearance of the average $\langle g(\vec{r})(\vec{\nabla}c')^2\rangle$ from the right hand side; the minimum value of the ratio $R \equiv \langle(\vec{\nabla}c')^2\rangle/\langle g\vec{\nabla}c'\rangle^2$ can be explicitly obtained, as we now show. The Euler-Lagrange equation associated with the minimization of R is

$$\nabla^2c'(\vec{r}) = \vec{B}\cdot\vec{\nabla}g', \quad \langle\vec{\nabla}c'\rangle = 0 \quad (2.4)$$

$$(\vec{B} \equiv \frac{\langle g'\vec{\nabla}c'\rangle}{\langle(\vec{\nabla}c')^2\rangle})$$

Equation (2.4) is readily solved, since the stochastic character of the problem appears only in the inhomogeneous term:

$$c'(\vec{r}) = \frac{1}{4\pi} \vec{B} \cdot \int \frac{\vec{\nabla} g'(\vec{r} + \vec{\rho})}{\rho} d^3\vec{\rho} \,. \tag{2.5}$$

The average $\langle(\vec{\nabla} c')^2\rangle$ therefore becomes

$$\begin{aligned}
\langle(\vec{\nabla} c')^2\rangle &= \frac{1}{16\pi^2} \iint [\vec{B}\,\vec{B} : \langle \vec{\nabla} g'(\vec{r} + \vec{\rho})\, \vec{\nabla} g'(\vec{r} + \vec{\rho}\,')\rangle] \\
&\qquad \times [\, \vec{\nabla} (\tfrac{1}{\rho}) \cdot \vec{\nabla} (\tfrac{1}{\rho'})] \, d^3\vec{\rho}\, d^3\vec{\rho}\,' \\
&= \frac{B^2}{48\pi^2} \iint \langle g'(\vec{r} + \vec{\rho})\, g'(\vec{r} + \vec{\rho}\,')\rangle \\
&\qquad \times [\, \vec{\nabla}\vec{\nabla} (\tfrac{1}{\rho}) : \vec{\nabla}\vec{\nabla} (\tfrac{1}{\rho'}) \times d^3\vec{\rho}\, d^3\vec{\rho}\,' \\
&= \frac{B^2}{3} \iint \langle g'(\vec{r} + \vec{\rho}) g'(\vec{r} + \vec{\rho}\,') \\
&\qquad \times \nabla^2 (\frac{1}{4\pi\rho}) \nabla^2 (\frac{1}{4\pi\rho'}) \, d^3\vec{\rho}\, d^3\vec{\rho}\,' \\
&= \frac{B^2}{3} \iint \langle g'(\vec{r} + \vec{\rho})\, g'(\vec{r} + \vec{\rho}\,')\rangle\, \delta(\vec{\rho})\, \delta(\vec{\rho}\,')\, d^3\vec{\rho}\, d^3\vec{\rho}\,' \\
&= \frac{B^2}{3} \langle (g')^2 \rangle = \frac{B^2}{3} \phi(1 - \phi)
\end{aligned} \tag{2.6}$$

A similar sequence of manipulations gives

$$|\langle g'\, \vec{\nabla} c' \rangle| = \frac{B}{3} \phi(1 - \phi) \,, \tag{2.7}$$

and we finally arrive at the Hashin-Shtrikman upper bound

$$D_e < \langle D \rangle - \frac{(D_\alpha - D_\beta)^2 \phi(1-\phi)}{3D_\alpha - (D_\beta - D_\alpha)\phi} \,. \tag{2.8}$$

A corresponding line of argument can also be applied to the lower bound in (1.14), resulting in the inequality

$$D_e > [\, \langle \frac{1}{D} \rangle - \frac{(D_\alpha - D_\beta)^2 \phi(1 - \phi)}{(\frac{3}{2} D_\alpha - (D_\alpha - D_\beta)(1 - \phi)) D_\alpha D_\beta} \,]^{-1} \,. \tag{2.9}$$

Hashin and Shtrikman were also able to construct isotropic two-phase geometries for which the bounds (2.8) and (2.9) are actually attained, thus demonstrating that these two inequalities provide the best bounds obtainable for an isotropic material if only the volume fractions of the two phases are known.

3. Bounds Involving Spatial Correlations of $g(\vec{r})$.

Information about $g(\vec{r})$ that goes beyond mere specification of its average value can be offered and used in many forms. We can for instance substitute into (2.2) a trial fluctuation distribution of the form [9]

$$c'(\vec{r}) = \int g'(\vec{r} + \vec{\rho})\, h(\vec{\rho})\, d^3\vec{\rho} \,, \tag{3.1}$$

where $h(\vec{\rho})$ reflects the influence upon c' at $\vec{r}$ exerted by the distribution of α and β phases in the vicinity of $\vec{r}$. For the case of an isotropic porous material $(D_\beta = 0)$, the result is

$$D_e < \phi D_\alpha \left[1 - \frac{|\int h(\vec{\rho})\vec{\nabla}s(\rho)d^3\vec{\rho}|^2}{\phi \iint h(\vec{\rho})h(\vec{\rho}):\vec{\nabla}\,\vec{\nabla}'G^*(\vec{\rho},\vec{\rho}\,')d^3\vec{\rho}\, d^3\vec{\rho}\,'}\right] \tag{3.2}$$

where $s(\rho) \equiv \langle g'(\vec{r})g'(\vec{r} + \vec{\rho})\rangle$, $G^*(\vec{\rho},\vec{\rho}\,') \equiv \langle g'(\vec{r})g'(\vec{r} + \vec{\rho})g'(\vec{r} + \vec{\rho}\,')\rangle$, and the gradients $\vec{\nabla}$ and $\vec{\nabla}'$ are to be taken with respect to $\vec{\rho}$ and $\vec{\rho}\,'$ respectively. The influence function $h(\vec{\rho})$ that will give the least upper bound on D_e must, in view of the isotropy of the material, have the form

$$h(\vec{\rho}) = \vec{k} \cdot \vec{\rho}\;\; H(\rho) \,, \tag{3.3}$$

where $\vec{k}$ is the unit vector in the direction of the mean concentration gradient, and $H(\rho)$ satisfies the integral equation

$$\int_0^\infty H(\rho')\, \Gamma\, (\rho,\rho')\, \rho'^2\, d\rho' = -\, \rho\, \frac{ds}{d\rho} \tag{3.4}$$

$$(\Gamma(\rho,\rho') \equiv \iint \vec{\rho} \cdot \vec{\rho}\,'\; \vec{\nabla} \cdot \vec{\nabla}'\; G^*(\vec{\rho},\vec{\rho}\,')\; d^2\vec{\Omega}')$$

($d^2\vec{\Omega}'$ is the solid angle element in the space of $\vec{\rho}\,'$) .

In contrast to (2.4), which it replaces, (3.4) cannot be solved in closed form; Beran [11] has suggested that the solution to (2.4) should nevertheless generate a valid upper bound on D_e, which amounts to setting $H(\rho)$ equal to $1/\rho^3$. This simplification (and consequent weakening) of (3.2) does not however remove the main obstacle to the wider application of this inequality: the lack of information about G^*. Until recently, the only explicit evaluation of the bound was for the special case of a so-called "symmetric" material [12], in which $\phi = 1/2$ and the statistics of the phase geometry is invariant with respect to interchange of the α- and β- phases. For that case, it can be shown [13] that $G^*(\vec{\rho},\vec{\rho}\,') = \phi s(|\vec{\rho} - \vec{\rho}\,'|)$, and (3.4) can then be solved to give $H(\rho) = 1/\rho^3$, i.e. Beran's approximation provides the optimal bound here. For an isotropic, symmetric material, with $D_\beta = 0$, the effective diffusion coefficient can not exceed $(1/3)D_\alpha$ (the Hashin-Shtrikman upper bound is $(2/5)D_\alpha$).

4. Bounds for Beds of Spherical Particles

Instead of a direct statistical description of the phase geometry, one may instead know something of the process used to generate that geometry. For example Weissberg [14] has studied diffusion in a porous medium generated by random placement of n spheres of radius $\underline{a}$ per unit volume, with no correlation between positions of different spheres; clusters of overlapping spheres are simply treated as particles of more complex shape. A simple trial fluctuation distribution for this model is

$$\vec{\nabla}c'(\vec{r}) = \vec{A} + \sum_i \vec{\nabla} f(\vec{r} - \vec{r}_i) , \tag{4.1}$$

where

$$f(\vec{\rho}) = \vec{k} \cdot \vec{\rho}\, F(\vec{\rho}) ,$$

$$\vec{A} = -n \int \vec{\nabla} f(\vec{\rho})\, d^3\vec{\rho} = -\frac{4\pi}{3} n \vec{k} \lim_{\rho\to\infty} (\rho^3 F(\rho)) ,$$

$\vec{r}_i$ is the position of sphere i, and $\vec{k}$ is once more the unit vector in the direction of the mean concentration gradient. Essentially (4.1) constructs the trial function as a superposition of contributions from individual spheres, with the constant vector $\vec{A}$ added on to ensure $\langle \vec{\nabla} c' \rangle = 0$ (requiring $\rho^3 F$ to vanish at infinity achieves the same purpose, but produces a much weaker bound). Weissberg now shows that the best choice for $f(\vec{\rho})$ is the concentration perturbation caused by an isolated impenetrable sphere set in a uniform gradient field. His final result is

$$D_e < D_\alpha \phi/(1 - \tfrac{1}{2} \ln \phi) \tag{4.2}$$

The trial function (4.1) can of course be used even if the positions of the spheres are not uncorrelated [15]. This results in bounds that are expressed in terms of distribution functions closely akin to the two- and three- particle distributions of statistical mechanics [8]. If $nQ(\vec{\rho})$ is the concentration of sphere centers at $\vec{r} + \vec{\rho}$ when $\vec{r}$ is known to be in the α phase, and if $n^2 R(\vec{\rho}, \vec{\rho}\,')$ is the analogous two-sphere distribution, then Weissberg's bound generalizes to

$$D_e < D_\alpha \phi \left[1 - \frac{n(\int (Q(\rho) - 1)\, \vec{\nabla} f(\vec{\rho}) d^3\vec{\rho})^2}{\int Q(\rho)(\vec{\nabla} f(\vec{\rho}))^2 d^3\vec{\rho} + n \iint R^*(\vec{\rho}, \vec{\rho}\,') \vec{\nabla} f(\vec{\rho}) \cdot \vec{\nabla} f(\vec{\rho}\,') d^3\vec{\rho}\, d^3\vec{\rho}\,'}\right], \tag{4.3}$$

$$(R^*(\vec{\rho}, \vec{\rho}\,') \equiv R(\vec{\rho}, \vec{\rho}\,') - Q(\rho) - Q(\rho') + 1)$$

with $f(\vec{\rho})$ once again to be chosen so as to minimize the right hand side. For randomly overlapping spheres, $Q(\rho)$ is 1 for $\rho > a$ and zero other wise, $R(\vec{\rho}, \vec{\rho}\,') = Q(\rho)\, Q(\rho')$, and $\phi = \exp(-\frac{4\pi}{3} n a^3)$; the optimum choice for $f(\vec{\rho})$ is $\vec{k} \cdot \vec{\rho}/\rho^3$, and substitution of all this into (4.3) recovers the inequality (4.2).

The inequalities (3.2) and (4.2) may be generalized to materials in which D_α and D_β are both finite, and it is then possible to obtain non-trivial lower bounds as well [16]. Recent calculations by Stell permit comparison of these results for the randomly overlapping spheres geometry. The outcome (see Figs. 2-4 of Stell's article in this volume) show that the considerable labor involved in doing the integrals in (3.2) and the corresponding lower bound does produce a narrower squeeze on D_e, even though Beran's choice for $h(\vec{\rho})$ was used; one would like to know whether optimization according to (3.4) is capable of producing significant improvement, particularly as D_α/D_β becomes large.

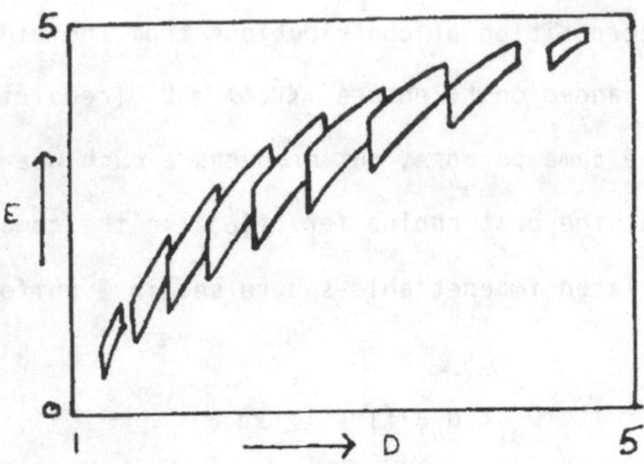

Fig. 1. Bounds on $D \equiv D_e/D_\beta$ and $E \equiv (d\, \ell n(D_e/D_\beta)/d\, \ell n\, D_\beta)$ at various volume fractions ϕ of α-phase for the case $D_\alpha/D_\beta = 5$ and $d\, \ell n(D_\alpha/D_\beta)/d\, \ell n\, D_\beta = 5$ (ref. 17). Each window represents the possible values of D and E for a particular choice of ϕ ; from left to right, $\phi = 0.1, 0.2, \ldots 0.9$.

---

5. Bounds Involving Knowledge of Other Bulk Properties

Yet a third way of characterizing a random material is in terms of measurements of other bulk properties. Here we are told nothing about the phase geometry as such, but are instead given, for example, the information that at temperature T , where the diffusion coefficients in the individual phase are $D_\alpha^{(1)}$ and $D_\beta^{(1)}$, the effective diffusion coefficient is $D_e^{(1)}$. What can be predicted about the value $D_e^{(2)}$ of the effective diffusion coefficient of the same material at another temperature T_2, where D_α and D_β take on different values $D_\alpha^{(2)}$ and $D_\beta^{(2)}$? Since the concentration and flux distributions at T_1 are legitimate trial functions for the variational bounds (1.14) at T_2 , it should be possible to use knowledge of $D_e^{(1)}$ to impose limits on $D_e^{(2)}$. Such arguments, when combined with applications of the Schwartz inequality along lines similar to those used in the derivation of (2.8), can produce a variety of useful inequalities that do not require knowledge of the actual phase geometry [17]. An example is shown in Fig. 1.

Bergman [20] has found a more direct route to this type of result. For a two-phase material, he first shows that D_e can be expressed in the form

$$D_e = D_\alpha \left[1 - (1 - \phi)(1 - \frac{D_\beta}{D_\alpha}) \int \frac{B(u)du}{1 - \xi(u)(1 - \frac{D_\beta}{D_\alpha})}\right] , \qquad (5.1)$$

where $D_\alpha > D_\beta$, and $B(u)$ and $\xi(u)$ are subject to the restrictions

$$B(u) < 0 \qquad 0 < \xi(u) < 1 \tag{5.2}$$

For an isotropic geometry, Brown's expansion (1.5) imposes the further conditions

$$\begin{aligned} \int B(u)\,du &= 1 \\ \int B(u)\,\xi(u)du &= \tfrac{1}{3}\,\phi\,, \end{aligned} \tag{5.3}$$

and maximizing the right hand side of (5.1) subject to (5.2) and (5.3) then gives $\xi(u) = \frac{1}{3}\phi$, which returns us to the Hashin-Shtrikman upper bound (2.8). Moreover specifying D_e for one or more D_α, D_β pairs provides additional constraints on $\xi(u)$ and $B(u)$, and an improvement in the upper bound. Bergman has worked out several examples of both upper and lower bounds along these lines, including cases involving three-phase materials.

On a more fundamental level, Bergman's work makes the connection between the variational approach discussed here and the Padé approximant methods used by Gordon and others in a variety of applications [19]. Bergman has also discussed the question of percolation and the approach to the percolation limit [20], an area where variational methods have been notoriously unhelpful.

6. Bounds for Many-Body Diffusion Problems

So far we have assumed a static phase geometry and no interaction between molecules of the diffusing species. However the basic diffusion problem is readily extended to include many-body Brownian motions, such as arise in the interdiffusion of two solute species α and β [21]. Let D_α, D_β be the diffusion coefficients of isolated α and β species, c_α, c_β their bulk concentrations, and $\vec{A}_\alpha$, $\vec{A}_\beta$ constant external forces acting on each α or β molecule, subject to the force balance restriction

$$\vec{A}_\alpha c_\alpha + \vec{A}_\beta c_\beta = 0 \tag{6.1}$$

In addition to the external fields, solute particles also experience intermolecular forces which can be written as gradients of a system potential function

$$\Phi(\vec{r}_1,\vec{r}_2,\ldots) = \sum_m \sum_{<n} \zeta_{mn}(r_{mn})\,, \tag{6.2}$$

where $\vec{r}_i$ is the position of particle i, $\vec{r}_{mn} \equiv |\vec{r}_m - \vec{r}_n|$, and the pair interaction potential $\zeta_{mn}(r)$ is $\zeta_{\alpha\alpha}(r)$, $\zeta_{\alpha\beta}(r)$, or $\zeta_{\beta\beta}(r)$, depending on the species of particles m and n.

An ensemble of such systems, all of the same composition, size, and temperature, may be characterized by a configurational density $\psi(\vec{r}_1,\vec{r}_2,\ldots,t)$ which replaces the concentration c in our earlier discussions ($\psi\, d^3\vec{r}_1, d^3\vec{r}_2\ldots$ is the fraction of ensemble members at time t in the volume element $d^3\vec{r}_1 d^3\vec{r}_2\ldots$ at the location $\vec{r}_1,\vec{r}_2,\ldots$). Under the action of the time- and position - independent external forces $\vec{A}_\alpha$, $\vec{A}_\beta$, the ensemble eventually reaches a steady state in which ψ becomes time-independent (we ignore accumulation or depletion at the boundaries of the system). If $|c_\alpha \vec{A}_\alpha| = |c_\beta \vec{A}_\beta|$ is sufficiently small, this steady state ψ will differ but little from the equilibrium distribution

$$\psi_e\,(\vec{r}_1,\vec{r}_2,\ldots) = \text{const. } e^{-\Phi/kT} \tag{6.3}$$

reached by the system in the absence of external forces, and the deviation $H \equiv \frac{\psi}{\psi_e} - 1$ then satisfies the multi-dimensional diffusion equation

$$\sum D_i\, \vec{\nabla}_i \cdot [\, \psi_e(\vec{\nabla}_i H - \vec{A}_i/kT)] = 0\,, \tag{6.4}$$

where $\vec{\nabla}_i$ is the gradient operator with respect to $\vec{r}_i$. The detailed boundary conditions on equation (6.4) are not important if the system is large enough, so long as they ensure that all gradients of H vanish on the average. This type of many-particle diffusion was first introduced by Debye and Hueckel and by Onsager [22] in connection with electrolytic conduction.

Just as in the case of simple diffusion, solving the steady state problem (6.4) is equivalent to the minimization of a functional,

$$S\{H\} \equiv \tfrac{1}{2} \sum_i D_i \int \psi_e\, [\, \vec{\nabla}_i H - \vec{A}_i/kT]^2\, d^3\vec{r}_i d^3\vec{r}_2\ldots d^3\vec{r}_N \tag{6.5}$$

where V is the 3-dimensional volume to which the N solute particles are confined. Also, S is once again related to an experimentally accessible characteristic of the transport induced by the forces $\vec{A}_\alpha$, $\vec{A}_\beta$, namely the mean speed u of an α - relative to a β - molecule:

$$u \equiv -\frac{1}{A} \sum_i D_i \vec{A}_i \cdot \int_{\vec{r}_1,\vec{r}_2..\varepsilon V} \psi_e \, (\vec{\nabla}_i H - \vec{A}_i/kT) d^3\vec{r}_1 d^3\vec{r}_2 \dots d^3\vec{r}_N \tag{6.6}$$

$$(A \equiv |Vc_\alpha \vec{A}_\alpha| = |Vc_\beta \vec{A}_\beta|)$$

If $\overline{H}$ is the true steady state deviation from equilibrium, the analog of (1.8) for this system is

$$S\{\overline{H}\} = \frac{1}{2kT} Au , \tag{6.7}$$

and for a trial deviation H^* we have therefore the inequality

$$S\{H^*\} \geq \frac{1}{2kT} Au , \tag{6.8}$$

The simplest choice for H^* is a superposition similar to (4.1):

$$H^* = \sum_i{}' \sum_j \xi_{ij} h(\vec{r}_i - \vec{r}_j) , \tag{6.9}$$

where ξ_{ij} is 1 if particle i belongs to the α species and particle j to the β species, and vanishes otherwise. The result of substituting (6.9) into (6.8) can be written in terms of the relative two- and three-particle distribution functions $g_{\alpha\beta}(r_{\alpha\beta})$, $g_{\alpha\alpha\beta}(r_{\alpha\beta}, r_{\alpha'\beta})$, $g_{\beta\beta\alpha}(r_{\beta\alpha}, r_{\beta'\alpha})$ of equilibrium statistical mechanics, all normalized to approach unity when <u>all</u> distances between the particles involved become large [8] (for example, $c_\alpha^2 c_\beta g_{\alpha\alpha\beta} \, d^3\vec{r}_\alpha \, d^3\vec{r}_{\alpha'} d^3\vec{r}_\beta$ is the probability that, at equilibrium in the absence of external forces, the three volume elements $d^3\vec{r}_\alpha$, $d^3\vec{r}_{\alpha'}$, and $d^3\vec{r}_\beta$ at $\vec{r}_\alpha \equiv \vec{r}_{\alpha\beta} + \vec{r}_\beta$, $\vec{r}_{\alpha'} \equiv \vec{r}_{\alpha'\beta} + \vec{r}_\beta$, and $\vec{r}_\beta$ are occupied respectively by an α- , a second α- , and a β- particle). After simplification, we obtain the inequality

$$kTc_\beta \frac{u - u_0}{A_\alpha} \leq - \frac{(D_\alpha c_\beta + D_\beta c_\alpha)^2 \, [\int g_{\alpha\beta}(\rho) \vec{\nabla} h(\vec{\rho}) d^3\vec{\rho}]^2}{(D_\alpha + D_\beta) \int g_{\alpha\beta}(\rho) \, (\vec{\nabla} h(\vec{\rho}))^2 \, d^3\vec{\rho} + \Gamma} , \tag{6.10}$$

$$\Gamma \equiv \iint [\, D_\alpha c_\beta g_{\alpha\beta\beta} \, (\vec{\rho},\vec{\rho}\,') + D_\beta c_\alpha g_{\alpha\alpha\beta}(\vec{\rho},\vec{\rho}\,')] \; \vec{\nabla} h(\vec{\rho}) \cdot \vec{\nabla} h(\vec{\rho}\,') d^3\vec{\rho} \, d^3\vec{\rho}\,'$$

$$u_0 \equiv \frac{1}{kT} (D_\alpha A_\alpha - D_\beta A_\beta)$$

which is essentially a generalization of (4.3). Indeed, if $\zeta_{\alpha\alpha}$ and $\zeta_{\beta\beta}$ vanish and $c_\alpha \ll c_\beta$, (6.10) begins to look very similar to Weissberg's bound (24):

$$u < u_0/(1 + c_\beta D_\alpha \gamma / D(D_\alpha + D_\beta)) , \qquad (6.11)$$

where $\gamma \equiv (1 + D_\beta/D_\alpha) \lim_{c_\beta \to 0} (d(u/u_0)/dc_\beta)$ depends only on the form of the pair potential $\zeta_{\alpha\beta}(r)$.

Before concluding, we should point out that variational inequalities have been developed for properties of random and many-particle systems other than effective diffusion coefficients or mobilities and their mathematical isomorphs. A partial listing is given in reference [23] through [29]. Much additional material is to be found in the book by Strieder and Aris [30].

References

1. J.B. Keller, "Conductivity of a Medium Containing a Dense Array of Perfectly Conducting Spheres or Cylinders or Non-Conducting Cylinders", J. Appl. Phys. 34 (1963) 991-993.

   J.L. Jackson and S.R. Coriell, "Effective Diffusion Constant in a Polyelectrolyte Solution", J. Chem. Phys. 38 (1963) 959-968.

   H. Levine, "The Effective Conductiviy of a Regular Composite Medium", J. Inst. Maths. Applics. 2 (1966) 12-28.

   D.R. McKenzie and R.C. Mc Phedran, "Exact Modeling of Cubic Lattice Permittivity and Conductivity", Nature 265 (1977) 128-129.

2. For a review see G.K. Batchelor "Transport Properties of Two-Phase Materials of Random Structure", Ann. Rev. Fluid Mech. 6 (1974) 227-255.

   P.H. Winterfeld, L.E. Scriven and H.T. Davis, "Percolation and Conductivity of Random Two-Dimensional Composites" J. Phys. C 14 (1981) 2361-2376.

3. W.F. Brown, "Solid Mixture Permittivities", J. Chem. Phys. 23 (1955) 1514-1517.

   S. Prager, W. Kunkin, and H.L. Frisch, "Variational Approach to the Dielectric Constant of a Polarizable Medium", J. Chem. Phys. 52 (1970) 4925-4930.

   W.E.A. Davies, "The Theory of Composite Dielectrics" J. Phys. D 4 (1971) 318-328.

4. For a review see M. Beran, Statistical Continuum Theories, (New York, Interscience, 1968).

5. Z. Hashin and S. Shtrikman, "A Variational Approach to the Theory of the Effective Magnetic Permeability of Multiphase Materials", J. Appl. Phys. 33 (1962) 3125-3131.

6. S. Prager, "Diffusion in Inhomogeneous Media", J. Chem. Phys. 33 (1960) 122-127.

7. J.C. Maxwell, A Treatise on Electricity and Magnetism, Vol. 1 (Oxford, Clarendon Press, first published 1873, 3rd edition 1891; reprinted 1954 by Dover, New York).

D.J. Jeffrey, "Conduction through a Random Suspension of Spheres", Proc. Roy. Soc. Lond. A. 335 (1973) 355-367.

8. See for example D.A. McQuarrie, Statistical Mechanics" (New York, Harper and Row, N.Y. 1976), Chap. 13.

9. S. Prager, "Diffusion and Viscous Flow in Concentrated Suspensions", Physica 29 (1963) 129.

10. Excellent discussions of reciprocal variation principles is to be found in J.L. Synge, The Hypercircle in Mathematical Physics, (Cambridge University Press, 1957) and in A.M. Arthurs, Complementary Variational Principles, (Oxford, Clarendon Press, London, 1970).

11. M.J Beran, "Use of the Variational Approach to Get Bounds for the Effective Permittivity in Random Media", Nuovo Cimento 38 (1965) 771-782.

12. W.F. Brown, "Dielectric Constants, Permeabilities, and Conductivities of Random Media", Trans. Soc. Rheology 9 part 1 (1965) 357-380; an interesting extension to non-symmetric media has been given by M.N. Miller, "Bounds for effective electrical, thermal, and magnetic properties of heterogeneous materials", J. Math. Phys. 10 (1969) 1988-2004.

13. H.L. Frisch, "Statistics of Random Media", Trans. Soc. Rheology 9, part 1 (1965) 293-312

14. H.L. Weissberg, "Effective Diffusion Coefficients in Porous Media", J. Appl. Phys. 34 (1963) 2636-2639.

15. H.L. Weissberg and S. Prager, "Viscosity of Concentrated Suspension", Trans. Soc. Rheology 9, part 1 (1965) 321-338 (discusses the viscosity analog of the diffusion problem treated here).

16. A.L. DeVera and W. Strieder, "Upper and Lower Bounds on the Effective Thermal Conductivities of a Random, Two-Phase Material", J. Phys. Chem. 81 (1977) 1783-1790.

17. S. Prager, "Improved Variational Bounds on Some Bulk Properties of a Two-Phase Random Medium", J. Chem. Phys. 50 (1969) 4305-4312.

18. D.J. Bergman, "Calculation of Bounds for Some Average Bulk Properties of Composite Materials", Phys. Rev. B 14 (1976) 4304-4312.

19. R.G. Gordon, "Error Bounds in Equilibrium Statistical Mechanics", J. Math. Phys. 9 (1968) 655-663.

G.A. Baker, Jr. Essentials of Pade Approximants, (New York, Academic Press, 1975).

20. D.J. Bergman, "Analytical Properties of the Complex Effective Dielectric Constant of a Composite Medium with Applications to the Derivation of Rigorous Bounds and to Percolation Problems", in Electrical Transport and Optical Properties of Inhomogeneous Media, J.C. Garland and D.B. Tanner, eds., AIP Conference Proceedings No. 40 (New York, American Institute of Physics, 1978), pp. 46-62.

21. G. Woodbury and S. Prager, "Brownian Motion in Many-Particle Systems", J. Chem. Phys. 38 (1963) 1446, and J. Am. Chem. Soc. 86, (1964) 3417.

22. P. Debye and W. Hueckel, "Zur Theorie der Elektrolye II. Das Grenzgesetz fuer die Elektrolytische Leiffaehigkeit", Physik. Z. 24 (1923) 185-206.

L. Onsager, "Zur Theorie der Elektrolyte", Physik Z. 27 (1926) 388.

23. S. Prager, "Viscous Flow Through Porous Media", Phys. Fluids 4 (1961) 1477-1482.

H.L. Weissberg and S. Prager, "Viscous Flow Through Porous Media III. Upper Bounds on the Permeability for a Simple Random Geometry", Phys. Fluids 13 (1970) 2958-2965.

24. Z. Hashin and S. Shtrikman, "A variational approach to the theory of the elastic behaviour of multiphase materials", J. Mech. Phys. Solids (1963) 127-140.

Z. Hashin, "Theory of mechanical behavior of heterogeneous media", Appl. Mech. Rev. 17 (1964) 1-9.

25. R.A. Reck and S. Prager, "Diffusion-Controlled Quenching at Higher Quenches Concentrations", J. Chem. Phys. 42 (1965) 3027-3032.

26. W. Strieder and S. Prager, "Knudsen Flow Through a Porous Medium", Physics of Fluids 11 (1968) 2544-2548, W. Strieder and C.Y. Shiau, "Surface Mobility in Transport Across a Porous Medium with Knudsen Diffusion in the Pores", J. Colloid Interface Sci. 51 (1975) 152-161.

F.G. Ho and W. Strieder, "A Variational Calculation of Effective Surface Diffusion Coefficient and Tortuosity", Chem. Eng. Sci. 36 (1982) 253-258.

27. J. Rotne and S. Prager, "Variational Treatment of Hydrodynamic Interaction in Polymers", J. Chem. Phys. 50 (1969) 4831-4837.

S. Prager, "Variational Bounds on the Intrinsic Viscosity", J. Phys. Chem. 75 (1971) 72-78.

M. Fixman, "Variational Bounds for Polymer Transport Coefficients", J. Chem. Phys. 78 (1983) 1588-1593.

M. Fixman, "Effect of Fluctuating Hydrodynamic Interaction", J. Chem. Phys. 78 (1983) 1588-1593.

28. G.H. Malone, T.F. Hutchinson and S. Prager, "Molecular Models for Permeation Through Thin Membranes", J. Fluid Mech. 65 (1974) 753-767. See also the article by Malone, Suh and Prager in this volume.

29. K. Schulgasser and Z. Hashin, "Bounds for Effective Permittivities of Lossy Dielectric Composites", J. Appl. Phys. 47 (1976) 424-427.

D.J. Bergman, "Exactly Solvable Microscopic Geometries and Rigorous Bounds for the Complex Dielectric Constnat of a Two-Component Composite Material", Phys. Rev. Letters 44 1285-1287 (1980); see also ref. 20.

30. W. Strieder and R. Aris, Variational Methods Applied to Problems of Diffusion and Reaction (Berlin, Springer Verlag, 1973).

ON THE WEDDING OF CERTAIN DYNAMICAL PROCESSES IN DISORDERED COMPLEX MATERIALS TO THE THEORY OF STABLE (LÉVY) DISTRIBUTION FUNCTIONS

Elliott W. Montroll and Michael F. Shlesinger

Institute for Physical Science and Technology
University of Maryland
College Park, MD 20742

and

La Jolla Institute
Post Office Box 1434
La Jolla, CA 92038

ABSTRACT

We review and comment on styles of applied mathematics before exhibiting our own in regard to relating the theory of Lévy's stable distributions to dynamic processes in complex disordered materials. Lévy's probability distributions have long tails, infinite moments and elegant scaling properties. Our first example connects intermittant currents in certain xerographic films to a Lévy distibution of waiting times for the jumping of charges out of a distribution of deep traps. We then extend our analysis from transport to electron-hole recombination reactions in amorphous materials. A Lévy distribution of first passage times appears both in this recombination problem as well as in the dielectric relaxation phenomena described by the Williams-Watts formula. Lastly, the most famous scaling problem, "1/f noise", is shown to be related to a log-normal distribution of relaxation times. We derive the log-normal distribution in a generic fashion and show it to be a limiting form of a Lévy distribution.

1. Some Remarks on a Certain Classical Style of Applied Mathematics

Hans, George, and Barry, we thank you for inviting us to participate in your important conference series that is attempting to apply mathematical ideas to significant technical problems, and we especially applaud (for reasons given below) your excellent choice of the subject of the meeting.

Among the many schools of applied mathematics a variety of styles have evolved. As an introduction to our report, we thought it useful to describe our style. Perhaps the style has been partly motivated by the fact that the senior of us (EWM) has in raising ten growing children been forced to make consulting a critical component of his scientific life.

We feel that it is important to have a cadre of applied mathematicians who operate in the manner of the old family doctor, the general practictioner whose door is open to receive all clients with a collection of numbers, or other mathematically describable objects that they wish to have put into some usable or understandable form. Such applied mathematicians must have a great (but not necessarily deep) storage of lore of old and new forms of mathematical science including some forgotten old relics of the 19th and 18th centuries; they should be expert in making rapid searches through the literature; they should be able to recall a wide variety of mathematical models that have been applied successfully and some unsuccessfully in a diversity of other sciences; and it is absolutely necessary that they have numerous learned friends who are easily available through the wonderful electronic data retriever, the telephone. Of course today everyone has some computer available from pocket size up.

What are some successful examples of this style of applied mathematics? We briefly relate one that is universally known and a second of almost as great importance but which is practically unknown to the mathematics and physics communities.

Our first success story relates how a hundred years of intense mathematical research on second order differential equations and thirty years of matrix theory provided the precise instrument for putting order into enormous lists of spectroscopic data that had been accumulating for about seventy-five years. On the mathematical side, Richard Courant provided a useful service of the publication of Courant-Hilbert's Methoden der Mathematischen Physik [1] in 1924 by giving a concise summary of many of the important results of the two fields. The physical side of the story has been written in dozens of places but our favorite rendition is the charming set of reminiscences [2] of Felix Bloch on his student days in the heroic period of the middle 1920's. As a young student in Peter Debye's Theoretical Physics Seminar at the ETH (Zürich) in 1926, he was an eyewitness to the birth of wave mechanics.

Niels Bohr had, through the old quantum theory, put a bit of order into some simple spectra but it was clear in the 1920's that a deeper theory was needed to give a real understanding to the subject. In 1926 Louis de Broglie wrote a remarkably imaginative and original Ph.D. thesis proposing that the motion of small matter particles has a wave character. Copies of the thesis were delivered to several important physicists including Albert Einstein and Peter Debye. Einstein sent a postcard to his friend Erwin Schrödinger (who was at University of Zürich and a participant of the Debye Seminar) suggesting that he might find the de Broglie thesis entertaining and provocative.

According to Bloch, at the end of a colloquium: "I heard Debye say something like: 'Schrödinger, you are not working now on very important problems anyway. Why don't you tell us something about that thesis of de Broglie, which seems to have attracted some attention'." With orders from the boss and the recommendation of Einstein, he hadn't much choice but to dutifully, eventually review the thesis. "So in one of the next colloquia, Schrödinger gave a beautifully clear account of how de Broglie associated a wave with a particle .... When he had finished, Debye casually remarked that he thought this way of talking was rather childish. As a student of Sommerfeld he had learned that to deal properly with waves, one had to have a wave equation... Just a few weeks later Schrödinger gave another talk at the colloquium which he had started by saying: 'My colleague Debye suggested that one should have a wave equation; well I found one!,'" Thus the Schrödinger equation was born.

The small band of theoretical physicsts at Zürich was on very friendly terms with the corresponding tribe of mathematicians. Schrödinger's mathematical colleagues H. Weyl and E. Fues are thanked for their advice at various points in the original papers and Courant-Hilbert is quoted far more often than any other reference source. The original Schrödinger papers (which are available in English in a Chelsea reprint [3]) are still a joy to read and are highly recommended to young applied mathematicians and theoretical physicists who wish to study the style of the Masters.

Unfortunately, when certain more complicated forces were added to atomic models, say electrical and magnetic fields to the hydrogen atom and anharmonic terms to the oscillator problem, Courant-Hilbert wasn't very helpful to Schrödinger. However, as did all good theoretical physicists of his time, Schrödinger had carefully studied Rayleigh's "Theory of Sound" and remembered Rayleigh's perturbation methods that applied well to his more difficult problems.

The finale of the story is, of course, that Schrödinger's wave mechanics combined with the mechanics of Heisenberg, Born, and Jordan based on matrix theory revolutionized our understanding of atoms, molecules, and condensed matter and gave us a completely new style of thought for the investigation of physical, chemical, and biological processes.

Our second success story, one practically never told, yielded our basic understanding of long distance radio propagation, a technology central to our modern life style. The mathematical background for the success was a hundred years of intense mathematical research on the theory of Bessel functions.

Our saga begins on the stormy winter day, December 12, 1901. Young Guilio Marconi had for some years been constructing higher powered wireless transmitters and larger antennas to the degree that he could transmit signals across the English Channel; then, the Irish Sea. The obvious next step was to span the Atlantic Ocean. At this point, his learned theoretical friends said: "Guilio, enough is enough! The inverse square law will kill you. Your dipole radiation source will be spread in all directions -- nothing will be observed at your Newfoundland Station." As a good experimentalist, he ignored the theoreticians' advice. When anomalously strong signals were received on December 12, 1901, Lord Rayleigh emphasized that plane wave transmission could not account for the observations and proposed that a diffraction mechanism must be used to explain how the wave followed the curvature of the earth [4].

Rayleigh's proposal led to numerous papers on this diffraction theory by the great theoreticians of the early 1900's: Rayleigh, Poincaré, Sommerfeld, Love, van der Pol, Nicholson, etc. They derived complicated Legendre-Bessel function series to describe the phenomenon--none of which exhibited the slightest agreement with the experimental observations. "Many years of work by some of the most distinguished of the world's mathematicians did not suffice to bring this apparently innocent problem beyond the stage at which Nicholson could say that it was the one in the whole field of mathematical analysis about which the most divergent views were held." [4]

The wisest of the theoreticians facing the problem was Oliver Heaviside (whom we consider, in agreement with Norbert Wiener, to have been one of the giants of applied mathematicians in the period 1880-1910). Oliver's father died when the boy was quite young. His formal education at the girls' school where his mother (a sister of Charles Wheatstone) taught was terminated at age 13 when he refused to further mingle with the girls. He became apprenticed to a telegrapher and studied telegraphy and electricity by himself. In the process he became a self-taught, sophisticated mathematician of a most imaginative and eccentric type, at constant odds with the systematic, rigorous, Cambridge Tripos-trained professionals.

Heaviside's response to the unexpected power in the signal of 12 December was clear and rapid. Since the tyranny of the inverse square law failed to emerge, he proposed that there must be a mirror "up there" to reflect the signals. No journals seemed to be interested in publishing his conjecture. By good fortune, he was writing an article on a slightly different topic for the Encyclopedia Britannica and freely appended the remark [5]: "There may possibly be a sufficiently conducting layer in the upper air. If so, the waves, so to speak, catch onto it more or less. Then guidance will be by sea on one side and the upper air on the other". Today we would consider this phenomenon as electromagnetic wave propagation in a naturally occurring wave guide. A similar remark was made by Kennelly. The idea was further developed by Eccles (1912) who named the conducting layer "Heaviside's Layer".

The surprise of 12 December 1901 immediately awakened the navies of the world to the potential of wireless communication. The U.S. Navy conducted experiments in 1909 and 1910 under the supervision of L.W. Austin of the National Bureau of Standards. "On the cruisers Birmingham and Salem, numerous qualitative observations of signal strength versus wave length were made in cooperation with the powerful land station at Brant Rock, Massachusetts, while the ships moved away from fixed stations" [4]. Antennas of varying height were also employed. Reams of data of varying reliability were collected until the situation became comparable to that of the early spectroscopists.

Two geniuses of data analysis, Austin and L. Cohen of the Naval Electrical Laboratory (now NRL) completely summarized the mess in the magic Austin-Cohen formula [4]

$$I_R = 4.25\ I_S\ (h_1 h_2/\lambda d)\ \exp\ (-\alpha d \lambda^{-1/2})$$

where: I_R = current in a 25 ohm receiving antenna of height h_2 kilometers, I_S = current in transmitting antenna of height h_1 kilometers, d = distance between transmitter and receiver in kilometers, λ = wave-length in kilometers, and finally $\alpha = 0.0015$ for transmission over sea water. Cohen seemed to have had the inspiration to put the $\lambda^{-1/2}$ in the exponent so that all the data finally fell into place.

The hero of our story, G.N. Watson, was a mathematician at University College, London, who wanted to do war work at the beginning of World War I, his choice of topic becoming the investigation of naval wireless propagation. He started his investigation by studying the various unsuccessful diffraction theories of his distinguished predecessors. In attempting to put order into their mathematics, he converted their slowly convergent series of combinations of Bessel functions and Legendre polynomials, which represented the electric field strength near the earth, into a rapidly converging series through the use of a transformation now known as the Watson transformation, thus solving the problem but not quite yielding results that checked with Austin-Cohen [6]. In the diffraction theory a $\lambda^{-1/3}$ appeared in the exponent and the various constants, especially α, were completely unrealistic. With this deeper understanding of the basic mathematics and having heard of the Heaviside mirror "up there", he was prepared to tackle the real problem of the propagation of a signal from a point dipole source through a spherical shell with conducting boundaries [7]. Every budding applied mathematician should study his magnificent <u>tour de force</u> in solving this problem by exploiting a hundred years of Bessel function research. Since there was no Courant-Hilbert compendium to guide him, he applied the art of a great applied mathematician of being able to make rapid searches through the literature.

Watson's remarkable mathematical skill finally produced the Austin-Cohen formula from a jungle of Bessel functions. His one unknown parameter, the height of Heaviside's layer, was chosen to be about 100 kilometers to yield the Austin-Cohen constant $\alpha = 0.0015$. The prediction was verified by measurements in 1925 by Breit and Tuve in Washington and Appleton in England.

The modern theory of long range radio propagation was thus the wedding of a hundred years of research on Bessel functions with reams of wireless data painfully taken by a group of naval enthusiasts at sea.

Watson also deserves special appreciation for patiently assembling his accumulated notes taken from a century's literature on Bessel functions and publishing them in his classic [8] "Theory of Bessel Functions" (1922) that became a bible for two generations of applied mathematicians. Incidentally, by the 1950's when scattering techniques supported critical experiments in high energy physics, a basic theory of the process emerged from the Watson transformation. The singularities in the Watson series were renamed "Regge poles", whose investigation became a great cottage industry among nuclear theorists. Watson is well known to condensed matter theorists through the three Watson triple integrals [9], special cases of lattice Green's functions that appear in spin wave theory, various phase transition models, lattice random walks, etc.

Enough of philosophy and stories; let us return to our own topic of research, which incidentally we pursued somewhat in the style described above.

Hans, George, and Barry, in the introductory sentence of this report we applauded your excellent choice of subject matter for this symposium. To show that you are in excellent company in the selection we quote a passage from a lectur given in 1978 by Victor Weisskopf [10] at an International Symposium on

Science, Invention, and Social Change in celebration of the 100th Anniversary of the founding of the General Electric Company by Thomas A. Edison. In his assignment to predict the course of physics in the 1980's, he included the statement: "It has been known for a long time that the deviations from ideal order in solid crystals are decisive for the properties of materials. Still we do not have a systematic way of describing grain boundaries, dislocations, and other lattice defects, or of predicting under what circumstances they occur. Furthermore, it was found that amorphous solids show surprising properties previously ascribed only to crystals. We begin to understand the reasons. Effects of random distributions of spins, of foreign atoms and the like are studied in greater detail and a theoretical treatment of such problems has become possible. To quote Phillip Anderson: 'The next decade is very likely to be the most "disorderly" decade in theoretical physics.'"

We are now experiencing an information and systems control revolution as well as one in the development of structural materials. With each phase of the evolution of the process there is also a cost revolution based upon more extensive use of disordered, amorphous, randomized materials. In the xerographic process the basic photoconductors are amorphous materials such as selenium and arsenic triselenide or amorphous polymeric mixtures. A critical phase of the production of integrated circuits is the processing required of the insulation between active elements as amorphous silicon oxides. Materials used in fiber optics and in dielectrics with special very high frequency response characteristics are generally glassy in nature or odd mixtures of exotic polymeric substances. Finally, the new structural materials tend to be derived from the polymer-plastic family; their use frequently demanding special responses to high frequency mechanical vibrations. The future solar cells will certainly be amorphous, random systems. These technological developments provide an important driving force behind the Anderson prediction.

While at first glance it would seem doubtful that any elegant sophisticated branch of mathematics could be of any application to the understanding of the crazy systems we have described, the main thrust of this paper is to show that the theory of stable (Lévy) distribution functions so intensely studied by some of the great mathematical analysts of the period 1930-1950 provides a key to understanding the characteristic behavior of these materials just as the theory of Bessel functions did for the understanding of long range radio propagation.

We start the technical aspect of this report with a very brief review of the theory of stable distributions. Then we proceed to connect it with the behavior of our subject materials.

2. A Summary of the Nature of Stable (Lévy) Distributions

An early problem considered by pioneering statisticians of the 18th century was to find the best fit of an equation to a set of observed data points. After some false starts, even among giants such as Lagrange and Laplace, Legendre (and several others as noted by C. Eisenhart [11]) they hit upon the method of least squares. Legendre's work seemed to be the most influential at the time. Laplace elaborated upon it, and finally in a discussion of the distribution of errors, Gauss emphasized the importance of the normal or Gauss distribution. A role of this function in probability theory had already emerged in De Moivre's [12] investigation of gambling games 250 years ago (1733), 76 years before the Gauss publication. In today's language he was the first to consider a simple form of the central limit theorem of probability, essentially observing that in a Bernoulli process the limit of the sum of a number of random variables has a normal distribution. These ideas were generalized by Laplace and became a primary branch of the theory of probability, after the innovations of the Russian School led by Chebyshev, Markoff, and Lyapunov in the early 1900's.

Laplace was a great exploiter of generating functions for the solution of complex probability problems. As the theory of Fourier series and integrals emerged in the early 1800's, he and his protégé Poisson made the natural next step of introducing those representations of probability distributions as new natural starting points for analysis, thus extending the power of the generating function techniques. The Fourier transforms of probability distributions are now called characteristic functions. Laplace seemed to be especially pleased in observing that the normal distribution was its own Fourier transform (or characteristic function):

$$\frac{1}{\pi} \int_0^\infty e^{-c\theta^2} \cos \theta\varepsilon \, d\theta = (4c\pi)^{-1/2} \exp\{-\varepsilon^2/4c\} . \tag{1}$$

In the early 1850's Laplace's former student Augustine Cauchy became interested in the theory of errors, extending the analysis to other powers than squares (say N-th powers), thus generalizing Gauss's distribution to a new one

$$f_N(\varepsilon) = \frac{1}{\pi} \int_0^\infty e^{-c\theta^N} \cos \theta\varepsilon \, d\theta , \tag{2}$$

expressed as a Fourier integral. He succeeded in evaluating the integral (in a non-Gauss case) only for $N = 1$, thus obtaining the famous Cauchy distribution

$$f_1(\varepsilon) = \left(\frac{1}{\pi}\right) \frac{k}{1 + k^2\varepsilon^2} \quad \text{with} \quad k = 1/c . \tag{3}$$

It was realized only much later that $f_N(\varepsilon)$ is nonnegative, and hence is a probability distribution, only when $0 < N \leq 2$ (F. Bernstein [14], 1919). With N in the appropriate range the harmless looking integrals (2) are functions of ε and N which are full of remarkable surprises.

After Cauchy there was a decline in mathematicians' interest in theory of probability, and especially in the use of characteristic functions; the main author of the period, Todhunter, was completely opposed to their use.

The theory of stable distribution, our basic mathematical topic, originated with Paul Lévy, whom his champion and self-appointed "student" Benoit Mandelbrot considers to be a shadowy figure of French mathematics in the first half of this century. His views of mathematics were generally somewhat orthogonal to those of many of his colleagues. To quote Mandelbrot [15] quoting von Neumann: "I recall John von Neumann saying in 1954, 'I think I understand how every other mathematician operates, but Lévy is like a visitor from a strange planet. He seems to have his own private methods of arriving at the truth which leaves me ill at ease.'"

As an example, when fashion sought the most general conditions for the validity of the Central Limit Theorem, he found simple exceptions to it. Consider the probability density function (pdf) of $s_2 = x_1 + x_2$, the sum of two random variables x_1 and x_2. Then, if $f_1(x_1)$ and $f_2(x_2)$ are the pdf's respectively of x_1 and x_2, the characteristic function of s_2 is

$$\langle \exp iqs_2 \rangle = \int_{-\infty}^{\infty} \int f_1(x_1) f_2(x_2) \exp iq(x_1 + x_2) dx_1 dx_2 = \tilde{f}_1(q)\tilde{f}_2(q) \tag{4}$$

$\tilde{f}_j(q)$ being the characteristic function of x_j. Furthermore, if

$$\tilde{f}_j(q) = \exp - a_j |q|^\alpha \tag{5}$$

for $0 < \alpha \leq 2$, then

$$\langle \exp iqs_2 \rangle = \exp\left(-(a_1 + a_2)|q|^\alpha\right) .$$

This has the same form as $\tilde{f}_j(q)$ but a_j has now been replaced by $(a_1 + a_2)$. In general, the characteristic function of the sum s_n of n random variables, each with its characteristic function of the form (5) is

$$\langle \exp iqs_n \rangle = \exp\left(-(a_1 + \ldots + a_n)|q|^\alpha\right). \tag{6}$$

The Fourier transform of (5) is our old friend (2) but (in new notation)

$$f(x,\alpha) = \frac{1}{2\pi} \int_{-\infty}^{\infty} \exp(-iqx - a|q|^\alpha) dq , \tag{7}$$

now called a Lévy or stable distribution [16,17]. The name stable has been assigned to these distributions because, if x_1 and x_2 are random variables having such a distribution, the linear combination $cx = c_1 x_1 + c_2 x_2$ has a similar distribution with the same α for any real values of the constants c, c_1 and c_2, when $c^\alpha = c_1^\alpha + c_2^\alpha$.

At first glance it might seem that other exponential forms of $f_j(q)$ and $\langle \exp iqs_2 \rangle$ might satisfy (4). However, the restriction that the Fourier transform of $\tilde{f}_j(q)$ must be a probability distribution, thus being everywhere non-negative and normalized, becomes very severe. In his investigation of the necessary and sufficient condition for a distribution to be stable, Lévy found that one more parameter γ could be included in (7) through a new factor in the exponential. The most general stable distribution for $\alpha \neq 1$ has the form (without loss of generality, in standard form, one chooses $\text{Re}(a) = 1$)

$$p(x,\alpha,\gamma) = \frac{1}{\pi} \text{Re} \int_0^\infty \exp\,(-iqx - q^\alpha \exp \frac{1}{2} i\pi\gamma)\, dq \quad \text{if} \quad x > 0 \tag{8a}$$

$$|\gamma| \leq \begin{cases} \alpha & \text{if } 0 < \alpha < 1 \\ 2-\alpha & \text{if } 1 < \alpha < 2 \end{cases} \tag{8b}$$

We omit comments here about the case $\alpha = 1$, which is more complicated and does not arise in any of our physical applications. When $x < 0$

$$p(-x,\alpha,\gamma) = p(x,\alpha,-\gamma) \ . \tag{9}$$

Let us return to our original stable distribution (7) . Lévy noted that the Gauss case $(\alpha = 2)$ is "singular" because for all other α , $0 < \alpha < 2$, $f(x,\alpha)$ has an inverse power tail for large $|x|$:

$$f(x,\alpha) \sim (\alpha a/\pi|x|^{\alpha+1})\, \Gamma(\alpha) \sin \pi\alpha/2 \quad \text{as} \quad |x| \to \infty \ . \tag{10}$$

Since this distribution and those of the x_j in (4) have no second moment, the second moment existence condition for the Gauss Central Limit Theorem is violated, allowing the possiblity of unusual results. It is this long tail which makes stable distributions interesting to us.

The distinctive features of $f(x,\alpha)$ become exaggerated as one lets α become very small, thus receding from the Gauss case $(\alpha = 2)$. The peak remains at the origin but becomes δ-function like since

$$f(0,\alpha) = (1/\pi\alpha)\, \Gamma(1/\alpha) \ . \tag{11}$$

At the same time, as $\alpha \to 0$, from (10) the tail of the distribution extends further.

The stable distributions (except for the Gauss case $\alpha = 2$) would seem to be monstrosities from the view of a statistician or anyone concerned with data analysis since they would correspond to measurements with no precision. According to (7) the mean value of x would be 0 . From (10), the second moment about the mean, which is usually taken to be an estimate of the accuracy, diverges, thus indicating a measurement of no precision.

Even though the stable distributions did not create much of a stir in the statistical world, their purely mathematical properties were sufficiently entertaining to attract the attention of some of the great mathematical analysts of the 1930's and 1940's; for example, A. Wintner [18], S. Bochner [19,20], W. Feller [21], B. Gnedenko and A. Kolmogorov [22], and others. Even G.H. Hardy includes an appendix involving (7) in his classic book on "Divergent Series. The inherent self-similarity of Lévy's stable distributions is related to fractals in Mandelbrot's well-known and magnificently illustrated book [15]. Deep results of the analysis become transparent when the random process is restricted to a lattice [32,33]. Feller devotes considerable space to stable distributions in Volume II of his pair of probability books, but the admission is made that their role in applied science seems to be almost nonexistent. The definitive book on the subject (which few physicists would find written in their style) by Gnedenko and Kolmogorov ends on a more optimistic note with the quotation: "It is probable that the scope of applied problems in which they play an essential role will become in due course rather wide." Our article should be viewed as a small step forward towards fulfilling their prophecy.

3. Intermittant Transport in Amorphous Films

Near the conclusion of Sir Neville Mott's 1977 Nobel Prize lecture, he made the following remark [23]: "I said I would end by talking about two old things. One, of course, is the use of amorphous selenium for office copying by the Xerox company--a multibillion dollar industry developed, as is so often the case, before anybody had tried to make theories of the processes involved. When the subject became fashionable all over the world, we found, of course, that the Xerox scientists knew a great deal about it; and their recent contributions, particularly on dispersive transport (Scher and Montroll [24], 1975), are of the highest importance." It is with this dispersive transport that we start our own story.

The inventor of the Xerox machine, Chester Carlson, who received his B.S. in physics at Cal Tech at a time during the great depression when the demand for practicioners of our science was essentially nil, eventually found a position as an assistant to a patent attorney in New York City. The mechanics of the job were not a derivative of those of young Einstein in the patent office at Berne. Carlson instead soon found that most of his time was devoted to copying by hand long passages from the patent literature in the New York City public library. He thought there must be a better way to achieve the end result. It was with his motivation, with his inventive genius, with the later aid of imaginative colleagues, with the foresight of the little Haloid Corporation in Rochester to take a risk in his process (after major electronic companies in the U.S. considered it hopeless), and finally with the organizational genius of the managers of Haloid (especially Joseph Wilson, Jr., and Sol Linowitz) that the Xerox Corporation was born and flourished. The story is well told by John Dessauer [25], who was one of the key early participants in the machine development process.

The basic process depends upon the properties of the photoconducting materials employed in the xerographic films. In the dark these films are insulators. However, a light flash may generate a sheet of electron-hole pairs near the surface of the film. The commercial application of this process exploits the fact that light reflected from the blank part of a document forms electron-hole pairs in the film in an image of the document.

As the Xerox Corporation prospered, it was natural for them to add a more basic research component to their laboratory to investigate the nature of the materials that served them so well. Since the charge transport of the holes and electrons in their amorphous films was presumed to depend on the mobility of the carriers such measurements were high on the laboratory's program.

One method of studying charge transport properties of amorphous films is through time-of-flight experiments, where a thin film of thickness L is placed between two transparent electrodes and through which a short laser pulse generates a sheet of electron-hole pairs in the film. Under the influence of the external electric field E of the electrodes, the charge pairs separate. In a-As_2Se_3

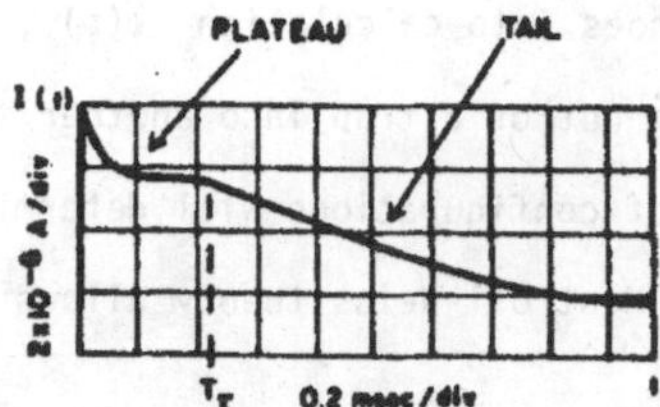

Figure 1. Current vs. time in a time-of-flight experiment of Scharfe in $a\text{-}As_2Se_3$ (from Scher and Montroll [24]).

only the holes propagate. The current $I(t)$ they generate to the surprise of the original investigators is that displayed in Fig. 1.

In a conventional transport theory the sheet of charges would move with a constant average velocity $v = \mu E$, where μ is the intrinsic mobility of the sample. The movement of the mobile charges will generate a constant current until the electrode of opposite polarity is reached. At that time, called the transit time T, the current would fall abruptly to zero because the opposite polarity electrode acts as an absorbing barrier. All of this is in marked contrast to Fig. 1. Again in a conventional theory, for time $t < T$, $|v| = L/T$ so $T^{-1} = \mu E/L$. In contrast for $a\text{-}As_2Se_3$

$$T^{-1} \sim (E/L)^{2.2} . \tag{12}$$

Any measured mobility would depend on E or L or both, a no-no in the conventional theory.

Scher and Lax [26] were successful in understanding a related problem concerning the a.c. conductivity in impurity bands by using the Montroll-Weiss [27] theory of continuous-time random walks (CTRW) to describe the hopping of localized charges governed by a distribution of waiting times between hops. Scher and Lax folded all the spatial and energetic disorder experienced by a charge into a distribution of waiting times $\psi(t)$ between hops on a perfect periodic lattice. Klafter and Silbey [28] later showed the procedure of Scher and Lax is rigorously correct. All the physics now goes into calculating $\psi(t)$, which can involve tunneling between sites or hopping out of a trap into another trap or into a band state, etc. The distribution of configurations will determine $\psi(t)$. However, once $\psi(t)$ is determined, the Montroll-Weiss theory allows for the calculation of all relevant quantities.

Montroll and Scher treated the electrical current in the time-of-flight experiments in amorphous films as arising from charges hopping on a periodic lattice with a bias towards an opposite polarity electrode. Given at $t = 0$ that the probability density for a hop in the time interval $(t, t + dt)$ is $\psi(t)dt$ and the probability for the hop displacement to be ℓ is $p(\ell)$, then Kenkre, Montroll and Shlesinger [29] have shown that for these conditions, the probability $P(\ell,t)$ for a random walker on a lattice to be at site ℓ at time t satisfies the generalized master equation (for simplicity we restrict our discussion to the one dimensional translationally invariant case)

$$\frac{dP(\ell,t)}{dt} = \int_0^t \phi(t - \tau)\left[-P(\ell,\tau) + \sum_{\ell'} P(\ell',\tau)p(\ell - \ell')\right]d\tau \tag{13}$$

where

$$\int_0^\infty \phi(t)\, e^{-st} dt \equiv \phi^*(s) = \frac{s\psi^*(s)}{1-\psi^*(s)}$$

relates the waiting time distribution and the kernel of the integral through their Laplace transforms. The mean current $\langle I(t)\rangle$ for a single charge is given by

$$\langle I(t)\rangle = e\frac{d}{dt}\langle \ell(t)\rangle = e\sum_{\ell} \ell\,\frac{dP(\ell,t)}{dt} \tag{14}$$

Here ℓ refers to fixed lattice sites and it is not affected by the time derivative. It can be shown that [30,31]

$$\langle \ell(t)\rangle = -i\bar{\ell}\, L^{-1}\left[\frac{\psi^*(s)}{s(1 - \psi^*(s))}\right] \tag{15}$$

where L^{-1} is the inverse Laplace transform, and $\bar{\ell} = \sum_{\ell} \ell p(\ell)$. If the first moment $\bar{t}$ of $\psi(t)$ is finite, then

$$\psi^*(s) = \int_0^\infty e^{-st}\,\psi(t)dt \sim 1 - s\bar{t} \quad \text{as} \quad s \to 0 . \tag{16}$$

This will lead to $\langle \ell(t)\rangle \sim \ell\overline{t}$ and thus a constant current, as is associated with traditional transport theory. This does not describe the experimental results in Fig. 1. To avoid this behavior, it is necessary that $\overline{t}$ be infinite. This is the case if (with $0 < \alpha < 1$)

$$\psi^*(s) = \exp(-\text{const. } s^\alpha) \sim 1 - \text{const. } s^\alpha \quad \text{as} \quad s \to 0 , \tag{17a}$$

which is the case if

$$\psi(t) \sim t^{-1-\alpha} \quad \text{as} \quad t \to \infty . \tag{17b}$$

Such a $\psi(t)$ may arise for systems whose random walkers encounter a distribution of trap depths whose mean release time is infinite. As time evolves, walkers will empty from shallow traps at a faster rate than from deeper traps. However, those leaving shallow traps will eventually fall into deeper traps. The current generated should decrease in time as more and more walkers become trapped for longer times and the trapped lifetime becomes longer than the duration of the experiment. When $\overline{t}$ is infinite, there is no characteristic time and intermittancies between jumps occur on all time scales. Jumps will occur in a self-similar pattern of bursts [32,33]. Such situations are not covered by Fermi's Golden Rule where a single rate constant is envisioned.

Using (17a) and (14) we find that

$$\langle \ell(t)\rangle \sim \sigma(t) = \langle[\ell(t) - \langle \ell(t)\rangle]^2\rangle^{1/2} \sim t^\alpha \tag{18}$$

and

$$\langle I(t)\rangle \sim \overline{\ell}\, t^{-1+\alpha} .$$

This analysis suggests that the data in Fig. 1 be replotted on a log-log scale, as is given in Fig. 2.

In the log-log plot we see that for $\alpha = 0.45$ the initial part of the curve (of slope -0.55) is well described. Also note how data from various experiments with different electric fields E and sample widths L fall on a single universal

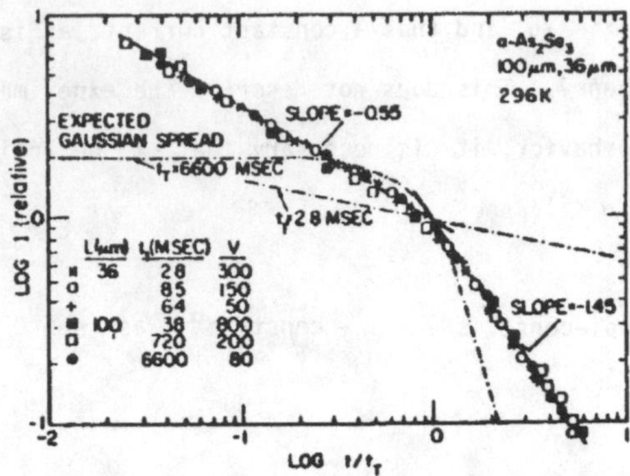

Fig. 2 A log-log plot of current vs. time data from various time-of-flight experiments in $a\text{-}As_2Se_3$ (from Pfister and Scher [34]).

In the log-log plot we see that for $\alpha = 0.45$ the initial part of the curve (of slope -0.55) is well described. Also note how data from various experiments with different electric fields E and sample widths L fall on a single universal curve. This is because for our scale invariant hopping the amount of fluctuation, $\sigma(t)$, per distance travelled $\langle \ell(t) \rangle$, is a constant (see Eq. (18)). Such universality is not found in standard photoconductors. To understand the transition of the curve from slope -0.55 to slope -1.45 it is necessary to take into account the absorbing electrode, as we proceed to do in section 4.

First, as will be shown in Section 5, $\psi^*(s)$ in (17a) is precisely the Laplace Transform of a Lévy distribution. It is the source of the self-similar intermittant transport leading to the new understanding of transport in certain amorphous materials.

4. Reactions

A. An Absorbing Plane Electrode

The change in slope of the current versus time curve at late times occurs because the electrode at $\ell = L$ acts as an absorbing boundary, so care must be exercised when calculating $P(\ell,t)$ not to include paths which reach site ℓ at time t, but have earlier passed through site L. This is accomplished with the equation

$$P(\ell,t) = P_0(\ell,t) - \int_0^t P_0(\ell - L, t - \tau) F(L,\tau) d\tau \qquad (19)$$

where P denotes the probability on the lattice with the absorbing boundary and P_o the probability on a perfect lattice without any boundaries, and $F(L,\tau)$ is the first passage time to site L . For a Levy $\psi^*(s)$ it can be shown that [24]

$$F(L,\tau) \sim \frac{\exp\{-([1-\alpha]/\alpha)(\alpha b/\tau^{\alpha})^{(1/[1-\alpha])}}{\tau[2\pi(1-\alpha)\{\tau^{\alpha}/(\alpha b)\}^{1/(1-\alpha)}]^{1/2}} \times \begin{cases} 1 & , \; b/\tau^{\alpha} \gg 1 \\ \tau^{-\alpha} + (\alpha/2(1-\alpha)) & , \; \frac{b}{\tau^{\alpha}} \ll 1 \end{cases} \tag{20}$$

where $b = L/\bar{\ell}$, and $\bar{\ell} = \bar{\ell}(E)$. We postulate the mean hop displacement to be linear in the electric field E . The transit time T can now be defined as $b = T^{\alpha}$ so that

$$\frac{1}{T} = \left(\frac{L}{E}\right)^{1/\alpha} . \tag{21}$$

For times less than the transit time, $F(L,t)$ is negligible, and at times greater than T , $F(L,t) \sim t^{-1-\alpha}$. This is the tail of a Levy distribution. Thus the behavior of $F(L,t)$ leads to [24,30,31]

$$\begin{aligned} \langle I(t)\rangle &\sim t^{-1+\alpha} \qquad b/t^{\alpha} \gg 1 \\ \langle I(t)\rangle &\sim t^{-1-\alpha} \qquad b/t^{\alpha} \ll 1 . \end{aligned} \tag{22}$$

The value $\alpha = 0.45$ consistently explains both the current data (Eq. (22)) and the mobility data (Eq. (21)). Montroll and Scher [30] first worked out the $\alpha = 1/2$ case while Shlesinger [31] provided the analysis for general α . Note that the sum of the slopes in (22) is always -2. The -2 law has been verifed for many organic and inorganic semiconductors [34].

B. Recombination

The above analysis need not be restricted to the specific geometry of the time-of-flight experiments. For example, Scher [35] uses a $\psi(t) \sim t^{-1-\alpha}$ to analyze the time dependent recombination of electron hole pairs generated from a laser pulse in a-S_i: H and a-As_2Se_3 . No external electric field is applied. The -2 law was again derived, but with not as sharp transition region as in the time-of-flight experiment where there was a well defined set of absorbers (a plane).

In general, if $N(t)$ is the number of reactants left at time t, then

$$N(t)/N(0) = \text{Probability (reactant's lifetime} > t) \\ = \int_t^\infty F(\tau)d\tau \tag{23}$$

with F the first passage time for a reaction. For the time-of-flight experiment, with $\alpha = 1/2$, Scher and Montroll [24] calculated that the number of charges which have not reached the absorbing barrier decays as $t^{-1/2}$. Shlesinger [36] considered a cell model with one charge hopping on a lattice with V sites including an absorbing site at the origin. For a non-exponential $\psi(t)$ this generalized Montroll's [37] model of exciton trapping in a photosynthetic unit. With $\psi(t) \sim t^{-1-\alpha}$ a first passage time $F(t)$ was calculated and leads to

$$N(t) \sim t^{-\alpha} \quad , \quad 0 < \alpha < 1 \ . \tag{24}$$

This result was later verified by Movaghar [38] using a master equation with site specific random absorption terms explicitly added, and by Scher [35] and by Blumen et. al. [39] who studied various approximations to the right-hand side of

$$\int_t^\infty F(\tau)d\tau = \sum_{n=0}^{\infty} \langle(1 - c)^{S_n}\rangle \ \psi_n(t) \quad , \tag{25}$$

where c is the trap concentration for a lattice, S_n is the number of new sites visited after n steps, and the brackets indicate an average over all trap positions.

C. **Reactions with a Log-Normal Waiting Time Distribution**

Reactions are a finer probe of $\psi(t)$ than transport. For example the diffusion constant D for a continuous time random walk depends on the ratio $\overline{\ell^2}/\overline{t}$. Many different $\psi(t)$ can have the same $\overline{t}$ and the same D but lead to different $N(t)$ in (23) since the whole $F(t)$ enters and not just $\overline{t}$.

Interesting cases are discovered by considering a theorem of Bedeaux, Lindenberg, and Shuler [40]. They proved for a continuous-time random walk with the moments $\langle t^n\rangle$ of $\psi(t)$, if

$$t^* = \sup_n \left[\frac{\langle t^n \rangle}{n!} \right]^{1/n} \tag{26}$$

is finite, then for times $t \gg t^*$ the walk behaves as if it were governed by a $\psi(t) = 1/\bar{t}\, \exp(-t/\bar{t})$. This leads to $N(t) \sim \exp(-(\text{constant})t/\bar{t})$. If $\psi(t)$ is such that $t^* = \infty$ then $P(\ell,t)$ will not approach that generated by an exponential $\psi(t)$. For example, the log-normal distribution

$$\psi(t) = \frac{1}{\sqrt{2\pi}\, t} \exp\{-\tfrac{1}{2}(\log t)^2\} \tag{27}$$

has $t^* = \infty$. We will show in the last section that such $\psi(t)$ appear naturally. The fact that this $\psi(t)$ may have the same first moment as an exponential $\psi(t)$ is insignificant. Reaction systems with t^* infinite have not been adequately studied.

5. The Williams-Watts Formula for Dielectric Relaxation

The frequency dependent dielectric constant $\varepsilon(\omega)$ of a material is given in terms of a relaxation function $\phi(t)$ by

$$\frac{\varepsilon(\omega) - \varepsilon_\infty}{\varepsilon_s - \varepsilon_\infty} = -\int_0^\infty e^{-i\omega t}\, [d\phi(t)/dt]dt \;, \tag{28}$$

the supscript s referring to static and ∞ to high frequency limits of $\varepsilon(\omega)$. Williams and Watts [41] have proposed the empirical form for $\phi(t)$ (with α and T constants for a given material)

$$\phi(t) = \exp\{-(t/T)^\alpha\} \tag{29}$$

from the examination of data from certain polymers. Later workers [42-44] found the form to be excellent for a very broad variety of materials. We now show that stable distributions appear naturally in two ways in the application of this form of $\phi(t)$.

The combination on the left hand side of (28) is generally written as $\varepsilon_n' - i\varepsilon_n''$. Then for example, the imaginary part of the normalized dielectric constant ε_n'' is given by

$$\varepsilon_n''(\omega) = \omega \int_0^\infty \phi(t)\cos\omega t\,dt$$
$$= x\int_0^\infty e^{-q^\alpha}\cos qx\,dq \quad (\text{with } x = \omega T) \tag{30}$$

in the Williams-Watts case. Hence $\varepsilon''(\omega)$ is precisely our old Lévy function (7) multiplied by πx.

The second way that stable distributions appear is somewhat subtler. The oldest form used for $\phi(t)$ is the exponential $\exp(-t/\tau)$ first proposed by Debye for small molecules. It has become traditional in more complex systems to express a general $\phi(t)$ as a linear combination of exponentials with different values of τ through the τ distribution function $\rho(\tau)$

$$\phi(t) = \int_0^\infty e^{-t/\tau}\rho(\tau)\,d\tau$$
$$= \int_0^\infty e^{-\mu t}\mu^{-2}\rho(1/\mu)d\mu\ , \tag{31}$$

where μ is a relaxation rate, $\mu = 1/\tau$. A distribution of relaxation rates $f(\mu)$ would then be related to $\rho(\tau)$ by

$$f(\mu) = \mu^{-2}\rho(1/\mu)\ . \tag{32}$$

Now let $\phi(t)$ be the Williams-Watts exponential (29). Then with $s = t/T$, making the substitution $x = \mu T$ in (32), and defining

$$\lambda_\alpha = f(x/T)\ , \tag{33}$$

Eq. (31) becomes

$$\exp(-s^\alpha) = \int_0^\infty \lambda_\alpha(x)\exp(-xs)dx\ . \tag{34}$$

The calculation of $\lambda_\alpha(x)$ as a purely mathematical problem by taking the Laplace inverse of (34) was first done by H. Pollard [45]. When $0 < \alpha < 1$ (the range of interest in dielectric relaxation processes)

$$\lambda_\alpha(x) = -\frac{1}{\pi}\sum_{\ell=0}^{\infty}\left(-\frac{1}{x^\alpha}\right)^\ell \frac{\sin\pi\alpha\ell}{x}\,\frac{\Gamma(\ell\alpha+1)}{\Gamma(\ell+1)}\ . \tag{35}$$

This particular sum is a special case of one in Feller [21, vol. II, p. 548, Eq. 6.8] in his section on stable distributions. Feller's expansion is a representation of the stable distribution function (8a), $p(x,\alpha,\gamma)$. In the special case (35), $\gamma = -\alpha$ so that

$$\lambda_\alpha(x) = p(x,\alpha,-\alpha) .$$

Hence the distribution of relaxation rates that yields the Williams-Watts relaxation function is precisely the stable distribution $p(x,\alpha,-\alpha)$.

The distribution $\lambda_\alpha(x)$ of (34) and (35) may also be identified with the same types of first passage time problems that are encountered in the Scher-Montroll [24] continuous-time random walk model of dispersive transport. Equation (35) was derived in [24, Appendix] as the distribution of the time for a random walker on a line to reach a point a distance x from the starting point for the first time when the pausing time distribution between steps has the asymptotic form $\psi(t) \sim t^{-1-\alpha}$. Montroll and Bendler [46] have made some suggestions concerning the dynamics of Williams-Watts relaxation based on the connection of (35) to first passage time distributions.

6. 1/f Noise and the Log Normal Distribution

We continue with our exposition of the scaling properties of Lévy distributions by considering the limit as $\alpha \to 0$; i.e. as the $t^{-1-\alpha}$ tail becomes a t^{-1} tail. In that regime the Lévy distribution is mimicked by a log normal distribution over a considerable portion of the sample variable range [48-49]. In this section we (i) give a simple derivation of this observation, (ii) show how a log normal distribution of relaxation times is a key to the understanding of 1/f noise, and (iii) derive the log normal distribution for the behavior of a broad class of complex dynamical systems.

A. Wintner [18] has derived the following convergent series expansion for the Lévy distribution (7) (for simplicity we set $a = 1$ in (7)) when $0 < \alpha < 1$ and $x > 0$ (when $x < 0$ one merely sets $f(x,\alpha) = f(-x,\alpha)$):

$$f(x,\alpha) = \frac{1}{\pi} \sum_{n=1}^{\infty} \frac{(-1)^{n+1}}{n!} \frac{\Gamma(1 + \alpha n)}{x^{1 + \alpha n}} \sin \frac{1}{2} \pi\alpha n . \quad (36)$$

as $\alpha \to 0$ this becomes (neglecting an expansion of the gamma function)

$$\begin{aligned} f(x,\alpha) &\sim \frac{1}{2} \alpha \sum_{n=1}^{\infty} \frac{(-1)^{n-1}}{(n-1)!x^{\alpha n+1}} = \frac{\alpha}{2x^{1+\alpha}} \exp - \frac{1}{x^{\alpha}}. \\ &= \alpha(2x^{1+\alpha})^{-1} \exp(-e^{-\alpha \log x}) \\ &= (\alpha/2x^{1+\alpha}) \exp (-1 + \alpha \log x - \frac{1}{2} \alpha^2 \log^2 x + O(\alpha^3)) \\ &\sim \frac{\alpha}{2xe} \exp - \frac{1}{2} \alpha^2 \log^2 x , \end{aligned}$$

neglecting terms of $O(\alpha^3)$ in the exponential. Hence $f(x,\alpha)$ has the form of log normal distribution. A longer but more careful treatment involving the expansion of the gamma function in (36) whould change $\log x$ to $(\log x) + \gamma$ in the final expression, with γ being Euler's constant ~ .577. As $\alpha \to 0$ the dispersion increases. By taking more terms in the small α expansion of (36) Montroll and Bendler [46] found a rapidly converging asymptotic series for $f(x,\alpha)$ for the regime $0 < \alpha < 1/4$.

We will now show how a log-normal distribution of relaxation times is related to the "1/f problem". Consider an autocorrelation function $A(t)$ for fluctuations. This may represent a velocity-velocity correlation, voltage-voltage correlation, water level-water level correlation, etc. The noise power spectrum $S(f)$ is defined as

$$S(f) = 4\text{Re} \int_0^{\infty} \exp(2\pi ift)A(t)dt . \quad (37)$$

It is found in many systems that $S(f)$ has a low frequency component

$$\lim_{f \to 0} S(f) \sim 1/f . \quad (38)$$

Note that, in this regime, the integrated power spectrum is independent of a rescaling $f \to f_0 \cdot f$:

$$\int_{f_1}^{f_2} \frac{df}{f} = \int_{f_1/f_0}^{f_2/f_0} \frac{df}{f} \quad (39)$$

Diverse examples of 1/f noise include contact noise in semiconductors, the height of the Nile, and voltage noise in squid nerve axons. As this phenomenon of 1/f noise is so ubiquitous, a generic explanation may exist.

Let us examine how 1/f noise can arise from a distribution of relaxation times. For a system with one relaxation time, τ, $A(t;\tau) = \exp(-t/\tau)$ and

$$S(f;\tau) = \frac{4\tau}{1 + (2\pi f\tau)^2} . \tag{40}$$

In a complex system it is likely that a distribution $\rho(t)$ of relaxation times exists; there is no loss of generality in assuming that $\rho(t)$ is normalized. Then

$$S(f) = \int_0^\infty \frac{4\tau\ \rho(\tau)}{1 + (2\pi f\tau)^2}\, d\tau . \tag{41}$$

Now suppose that $\rho(\tau)$ is a log normal distribution with dispersion σ. Then if we set $\rho(\tau) \equiv g(\tau/\langle\tau\rangle)$ with

$$g(\tau/\langle\tau\rangle)d\tau = \frac{\exp - (\log \tau/\langle\tau\rangle)^2/2\sigma^2}{(2\pi\sigma^2)^{1/2}}\ \frac{d(\tau/\langle\tau\rangle)}{\tau/\langle\tau\rangle}$$

and write $S(f) \equiv S(\omega,\sigma)$, $\omega = 2\pi f\langle\tau\rangle$ and $z = \tau/\langle\tau\rangle$

$$S(\omega,\sigma)/4\langle\tau\rangle = \int_0^\infty \frac{\exp[-(\log z)^2/2\sigma^2]dz}{(2\pi\sigma^2)^{1/2}\ [1 + \omega^2 z^2]} \tag{42}$$

In the large σ regime we find

$$S(\omega,\sigma)/4\langle\tau\rangle \sim (2\pi\sigma^2)^{-1/2} \int_0^\infty \frac{dz}{(1 + \omega^2 z^2)} = \frac{1}{2\omega\sigma}\left(\frac{\pi}{2}\right)^{1/2} \tag{43}$$

our required $1/\omega$ form.

Montroll and Shlesinger [28] have made a detailed analysis of (42) to find

$$S(\omega,\sigma)/4\langle\tau\rangle = \frac{1}{2\sigma\omega}\left(\frac{\pi}{2}\right)^{1/2} \exp\left[-\frac{1}{2}\left(\frac{\log\omega}{\sigma}\right)^2\right] \left\{ \left[1 - \frac{1}{2}\left(\frac{\pi}{2\sigma}\right)^2 + \frac{5}{8}\left(\frac{\pi}{2\sigma}\right)^4 + ..\right] \right.$$

$$+ \frac{\pi^2}{2}\left(\frac{\log\omega}{2\sigma^2}\right)^2 \left[1 - \frac{5}{2}\left(\frac{\pi}{2\sigma}\right)^2 + \frac{61}{8}\left(\frac{\pi}{2\sigma}\right)^4 + ...\right] \tag{44}$$

$$\left. + \frac{5\pi^4}{24}\left(\frac{\log\omega}{2\sigma^2}\right)^4 \left[1 - \frac{61}{10}\left(\frac{\pi}{2\sigma}\right)^2 + \frac{277}{8}\left(\frac{\pi}{2\sigma}\right)^4 + ...\right] + ... \right\}$$

so that if $(\sigma^{-1} \log \omega)$ is small (42) reduces to (43); the larger σ, the larger the frequency range that $S(\omega,\tau)$ has its $1/\omega$ noise character.

Our problem now is to find a generic model for a complex system that yields a log normal relaxation time distribution with a large dispersion.

We derive a log normal distribution in a general manner by considering a complex process whose successful completion requires the successful completion (in sequence or simultaneously) of n dependent subtasks [49]. The probability P of success for the primary task is $P = p_1 p_2 p_3 \cdots p_n$, where p_j is the probability of success of the j-th subtask). Hence

$$\log P = \log p_1 + \log p_2 + \log p_3 + \ldots + \log p_n . \tag{45}$$

Since the p_i are independent random variables so are the $\log p_i$. If the appropriate moments exist for $\log p_i$ and n is "large", then the central limit theorem is applicable and $\log P$ converges to a Gauss distribution and P to a log-normal distribution. This argument was first given by W. Shockley [50] to interpret the observation that the productivity of scientists publishing research papers has a log normal distribution. Shockley's data was taken from publication records of scientists at the Brookhaven National Laboratories. In that case the p_j's were probabilities of attributes necessary for the publication of a paper; p_1 being the probability of having an idea for investigation, p_2 that of having the competence to pursue the investigation, p_3 that of obtaining some interesting results, etc.

The time required for the occurrence for an event with probability P is proportional to $1/P$. As $\log p_i^{-1} = -\log p_i$ application of the central limit as theorem as before establishes that when n is large $1/P$ has a log-normal distribution. Let σ^2 be the dispersion of $(\log 1/p_j)$. Then the dispersion σ^2 of the sum in (45) is $\sigma^2 = \sigma_1^2 + \sigma_2^2 + \ldots + \sigma_n^2 = n\bar{\sigma}^2$, $\bar{\sigma}^2$ being the mean of the σ_j^2. Hence the larger the number of subtasks n required for the completion of the main task of our complex process, the larger is σ in (44), and thus the larger will be the frequency range in which the noise $S(\omega,\sigma)$ is mimicked by $1/\omega$ distribution.

Our basic message is that the log-normal is the generic distribution for "geometric mean"; that it has the required scale invariant nature to lead to 1/f noise; and that it is the limiting form of a Lévy distribution as $\alpha \to 0$.

Acknowledgement: This research was partially supported by the Defense Advanced Research Projects Agency.

References

1. R. Courant and D. Hilbert, Methoden der mathematischen Physik (Berlin, Springer, 1924).
2. F. Bloch, Physics Today, December 1976, p. 23.
3. E. Schrodinger, Collected Papers on Wave Mechanics (Blackie, London 1927; Reprint Edition, Chelsea, N.Y. 1978).
4. For survey of early work on long range radio propagation see H.R. Mimno Rev. Mod. Phys. 9, 1 (1937).
5. O. Heaviside, Encylopedia Britannica, Tenth edition vol. 33.
6. G.N. Watson, Proc. Roy. Soc. 95, 83 (1919).
7. G.N. Watson, Proc. Roy. Soc. 95, 546 (1919).
8. G.N. Watson, Theory of Bessel Functions (Cambridge Univ. Press, 1922).
9. G.N. Watson, Quart. J. Math. Oxford 10, 266 (1939).
10. V. Weisskopf, in proceedings of General Elective International Symposium on Sciences, Invention, and Social Change, p. 19, Albert Rosenfeld (editor), Schenectady and Albany, N.Y. (1978).
11. C. Eisenhart, J. Wash. Acad. of Sci. 54, 24 (1964).
12. A. De Moivre, Doctrine of Chances (1756 edition, this has recently been reprinted by Chelsea Press, 1967).
13. A. Cauchy, Compte Rendus, 37, 198 (185). Also in Oeuves Completes ser. 1, 12, p. 94.
14. F. Bernstein, Math. Ann. 79, 265 (1919).
15. B.B. Mandelbrot, The Fractal Geometry of Nature (W.H. Freeman, San Francisco, 1982).
16. P. Lévy, Calcul des probabilites (Gauthier-Villars, Paris, 1925).
17. P. Lévy, Théorie de l'addition des variables aléatoires (Gauthier-Villars, Paris, 1937).
18. A. Wintner, Duke Math. J. 8, 678 (1941).
19. S. Bochner, Duke Math. J. 3, 726 (1937).

20. S. Bochner, Lectures on Fourier Integrals (Princeton University Press, 1959).

21. W. Feller, An Introduction to Probability Theory and its Applications. Vol. II. (John Willey & Sons, New York) (1966).

22. B.V. Gnedenko and A.N. Kolmogorov, Limit Distributions for Sums of Independent Random Variables, Addison-Wesley, Reading, Mass. (1954).

23. N.F. Mott. Rev. Mod. Phys. 50 203, (1978).

24. H. Scher and E.W. Montroll, Phys. Rev. B12, 2455, (1975).

25. John H. Dessauer, My Years with Xerox, (Woodhill Pub). 1979.

26. H. Scher and M. Lax, Phys. Rev. B7, 4491 (1973).

27. E.W. Montroll and G.H. Weiss, J. Math. Phys 6, 167 (1965).

28. J. Klafter and R. Silbey, Phys. Rev. Lett. 44, 55 (1980).

29. V.M. Kenkre, E.W. Montroll, and M.F. Shlesinger, J. Stat. Phys. 9, 45 (1973).

30. E.W. Montroll and H. Scher, J. Stat. Phys. 9, 101 (1973)

31. M.F. Shlesinger, J. Stat. Phys. 10, 421 (1974).

32. M.F. Shlesinger and B.D. Hughes, Physica A109 (1981).

33. M.F. Shlesinger and E.W. Montroll (this volume).

34. G. Pfister and H. Scher, Phys. Rev. B15, 2067 (1977).

35. H. Scher, J. de Physique (Paris) Colloq. 42, C4, 547 (1981).

36. M.F. Shlesinger, J. Chem. Phys. 70, 4813 (1979).

37. E.W. Montroll, J. Math. Phys. 10, 753 (1969).

38. B. Movaghar, J. Phys. C., Solid State Phys. 13, 4915 (1980).

39. A. Blumen, J. Klafter, and G. Zumofen, Phys. Rev. B27, 3429 (1983).

40. D. Bedeaux, K. Lindenberg, and K.E. Shuler, J. Math. Phys. 12, 2116 (1971).

41. G. Williams and D. Watts, Trans. Faraday Soc. 66, 80 (1970).

42. C.T. Moynihan, L.P. Boesch, and N.L. La Berge, Phys. Chem. Glasses 14 122 (1973).

43. K.L. Ngai and C.T. White, Phys. Rev. B.20 2475 (1979).

44. C.P. Lindsey and G.D. Patterson, J. Chem. Phys. 73, 3348 (1980).

45. H. Pollard, Bull. Am. Math. Soc. 52, 908 (1946).

46. E.W. Montroll and J.T. Bendler, J. Stat. Phys. (in press).

47. Sixth Int. Conf. on Noise in Physical Systems. P.H.E. Meijer, R. Mountain, and R.J. Soulen, Jr. eds., (National Bureau of Standards, Washington, D.C.) Special Pub. 614.

48. E.W. Montroll and M.F. Shlesinger, J. Stat. Phys. 32, 209 (1983).

49. E.W. Montroll and M.F. Shlesinger, Proc. Nat. Acad. Sci. (USA) 79, 3380 (1982).

50. W. Shockley, Proc. IRE 45 279 (1957).

FRACTAL STOCHASTIC PROCESSES: CLUSTERS AND INTERMITTANCIES

Michael F. Shlesinger and Elliott W. Montroll
Institute for Physical Science and Technology
University of Maryland
College Park, Maryland 20742

and

La Jolla Institute
P.O. Box 1434
La Jolla, California 92038

ABSTRACT

Many scaling relations for complex systems in the physical sciences involve non-integer exponents. We list several examples all of which may not be well known. We interpret non-integer exponents as indicating singularities arising from a long tailed probability distribution governing the physical observables. If the first appropriate moment of the probability distribution diverges, then no scale exists in which to gauge measurements and phenomena occur on all scales. Self-similar fractals, non-differentiability, and also non-integer exponents will arise. Random walk examples are presented where the above characeristics appear simply and naturally. The analysis provides a generalization of Weierstrass' continuous, but nowhere differentiable function. Lastly, the Riemann Hypothesis is recast in a random walk framework.

1. Introduction

When one begins the study of physics, one invariably discovers simple laws such as momentum conservation (mv = constant) and energy conservation ($\frac{1}{2}mv^2$ = constant). One also encounters many well-defined concepts, such as the mobility μ relating the drift velocity v of a conductance electron in a metal to the applied electric field E : $v = \mu E$. Simple relations are ubiquitous, e.g. the mean square displacement of a random walker after N steps is proportional to N ; the vibrational density of states $\rho(\omega)$ of a D dimensional crystal at low frequencies scales as ω^{D-1}, etc. Everywhere exponents seem to be integers and quantities are well-defined.

When one deals with actual systems which are sufficiently complex, many times the best description is in terms of probability distributions. A Poisson distribution is characterized by its first moment, and a Gaussian distribution by its first two moments. When distributions have long enough tails, the first few moments fail to exist. Distributions with infinite moments will lead to physics described by non-integer exponents and to surprises which counter our intuition. We list some examples where non-integer exponents abound.

2. Surprises in Experimental Physics

A. High Velocity Impacts [1,2]

An important materials science problem is the effect of high velocity micrometeorites (mass m, velocity v) on spacecraft. As a result of a collision many interesting phenomena occur: deformation of the meteor (projectile), crater formation on the surface of the spacecraft (target), shock wave generation, melting, and crack propagation. One would suspect that the volume of the crater formed by the collision would be proportional to the kinetic energy of the meteor ($\frac{1}{2}mv^2$) leading to the radius of the crater R scaling as $R^3 \sim v^2$. Momentum conservation is not considered because target material is thrown backwards during the crater formation thus enhancing the forward momentum. The surprise is that

$$R \sim v^{0.58} \tag{1}$$

which implies that

$$mv^{1.74} = \text{constant.} \tag{2}$$

B. Charge Transport in Amorphous Films [3-5]

In the xerographic process a flash of light creates a layer of electron-hole pairs in an amorphous film. The charges separate under the influence of an external electric field E. If the film is of depth L, and there exists a mean transit time T for charges to cross the film, then one expects that $L/T = \mu E$. Instead one finds (for a-As_2Se_3) that

$$T^{-1} \sim (E/L)^{2.2} = (E/L)^{1/.45} \tag{3}$$

for a wide variety of E and L values. The current I(t) generated by the charge movements is found to have two regimes. In the early phase

$$I(t) \sim t^{-1 + .45} \tag{4a}$$

and in the late phase (caused by absorption at the far surface)

$$I(t) \sim t^{-1 - .45} . \tag{4b}$$

C. **Polymers** [6]

1. The world of polymers is full of fractional exponents. The mean square displacement $\langle R^2(N) \rangle$ of a self-avoiding random walk (SAW) after N steps can be used to model the end-to-end distance of a polymer in a good solvent. In 3D

$$\langle R^2(N) \rangle \sim N^{\gamma} \tag{5}$$

with $\gamma \sim 1.2$. Mandelbrot [7] has suggested that $\gamma = 2/\mu$ where μ is the "fractal" dimension of the SAW and is [8] approximately $\ln 6/\ln 3 \sim 1.6$.

2. The vibrational density of states $\rho(\omega)$ of certain proteins is found at low frequencies to be [9]

$$\rho(\omega) \sim \omega^{.65 \pm .04} . \tag{6}$$

This is consistent with the protein having a 3D SAW shape with $\rho(\omega) \sim \omega^{\mu-1}$, in contrast to the Debye law $\rho(\omega) \sim \omega^{D-1} = \omega^2$ in 3D.

3. The reptation time τ_R for polymer melts is defined as the cross-over time from rubber-like to liquid-like behavior. It is found to scale with polymer mass M as

$$\tau_R \sim M^{3.3} \ . \tag{7}$$

A simple diffusion model [6] for the motion of constrained polymer chains would yield $\tau_R \sim M^3$.

4. The dielectric response of polymers (as well as many other materials) is governed by non-integer exponents. The frequency-dependent dielectric constant $\varepsilon(\omega)$ is given by

$$\frac{\varepsilon(\omega) - \varepsilon_\infty}{\varepsilon_\infty - \varepsilon_0} = -\int_{-\infty}^{\infty} e^{-i\omega t} \frac{d\phi(t)}{dt} \, dt \ . \tag{8}$$

Williams and Watts [10] find that a good form for the response function is

$$\phi(t) = Ae^{-Bt^\alpha} \ , \quad 0 < \alpha < 1 \ . \tag{9}$$

D. **Income Distributions [11]**

Non-integer exponents are not restriced to physics. Income distributions are log-normal over a wide range. However, the last percent, or so, of a population follows Pareto's law

$$\text{Probability(Income} > x) \sim x^{-1.6} \tag{10}$$

Many more examples exist. We have chosen illustrative examples some of which may not be well known.

3. From Non-integer Exponents to Self-similarity

Integer exponents can usually be traced back to the analytic behavior of an appropriate function which can be expanded in a Taylor series. Non-integer exponents imply the presence of singularities and the breakdown of a Taylor series due to the divergence of a coefficient. Our main theme is that singularities and thus non-integer exponents arise in complex systems because they exhibit randomness on many scales. As an example, consider a probability $\psi(t)dt$ for an event (electron hopping, dislocation movement, crack propagation, etc.) to occur in the time interval $(t,t + dt)$. Its Laplace transform is defined by

$$\psi^*(s) \equiv \int_0^\infty e^{-st}\ \psi(t)dt\ . \tag{11}$$

If all the moments $\langle t^n \rangle$ of $\psi(t)$ exist, then

$$\psi^*(s) = 1 + \sum_{n=1}^{\infty} \frac{(-s)^n}{n!}\ \langle t^n \rangle \tag{12}$$

and times will be measured in units of $\langle t \rangle$. If however, $\langle t \rangle$ is infinite, i.e.,

$$\langle t \rangle = \int_0^\infty t\psi(t)dt = -\frac{\partial}{\partial s}\psi^*(s = 0) = \infty$$

then $t/\langle t \rangle$ is not defined, and $\psi^*(s)-1$ is not of order s at $s = 0$. For example if at long times $\psi(t) \sim ct^{-1-\alpha}$, $0 < \alpha < 1$, then [3-5,12] $\psi^*(s) = 1 - c's^\alpha + (s^\alpha)$. Effectively t is replaced by t^α in all dimensional analysis. This is the underlying cause for the non-integer exponents in Eqs. (3) and (4). Such $\psi(t)$ arise from a distribution of deep traps which can capture charges in amorphous materials. There is no average time in this problem, the duration between events occurs on all time scales. A self-similar set of bursts of events has this property.

Systems with inherent scaling are most naturally described by scaling equations and not by differential equations. In fact, the appropriate functions may be nowhere differentiable. Scaling brings us into the province of the renormalization group (RG). Let us consider how the free energy F of a system scales near a critical point [13].

The free energy is a function of a scalar or vector parameter K, which describes interactions over the fundamental length scale of the problem (typically a lattice spacing). Under RG transformation, the length scale of the interaction is changed, and K is mapped to a new value K'. Nontrivial fixed points $K = K'(K)$ of the transformation correspond to thermodynamic critical points, at which interactions occur over all length scales and at which the free energy F is nonanalytic. Regarding F as a function of the scaling variable u measuring distance from such a fixed point, one finds that [13]

$$F(u) = \ell^{-D}F(\lambda u) + G(u)\ . \tag{13}$$

Here D is the dimensionality of the system, λ is a relevant eigenvalue of the RG transformation ($\lambda > 1$) and G is a function which is analytic at $u = 0$. The parameter u describes interactions in the original system, while λu describes interactions on the transformed system. Let us iterate Eq. (13) n times and then let $n \to \infty$, to obtain

$$F(u) = \lim \left[\ell^{-nD} F(\lambda^n u) + \sum_{j=0}^{n-1} \ell^{-jD} G(\lambda^j u)\right] . \quad (14)$$

It is usually assumed that the first term on the right hand side vanishes as $n \to \infty$, and thus all singularities in $F(u)$ must shift to the term involving the summation over G. The singular part of any solution of Eq. (13) is

$$F_{sing}(u) = A(u) \; |u|^{D \, \ell n \, \ell / \ell n \, \lambda} \quad (15a)$$

where $A(u)$ is oscillatory in $\ell n \, u$ with period $\ell n \, \lambda$, i.e.

$$A(u) = A(\lambda u) = \sum_n A_n e^{2\pi i n \ell n \, u / \ell n \, \lambda} . \quad (15b)$$

In the next section we introduce fractal random walk processes which produce non-integer exponents naturally in the form of fractal dimensions, and integral transforms of probability distributions exhibit precisely the scaling of Eqs. (13)-(15).

However, first we describe what "fractals" are by way of a few simple examples. The word "fractal" was coined by Mandelbrot [7] to describe geometric shapes with features on many length scales or events occuring over many time scales. Consider first the example of a square of side N. It can be decomposed into N^2 self-similar squares each with side 1 (N^{-1} times smaller than the side of the original square). The topological dimension is, of course, 2 which we choose to write as

$$D = \frac{\log(\text{no. of small squares in a big square})}{\log(\text{side scale factor})} = \frac{\log N^2}{\log N} = 2 . \quad (16)$$

Next consider the construction of Cantor's triadic set. A line of length 1 has its middle third cut out, leaving 2 lines each of length 1/3. The same procedure is now applied to the lines producing 2^2 lines, each of length $(1/3)^2$. Each stage of iteration is self-similar. After an infinite number of iterations only a Cantor set of points remains. This set has measure zero and a topological dimension of zero, but it is closed, every point is a limit point, and it can be put into in a one-to-one correspondence with the real numbers. In analogy with Eq. (16) (since the set of points is self-similar over an infinite number of scales) we ascribe to the Cantor set the fractal (Hausdorff-Besicovitch) dimension $F = \log 2/\log 3$. In general, the fractal dimension of a set is never less than the topological dimension. As another example, the trajectory of Brownian motion in two dimensions is of topological dimension one, but has a fractal dimension of two (eventually the trajectory can fill the plane).

4. Fractal Random Walks

A. Gaussian Random Walks [14]

Consider a random walker starting at the origin of a one-dimensional lattice with unit spacing between lattice sites. Let the walker jump a distance ℓ $(\ell = \pm 1, \pm 2, \ldots)$ with probability $p(\ell)$. The Fourier transform of $p(\ell)$ is called the structure function and denoted by $\tilde{p}(k) \equiv \sum_{\ell=-\infty}^{\infty} e^{ik\ell} p(\ell)$. Let $P_n(\ell)$ be the probability that the walker reaches site ℓ after n jumps. Then

$$P_{n+1}(\ell) = \sum_{\ell'=-\infty}^{\infty} P_n(\ell') p(\ell - \ell') . \tag{17}$$

Since Eq. (17) is in the form of a convolution, it may readily be Fourier transformed to give

$$\tilde{P}_{n+1}(k) = [\tilde{p}(k)]^{n+1} \tilde{P}_0(k) . \tag{18}$$

In this simple example the complete information of the random walk is contained in $\tilde{p}(k)$ and the initial condition $P_0(\ell)$. Inverse Fourier transforming Eq. (18) gives a Gaussian distribution for $P_n(\ell)$, i.e.

$$P_n(\ell) \sim (2\pi\langle\ell^2\rangle n)^{-1/2} \exp\left\{- \frac{\ell^2}{2\langle\ell^2\rangle n}\right\} \tag{19}$$

where $\langle \ell^2 \rangle = \sum_{\ell} \ell^2 p(\ell)$, provided $\langle \ell^2 \rangle < \infty$.

B. **The Weierstrass Random Walk** [15]

Gaussian behavior can be avoided by choosing $\langle \ell^2 \rangle$ to be infinite, which is the case for

$$p(\ell) = \frac{n-1}{2n} \sum_{j=0}^{\infty} [\delta_{\ell, b^j} + \delta_{\ell, -b^j}]\, n^{-j} \tag{20}$$

when $n > 1$ and $b^2 > n$, since

$$\langle \ell^2 \rangle \equiv \sum_{\ell} \ell^2 p(\ell) = \frac{n-1}{n} \sum_{j=0}^{\infty} (b^2/n)^j \, .$$

This random walker will take jumps of all orders of magnitude, but with each higher order of magnitude occuring with an order of magnitude less probability. On the average about n jumps of unit length occur forming a cluster of sites visited, before a jump of length b occurs and the trajectories again cluster with about n nearest-neighbor jumps. On the average about n such clusters (each with n points visited) are formed before a jump of length b^2 occurs, etc. One expects the set of points visited to have a fractal dimension of $\log n/\log b$. The fractal dimension appears naturally in the Fourier representation of $p(\ell)$, i.e. in $\tilde{p}(k)$, where from Eq. (20)

$$\tilde{p}(k) = \frac{n-1}{n} \sum_{j=0}^{\infty} n^{-j} \cos(b^j k) \, . \tag{21}$$

When $b > n$ this is precisely Weierstrass' example of a continuous, but nowhere differentiable function. The Weierstrass function can be considered to be a functional form of the self-similarity in a Cantor set. It is also the real part of a lacunary Taylor series, $\sum n^{-j} z^{b^j}$, which cannot be analytically continued beyond its radius of convergence $|z| = 1$. In addition $\tilde{p}(k)$ satisfies the scaling equation (with m any positive integer),

$$\tilde{p}(k) = n^{-m}\, \tilde{p}(b^m k) + \frac{n-1}{n} \sum_{j=0}^{m-1} \cos(b^j k) \tag{22}$$

in analogy with the free energy scaling of Eq. (14). Note how by definition any singularities of $\tilde{p}(k)$ shift from the first term on the right hand side, to the second term as $m \to \infty$. An alternative expression for $\tilde{p}(k)$ can be obtained by writing $\cos(b^n k)$ in terms of its inverse Mellin transform, i.e.,

$$\tilde{p}(k) = \frac{1}{2\pi i}\ \frac{n-1}{n} \sum_{j=0}^{\infty} n^{-j} \int_{c-i\infty}^{c+i\infty} \cos(\pi s/2)\ \Gamma(s)|k|^{-s}\ b^{-sj} ds \qquad (23)$$

for $0 < c = \mathrm{Re}\ s < 1$. Interchanging the summation and integral yields

$$2\pi i\ \tilde{p}(k) = \int_{c-i\infty}^{c+i\infty} \frac{|k|^{-s}\ \Gamma(s)\cos(\pi s/2)}{1 - n^{-1}\ b^{-s}}\, ds\ . \qquad (24)$$

The integrand has poles at $s = 0,-2,-4,\ldots$ from the cosine-gamma factor, and poles from the denominator at

$$s = -(\ell n\ n/\ell n\ b) \pm 2\pi i j/\ell n\ b \quad (j = 0,1,2,\ldots)\ .$$

Translating the contour to $\mathrm{Re}\ s = -\infty$ and taking account of the residues at the poles crossed, we arrive at

$$\tilde{p}(k) = 1 + |k|^{\alpha}\ Q(k) + \frac{n-1}{n} \sum_{j=1}^{\infty} \frac{(-1)^j k^{2j}}{(2j)!\,[1-b^{2j}/n]} \qquad (25)$$

with
$$Q(k) = \frac{n-1}{n\ \ell n\ b} \sum_{j=-\infty}^{\infty} \Gamma(q)\ \cos(\frac{\pi q}{2})\exp(-2\pi i j\ \ell n|k|/\ell n\ b)\ ,$$

$$q = -(\ell n\ n/\ell n\ b) + 2\pi i j/\ell n\ b$$

and

$$\alpha = \ell n\ n/\ell n\ b\ .$$

Note that $Q(k)$ is oscillatory, being periodic in $\ell n|k|$ with period $\ell n\ b$ (in analogy to the free energy oscillations of Eq. (15)). The quantity α is the fractal dimension of the set of sites visited and is less than 2 if $b^2 > n$. Actually in 1D we need $\alpha < 1$ to insure that not all sites on the lattice are visited. In higher dimensions one only needs $\alpha < 2$. If $\langle \ell^2 \rangle$ is finite, the pole at $s = -2$ dominates and the k^2 term precedes the k^{α} term in Eq. (25).

C. The Fractal Rayleigh-Pearson Random Walk [16,17]

Our Weierstrass random walk can be readily generalized to jumps on n-dimensional Bravais lattices. Non-lattice random walks offer another type of generalization. The first explicit mention of "random walks" was in a query to Nature by Pearson [18] concerning the behavior of a walker in the plane jumping fixed lengths, but in random directions. The problem, as it turned out, had already been solved by Lord Rayleigh [19] who pointed out that it was equivalent to the composition of equal amplitude waves with a random distribution of phases [20]. We now present the fractal version of the Rayleigh-Pearson random walk.

In D dimensions consider the spherically symmetric jump displacement probability density

$$p(\vec{x}) = \{A_D \, |\vec{x}|^{D-1}\}^{-1} \, f(|\vec{x}|) \qquad (26)$$

where $A_D = 2\pi^{D/2}/\Gamma(D/2)$ (the surface area of a unit hypersphere) and $f(|\vec{x}|)$ is the probability distribution for a step of length $|\vec{x}|$. In analogy with the Weierstrass random walk we choose

$$f(|\vec{x}|) = \frac{n-1}{n} \sum_{j=0}^{\infty} n^{-j} \, \delta(|\vec{x}| - b^j) \, . \qquad (27)$$

Fourier transforming $p(\vec{x})$, and performing the radial and angular integrations we obtain a generalization of the Weierstrass structure function,

$$\tilde{p}(\vec{k}) = \frac{n-1}{n} \sum_{j=0}^{\infty} n^{-j} \, \Gamma(\tfrac{D}{2}) (\tfrac{1}{2} |\vec{k}| b^j)^{1-\frac{D}{2}} \, J_{\frac{D}{2}-1}(|\vec{k}| b^j) \qquad (28)$$

where J is a Bessel function. The structure function satisfies the scaling equation

$$\tilde{p}(\vec{k}) = n^{-1} \tilde{p}(b\vec{k}) + \frac{n-1}{n} \Gamma(\tfrac{D}{2}) (\tfrac{1}{2} |\vec{k}|)^{1-\frac{D}{2}} J_{\frac{D}{2}-1}(|\vec{k}|) \qquad (29)$$

with the solution [16]

$$\tilde{p}(\vec{k}) = \frac{n-1}{n} \sum_{j=0}^{\infty} \frac{\Gamma(D/2)(-1)^j (\tfrac{1}{2}|\vec{k}|)^{2j}}{j!\,\Gamma(j + D/2)(1-b^{2j}/n)}$$

$$+ \frac{n-1}{2n \ln b} (\tfrac{1}{2}|\vec{k}|)^{\alpha} \Gamma(D/2) \sum_{m=-\infty}^{\infty} \frac{\Gamma(-\alpha/2 + m\pi i/\ln b)}{\Gamma(D/2 + \alpha/2 - m\pi i/\ln b)} \exp\left[-\frac{2m\pi i \ln(\tfrac{1}{2}|\vec{k}|)}{\ln b}\right] \qquad (30)$$

where $\alpha = \ln n/\ln b$. If $\alpha > 1/2(3 - D)$ this structure function is differentiable for $|\vec{k}| > 0$. At $\vec{k} = \vec{0}$, $\tilde{p}(\vec{k}) - 1 = O(|\vec{k}|^{\alpha})$ which ensures differentiability if $\alpha > 1$. However since the sum over m in Eq. (30) contains oscillatory terms qualitatively similar to the Weierstrass case (Eq. (25)) we leave as an open question whether or not a derivative of $\tilde{p}(\vec{k})$ will result in a continuous, but nowhere differentiable function.

D. The Riemann-Möbius Random Walk: On the Riemann Hypothesis

Various important number theoretic functions can be used to generate random walks [17]. For example, in one dimension, consider the following normalized structure function

$$\tilde{p}(k) = y(\zeta(\alpha) - \zeta^{-1}(\alpha))^{-1} \sum_{n=1}^{\infty} \frac{1-\mu(n)}{n^{\alpha}} \cos(nk) + (1-y)(\zeta(\alpha)+\zeta(\alpha)^{-1})^{-1} \sum_{n=1}^{\infty} \frac{1+\mu(n)}{n^{\alpha}} \cos(nk) , \tag{31}$$

where ζ is the Riemann zeta function, $1 < \alpha < 3$, and $0 \leq y \leq 1$, and μ is the Möbius function, i.e., $\mu(n) = (-1)^{\ell}$ if n is the product of ℓ distinct primes, and $\mu(n) = 0$ otherwise. Rewriting the cosine in terms of its inverse Mellin transform and then performing the summation in Eq. (31) leads to

$$\tilde{p}(k) = \frac{1}{2\pi i} \int_{c-i\infty}^{c+i\infty} k^{-s}[(\frac{y}{\zeta^2(\alpha)-1} + \frac{1-y}{\zeta^2(\alpha)+1})\zeta(\alpha + s) - (\frac{y}{\zeta^2(\alpha)-1} - \frac{1-y}{\zeta^2(\alpha)+1}) \frac{1}{\zeta(\alpha+s)}]\cos(\frac{\pi s}{2})\Gamma(s)ds . \tag{32}$$

For $\alpha \to 1$, $\zeta(\alpha) \to \infty$ and the integrand approaches

$$\frac{1}{2\pi i} \zeta^{-2}(\alpha) [\zeta(\alpha + s) + (1 - 2y)\zeta^{-1}(\alpha + s)]k^{-s} \cos(\frac{\pi s}{2})\Gamma(s) . \tag{33}$$

The cosine-gamma combination has poles at $s = 0,-2,-4,\ldots$ and $\zeta(\alpha + s)$ has a pole at $s = 1 - \alpha$. All other poles arise from zeros of the Riemann zeta function. An infinite number of zeros exist including the trivial zeros of $\zeta(s)$ at $s = -2,-4,-6,\ldots$. Riemann hypothesized in 1859 that all complex zeros of $\zeta(s)$ lie on the line $\mathrm{Re}\ s = 1/2$. This statement has never been proved or disproved.

From Eq. (34) we immediately conclude that

i) all zeros of $\zeta(s)$ occur in complex conjugate pairs since $\tilde{p}(k)$ is real.

ii) Real part of zeros of $\zeta(s) < 1$, or else the $(1-2y)\zeta^{-1}(\alpha + s)$ term could dominate the $\zeta(\alpha + s)$ term in the small k expansion of $\tilde{p}(k) - 1$. If this were true then the sign of the first term in the expansion depends on whether y is greater or less than $1/2$; since $\tilde{p}(k) < 1$, this cannot occur. Thus the $\zeta^{-1}(\alpha + s)$ term must not dominate the $\zeta(\alpha + s)$ term.

Facts i) and ii) are well known. We have extracted them from probabilistic arguments. In our example, if the Riemann Hypothesis were false, say a zero of $\zeta(s)$ appears with a real part equal to 3/4, no laws of probability would be violated. The prime number theorem is equivalent to showing that ii) is in fact a strict inequality. Perhaps our type of approach will eventually yield new results on the nature of Riemann's Conjecture.

5. Fractal Time [12]

Our examples have generated random walk paths which are clustered in space. No regard has yet been given to the time between events. A simple probability density $\psi(t)$ governing the time between events is the exponential,

$$\psi(t) = \lambda e^{-\lambda t} \tag{34}$$

which generates a Poisson distribution for the number of events in a time interval of length t. Our Weierstrass-like fractal form accounting for all time scales is

$$\psi(t) = \frac{1-a}{a} \sum_{j=0}^{\infty} a^n \lambda^n e^{-\lambda^n t}, \quad (a, \lambda < 1) \tag{35}$$

The evolution of this stochastic process is governed by a set of competing rates λ^n occuring with a probability proportional to a^n. Events occur in bursts of self-similar clusters. The scaling is most easily exhibited after Laplace transforming over time to obtain (in analogy to the free energy scaling in Eq. (14))

$$\psi^*(s) = \frac{1-a}{a} \sum_{j=1}^{\infty} \frac{(a\lambda)^j}{\lambda^j + s} = a\psi^*(s/\lambda) + \frac{1-a}{1+s/\lambda} \, . \qquad (36)$$

An explicit representation of $\psi^*(s)$ is

$$\psi^*(s) = 1 + s^\alpha K(s) + \frac{1-a}{a} \sum_{j=1}^{\infty} \frac{(-1)^j a \, s^j}{\lambda^j - a} \qquad (37)$$

where

$$K(s) = \frac{a-1}{a \, \ln \lambda} \sum_{j=-\infty}^{\infty} \frac{\pi a \lambda^x}{\sin(\pi x)} \exp(-2\pi i j \, \ln s/\ln \lambda) \quad ,$$

$$x = -\alpha + 2\pi i j/\ln \lambda$$

and

$$\alpha = \ln a/\ln \lambda \, .$$

When $0 < \alpha < 1$, the mean time $\langle t \rangle$ between events is infinite (see Eq. (12)) and fractal behavior ensues.

6. Conclusions

We have constructed stochastic processes which have inherent self-similar scaling in space or time, and can be used to model complex inhomogeneous systems which exhibit spatial or temporal features over many scales. As mathematicians we have categorized possible behaviors in terms of a fractal dimension. As physicists we search for the basic interactions that will lead to such fractal behavior. If the interaction is known, then in principle an expression for the fractal dimension can be derived. If the interaction depends on external variables such as temperature, pressure, etc. then so should the fractal dimension in a testable manner.

The long tailed probability distributions which generate our fractal stochastic processes appear naturally in mathematics as the limiting form for the distribution of a sum of identically distributed random variables with infinite second (and/or lower moments). The theory of these exceptions to the Central

Limit Theorem was developed by P. Lévy in the 1920's although the best known example is the Cauchy distribution $f(x) = [\pi(1 + x^2)]^{-1}$, which was discovered in this context in 1853. The Fourier transforms of these limit distributions of Lévy are non-analytic and involve non-integer exponents. The world of clusters and intermittancies is the result of these non-integer exponents. Our article is a celebration of the genius of Lévy for bringing the ingredients of this world into mathematical being.

ACKNOWLEDGEMENT

This research was partially supported by the Defense Advanced Research Projects Agency. We thank B.D. Hughes for many useful and interesting discussions.

REFERENCES

1. J.K. Dienes and J.M. Walsh in High Velocity Impact Phenomena, R. Kinslow, ed. (Academic Press, NY, 1979).

2. B.J. West and M.F. Shlesinger, J. Stat. Phys. 30, 547 (1983).

3. E.W. Montroll and H. Scher, J. Stat. Phys. 9, 101 (1973).

4. M.F. Shlesinger, J. Stat. Phys. 10, 421 (1974).

5. H. Scher and E.W. Montroll, Phys. Rev. B12, 2455 (1975).

6. P.G. deGennes, Scaling Concepts in Polymer Physics, (Cornell U., Ithaca, 1979).

7. B.B. Mandelbrot, The Fractal Geometry of Nature, (W.H. Freeman, San Francisco, 1982).

8. M.F. Shlesinger, J. Chem. Phys. 78, 416 (1983).

9. H.J. Stapelton, J.P. Allen, C.P. Flynn, D.G. Stinson and S.R. Kurtz, Phys. Rev. Lett. 45, 1456, (1980).

10. G. Williams and D.C. Watts, Trans. Faraday. Soc. 66, 80 (1970).

11. E.W. Montroll and M.F. Shlesinger, Proc. Nat. Acad. Sci. (USA) 79, 3380 (1982).

12. M.F. Shlesinger and B.D. Hughes, Physica A109, 597, (1981).

13. The Niemeijer and J.M.J. van Leeuwen, in C. Domb and M.S. Green (eds.), Phase Transitions and Critical Phenomena, Vol. 6 (Academic Press, London, 1976).

14. E.W. Montroll and G.H. Weiss, J. Math. Phys. 6, 167 (1965).

15. B.D. Hughes, M.F. Shlesinger, and E.W. Montroll, Proc. Natl. Acad. Sci. (USA) 78, 3287 (1981).

16. B.D. Hughes, E.W. Montroll, and M.F. Shlesinger, J. Stat. Phys. 28, 111 (1982).

17. B.D. Hughes, E.W. Montroll, and M.F. Shlesinger, J. Stat. Phys. 30, 273 (1983).

18. K. Pearson, Nature, 72, 294 (1905).

19. Lord Rayleigh, Phil. Mag. 10, 73 (1880).

20. Lord Rayleigh, Nature, 72, 318 (1905).

PERCOLATION AND FRACTAL PROPERTIES OF THIN GOLD FILMS

R. F. Voss, R. B. Laibowitz, and E. I. Alessandrini

IBM Thomas J. Watson Research Center

Yorktown Heights, NY 10598

ABSTRACT

Transmission electron micrographs of thin evaporated gold films with thickness varying from 6 to 10 nm were analyzed by computer. The films cover the range from electrically insulating to conducting and thus span the 2D percolation threshold. The computer analysis allows the direct comparison of actual geometric cluster statistics with both the scaling theory of percolation and Mandelbrot's fractal geometry. We find that Au-Au and Au-substrate interactions set a natural correlation length of order 10nm. Small clusters are dominated by these effects and have simple almost-circular shapes. At larger scales, however, the irregular connected clusters are ramified with a perimeter linearly proportional to area. Near the percolation threshold the large scale power-law correlations and area distributions are consistent with the scaling theory of 2nd order phase transitions. In the fractal interpretation, we demonstrate that the boundary of all clusters is a fractal of dimension D=2 while the largest cluster boundary has a fractal dimension $D_c \approx 1.9$. Moreover, many of the usual analytic scaling relations between universal exponents are shown to have fractal geometric basis.

1. Introduction

Thin film configurations of metal-insulator mixtures are technologically important and have been extensively investigated. Changes in the conductivity properties as the material composition is varied have been interpreted in terms of percolation [1]. The conductivity measurements are, however, complicated by the presence of other conductivity mechanisms not related to percolation. Leakage through the insulator or quantum mechanical tunneling between neighboring islands smear the percolation transition and give a temperature dependence to the conductivity. Although detailed structural micrographs of composite materials have been available for many years, there has traditionally been little work directly comparing geometric properties with percolation.

In the idealized percolation problem, a random mixture of conducting (fraction p) and insulating (fraction 1-p) material abruptly exhibits long-range conduction at a critical concentration $p=p_c$. This formulation [2] is mathematically equivalent to a second order phase transition [3] and is ideally suited to computer modeling [1,4-8]. The rich variety of universal "scaling" behavior near p_c is associated with the irregular shapes of the connected clusters. Until recently [9-11], however, statistical studies of cluster geometry were limited to computer simulations. In this paper, on the other hand, we use computer analysis of digitized micrographs to directly examine the cluster geometry of thin gold films near p_c.

2. Experiment

The thin Au films were made at room temperature by electron beam evaporation onto special insulating substrates developed for transmission electron microscopy (TEM) [12]. Most scanning electron microscopes (SEM) do not have sufficient resolution to study the gold clusters. The substrates consisted of a layer of amorphous Si_3N_4 deposited on a standard single-crystal Si wafer. The Si backing was then etched away from a portion of the wafer leaving a 30 nm thick Si_3N_4 window supported by a Si frame. This window structure allowed the transmission of the electron beam for the TEM studies. A series of substrates was mounted in a vacuum chamber with a base pressure of $2x10^{-7}$ Torr. An electron beam was used to heat Au metal and cause deposition of Au atoms on the insulating substrate. A moving shutter blocked successive samples from the Au and produced a series with nominal thicknesses ranging from about 6 to 10 nm that varied from electrically insulating to conducting.

Transmission electron micrographs were taken of each of the samples at varying magnifications. A typical micrograph is shown in Fig. 1(a). The light areas are metal and the dark background is the amorphous substrate. The complicated irregular shapes of the connected clusters are readily visible. The structure within each cluster is due to different crystallographic orientations of the individual Au grains that form a cluster. The micrographs were digitized with a scanning densitometer (typically on a 512 by 512 grid). The computer controlled densitometer moved the TEM negatives between a point light source

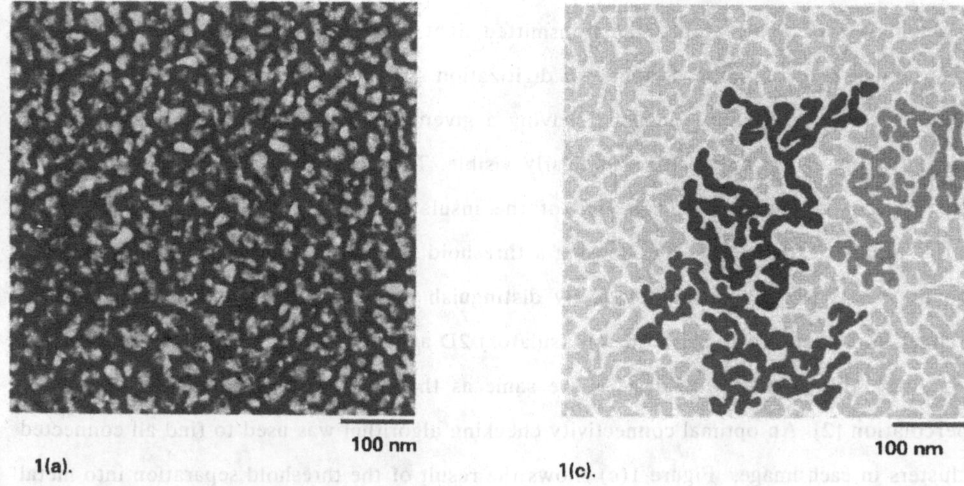

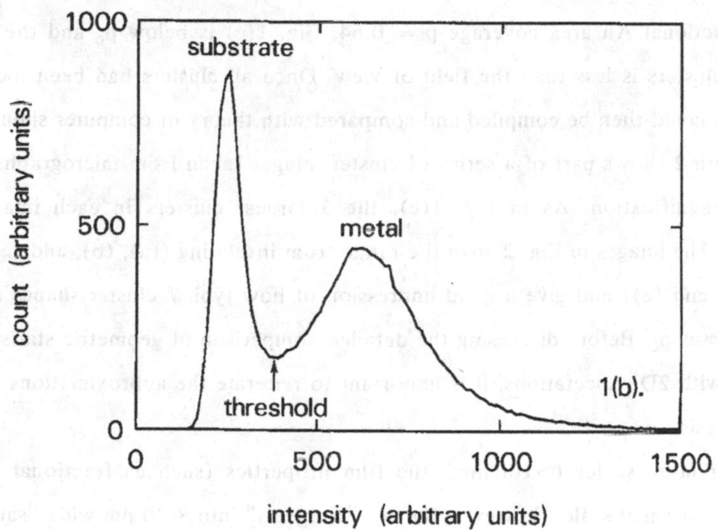

Figure 1. (a) Sample TEM micrograph. (b) Histogram of the number of pixels in (a) having a given transmitted light intensity. (c) Connectivity analysis of Fig. 1 with p=0.64. The three largest clusters are shown in dark shades.

and microscope and recorded the transmitted light intensity at each point in the image. Each image required about 8 hours for digitization. Figure 1(b) shows a histogram of the number of pixels (picture elements) having a given intensity of transmitted light for the image in Fig. 1(a). Two peaks are clearly visible. The peak on the left, at low intensity, corresponds to the dark background of the insulating substrate. The one on the right corresponds to the metal. By choosing a threshold intensity at the minimum between the peaks it was possible to automatically distinguish metal from insulator in the digitized images and create a two-state (metal-insulator) 2D array that approximated the actual film.

From this point, the analysis is the same as that used with computer simulations of percolation [2]. An optimal connectivity checking algorithm was used to find all connected clusters in each image. Figure 1(c) shows the result of the threshold separation into metal and insulator and the connectivity checking for the image in Fig. 1(a). The 3 largest clusters are seen in dark shades while the smaller clusters are seen in a uniform light gray. At a fractional Au area coverage $p = 0.64$, Fig. 1(c) is below p_c and the extent of the largest clusters is less than the field of view. Once all clusters had been located, detailed statistics could then be compiled and compared with theory or computer simulations.

Figure 2 shows part of a series of cluster images taken from micrographs at somewhat lower magnification. As in Fig. 1(c), the 3 largest clusters in each image are shown darkest. The images in Fig. 2 span the range from insulating ((a), (b), and (c)) to conducting ((d) and (e)) and give a good impression of how typical cluster shapes and sizes vary as p crosses p_c. Before discussing the detailed comparison of geometric statistics from such images with 2D expectations, it is important to reiterate the approximations being made in our analysis.

Over large scales (>100 nm) the film properties (such as fractional coverage) are uniform. At small scales, however, the Au "congeals" into 8-20 nm wide "sausages" with a 4-8 nm spacing [13]. The transition from insulator to metal in the films as more Au is added is actually a 3 dimensional process complicated by surface tension and substrate wetting effects. Although it is likely that the initial distribution of where Au atoms first hit the substrate is completely random, the atoms can move once on the surface. There is a tendency for the Au to form droplets on the surface. In fact, at elevated temperatures (around 550 C) the Au forms isolated round droplets rather than the irregular clusters shown in Figs. 1 and 2. The small scale roundness (of order 10 nm) in Fig. 2 is indicative of the beginnings of this process in our room temperature samples. The nominal thickness

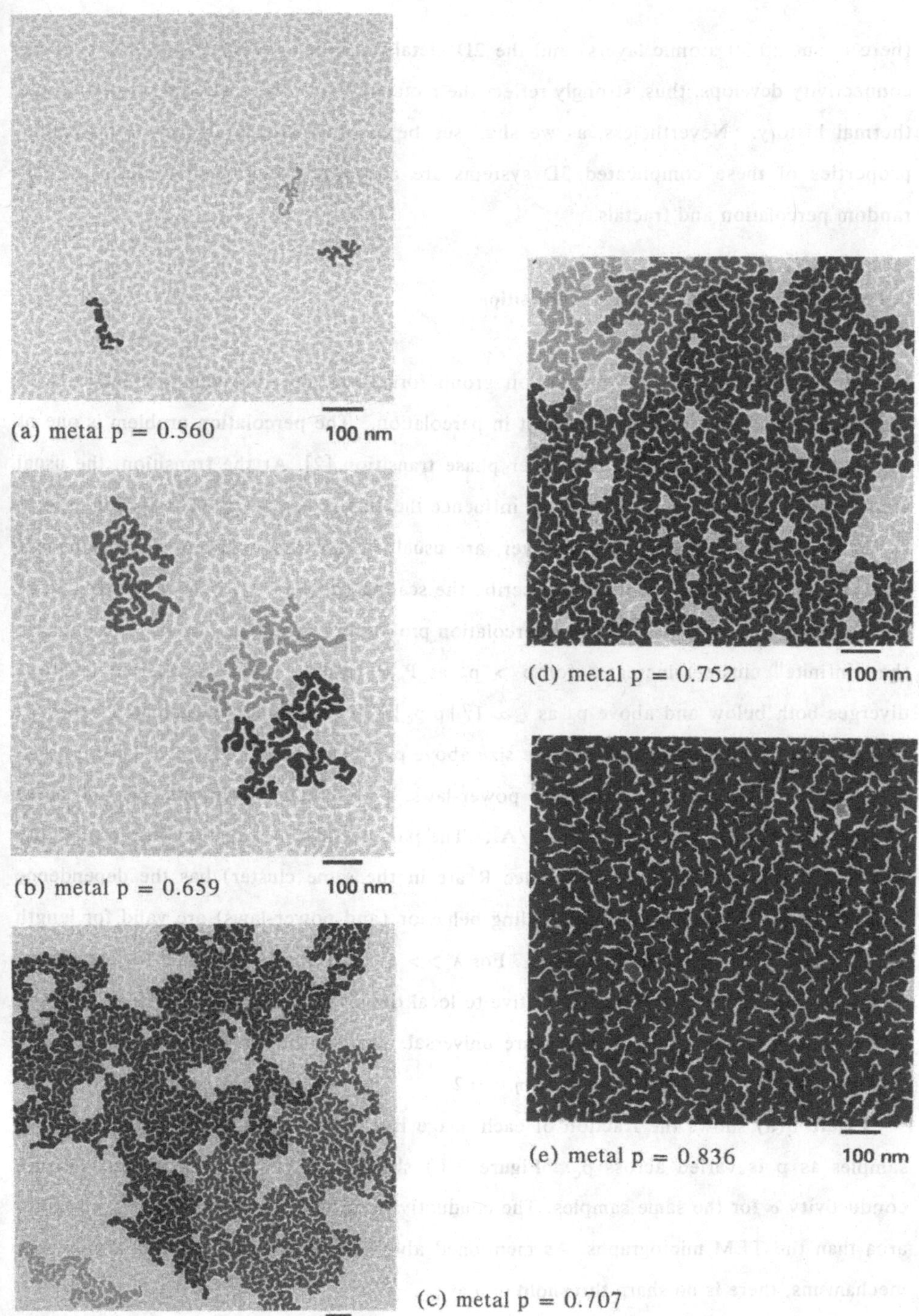

Figure 2. Cluster analysis from a series of Au films that span the percolative transition. In each case, the three largest clusters are shown in dark shades.

(here about 20-30 atomic layers) and the 2D metal fraction ($\approx$0.74) at which long range connectivity develops, thus, strongly reflect the metal and substrate material properties and thermal history. Nevertheless, as we shall see below, the large scale geometric scaling properties of these complicated 3D systems are adequately described in terms of 2D random percolation and fractals.

3. Percolation as a 2nd order phase transition

Recent advances in renormalization group formalism and the general popularity of scaling behavior have rekindled interest in percolation. The percolation problem is one of the simplest realizations of a 2nd order phase transition [2]. At the transition, the usual analytic scaling behavior is expected to influence the cluster geometry. The scaling aspects of 2nd order phase transitions, however, are usually presented in terms of the alphabet soup of "universal" exponents that describe the scaling (power-law) behavior at (or near) p_c. In terms of percolation, P, the "percolation probability" (the fraction of the sample in the "infinite" cluster) increases for $p > p_c$ as $P \propto (p-p_c)^\beta$. The correlation length ξ diverges both below and above p_c as $\xi \propto 1/|p-p_c|^\nu$. The average cluster area below p_c (which corresponds to the average hole size above p_c) diverges as $\langle A \rangle \propto 1/|p-p_c|^\gamma$. At p_c the interesting distributions become power-laws. The number of clusters (per unit area) having area A has the form $n(A) \propto 1/A^\tau$. The pair correlation function (the probability that two points separated by a distance R are in the same cluster) has the dependence $G(R) \propto 1/R^\eta$. Away from p_c the scaling behavior (and power-laws) are valid for length scales λ up to the correlation length ξ . For $\lambda >> \xi$ the properties become homogeneous. Although the exact value of p_c is sensitive to local details (square or triangular lattice, site or bond percolation), the exponents are universal. In 2D they have the values [2] $\beta \approx 0.14$, $\nu \approx 1.35$, $\gamma \approx 2.4$, $\tau \approx 2.05$, and $\eta \approx 0.2$.

Figure 3(a) shows the fraction of each image in the largest cluster for a series of Au samples as p is varied across p_c. Figure 3(b) shows the measured room temperature conductivity σ for the same samples. The conductivity measurements sample a much larger area than the TEM micrographs. As mentioned above, however, due to other conduction mechanisms, there is no sharp threshold in σ at p_c.

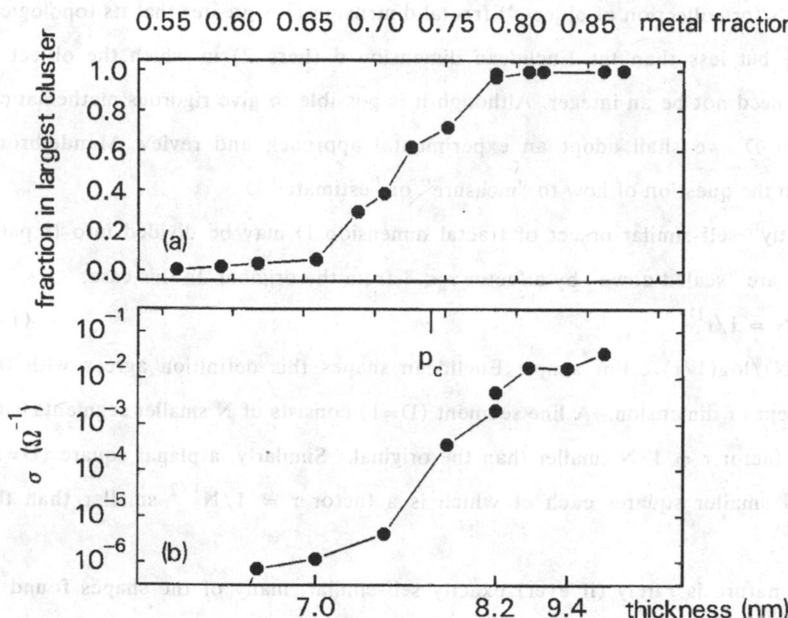

Figure 3. (a) Fraction of image in the largest cluster as p is varied. (b) Measured room temperature conductivity.

4. Fractal dimensions

The scaling theory of percolation provides a mature, analytic treatment of percolation as a 2nd order phase transition [2]. Figure 2, however, which shows the changing irregular shapes across the percolation transition, demonstrates the importance of geometrical concepts. The Au clusters in Figs. 1 and 2 are irregularly shaped, "stringy" or "ramified", and certainly not describable in terms of the familiar Euclidean shapes (circles, triangles, etc.). It is Mandelbrot's fractal geometry [14] that provides an alternate geometric formalism for describing the cluster shapes and understanding their scaling behavior. The concept of a fractal dimension D is finding wide use in many areas of science and fractals have provided specific geometric models for analytic calculations [4,15]. On a geographic scale, a fractal dimension can be deduced from the power-law scaling of the island size distribution on the earth [14], the changes in measured coastline length with yardstick size, or the area-perimeter relation of cloud and rain areas [16]. We shall apply these fractal geometric concepts to the Au clusters on the scale of 1-1000 nm and discuss their connection with scaling theory [2].

An object's (or collection of objects') fractal dimension D is greater that its topological dimension D_T but less than the Euclidean dimension d (here 2) in which the object is embedded. D need not be an integer. Although it is possible to give rigorous mathematical definitions for D, we shall adopt an experimental approach and review Mandelbrot's answer [14] to the question of how to "measure" or "estimate" D.

An "exactly" self-similar object of fractal dimension D may be divided into N parts each of which are "scaled down" by a factor $r < 1$ from the original. In this case,

$$N = 1/r^D \tag{1}$$

or $D = \log(N)/\log(1/r)$. For simple Euclidean shapes this definition agrees with the intuitive concept of dimension. A line segment (D=1) consists of N smaller segments each of which is a factor $r = 1/N$ smaller than the original. Similarly, a planar square (D=2) consists of N smaller squares each of which is a factor $r = 1/N^{1/2}$ smaller than the original.

Although nature is rarely (if ever) exactly self-similar, many of the shapes found in nature are statistically self-similar. A small portion looks like, but not exactly like, a scaled down version of the whole. Thus, $N(\lambda)$, the average number of elements having a characteristic length scale λ varies as

$$N(\lambda) \propto 1/\lambda^D. \tag{2}$$

For example, $L(\lambda)$, the length of a coastline when measured with a ruler of size λ varies as $L(\lambda) = \lambda N(\lambda) \propto 1/L^{D-1}$. Experimentally on the earth, $L(\lambda) \approx 1/\lambda^{0.2}$ corresponding to D ≈ 1.2.

D can also be estimated from the "mass" or area distribution. M(R) is the average amount of an object within a radius R (in d-space) of some point in the object. As R is varied

$$M(R) \propto R^D. \tag{3}$$

One again this definition agrees with intuition for simple Euclidean shapes such as lines where $M(R) \propto R^1$ and planes where $M(R) \propto R^2$.

So far, we have considered ways of estimating D for an individual object. D, however, can also characterize a distribution of objects. If $N(\xi > L)$ is the number of objects with a characteristic size $\xi >$ some specific value L, then

$$N(\xi > L) \propto 1/L^D\ . \tag{4}$$

In the plane (d=2), the fractal dimension D characterizing the boundaries of a collection of objects can also be deduced from perimeter P vs area A scaling. At a fixed resolution

λ, an object of characteristic size ξ has area $A \approx \xi^d$ while, from (2), the perimeter $P \approx \lambda N(\lambda) \approx \lambda(\xi/\lambda)^D$ so

$$P \propto A^{D/d}. \qquad (5)$$

This relation has been used by Lovejoy [16] to estimate D from P vs A scaling of rain and cloud areas taken from radar and satellite data.

5. Percolation, fractal dimensions, and scaling in the Au clusters

In spite of initial confusion in the application of fractal concepts to percolation [17] (and the proliferation of excess "dimensions"), there is now agreement that percolation cluster boundaries are actually characterized by two fractal dimensions D and D_c. The collection of ALL cluster boundaries at any p has $D = d$ (here, $d = 2$). At p_c the largest cluster extends across the sample but does not "fill" the plane. Its boundary has a fractal dimension $D_c \approx 1.9$ that directly reflects the universal exponents.

D=2 is an almost trivial consequence of the fact that at large scales the film properties become uniform. All points are close to a metal-insulator boundary and (like a Peano curve) the boundaries "fill" the plane. The D=2 relation, however, has a non-trivial reflection in the P vs A scaling. As first suggested by numerical simulation [6] and later proven rigorously [18] for percolation clusters, $P \propto A$ for large A at all p. As discussed above, in the fractal interpretation, $P \propto A^{D/d}$ leading to D=d=2. Figure 4 shows a scatter plot of P (defined as the number of unoccupied sites on the digitized grid adjacent to a given cluster) as a function of A for our films. Each point represents one cluster. The smallest clusters ($A < 250$ nm^2) have simple almost circular shapes and show $P \approx A^{1/2}$ corresponding to line-like boundaries with $D \approx 1$. For $A > 600$ nm^2, on the other hand, $P \propto A$ both above and below p_c. The crossover between $P \propto A$ and $P \propto A^{1/2}$ is the same for all p and is related to the Au-substrate interaction. Figure 2 shows a clustering of the largest Au clusters and it is plausible that large clusters grow by developing small connecting necks to their neighbors. Thus, the larger clusters are a weak connection of N smaller clusters (each of which has perimeter P_0 and area A_0) giving a total $P \approx NP_0$, $A \approx NA_0$, and $P/A \approx$ constant.

The scaling theory of percolation [2] is based upon computer simulation and renormalization group arguments. Typical studies center around the quantity n(A), the average number (per lattice site) of clusters of area A as a function of p. As discussed above, at

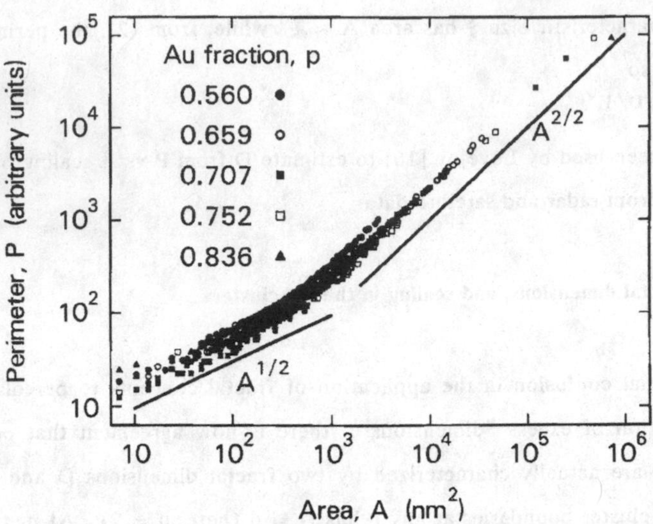

Figure 4. Scatter plot of the perimeter P vs area A for the gold clusters from the samples shown in Fig. 2. Each cluster produces one point. The solid line shows the expected P ∝ A behavior at large A.

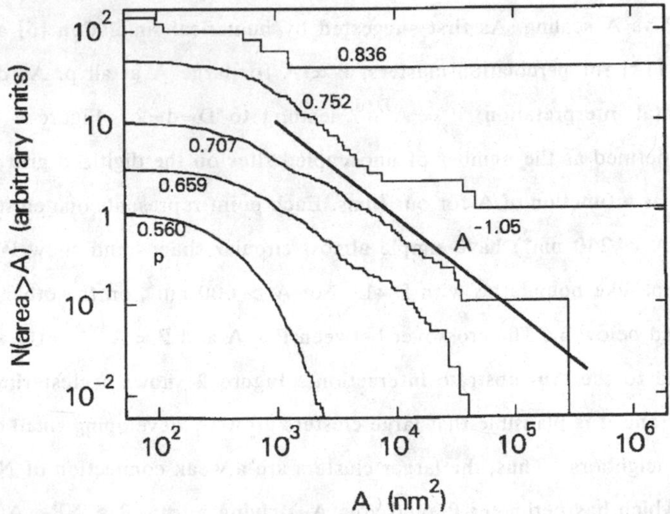

Figure 5. The cumulative distribution N(area>A) vs A at different fractional Au coverages p. The solid line shows the expected $1/A^{1.05}$ dependence at p_c.

p_c there is no characteristic size scale for the clusters and $n(A) \propto 1/A^{\tau}$. Similarly, the Korčak-Mandelbrot law [14] for the distribution of islands on the earth's surface states that the number of islands with area greater than some size A, $N(\text{area}>A) \propto A^{-B}$. Mandelbrot [14] explains this law using the fractal concepts discussed above. For a scaling system of islands in which the combined coastlines have a fractal dimension D, the number of islands having a characteristic size scale $\lambda >$ some length L is given by (4). The area of an individual island is given by $A \approx \lambda^2$ and the cumulative area distribution becomes

$$N(\text{area}>A) \propto 1/A^{D/2} \tag{6}$$

in agreement with empirical findings.

For percolation clusters at p_c, however, the area distribution involves both fractal dimensions D and D_c. D characterizes the size distribution of all clusters according to (4) while D_c characterizes the distribution within a given cluster. For length scales up to its size λ, the area or "mass" of an individual cluster is given by (3) as

$$A \propto \lambda^{D_c} \tag{7}$$

and the cumulative area distribution becomes

$$N(\text{area}>A) \propto 1/A^{D/D_c}. \tag{8}$$

Since $N(\text{area}>A)$ is simply the integral of $n(A)$ above, scaling theory (with $n(A) \propto 1/A^{\tau}$) corresponds to $N(\text{area}>A) \propto 1/A^{\tau-1}$ and we have the first relation between D_c and the universal exponents,

$$D_c = D/(\tau-1) = d/(\tau-1) \approx 1.9. \tag{9}$$

Figure 5 shows $N(\text{area}>A)$ vs A for some of the Au films studied. The use of the cumulative distribution $N(\text{area}>A)$ rather than $n(A)$ eliminates any problems with histogram bin sizes. For p well below p_c, $N(\text{area}>A)$ decays rapidly for A beyond some characteristic (maximum) size. For p well above p_c, $N(\text{area}>A)$ becomes independent of A at large A. Near p_c, however, $N(\text{area}>A)$ decays with the expected $1/A^{\tau-1} \approx 1/A^{1.05}$ dependence for a large range of A.

Many important physical properties of a system are determined by the pair correlation function [3] G(R). For percolation, G(R) is the probability that two points separated by a distance R are in the same cluster. At the critical point G(R) takes the power-law form $G(R) \propto 1/R^{\eta}$ and η can be related to D and D_c. We first consider the conditional correlation function for a single cluster. If the origin is in a specific cluster, then $G_c(R)$ is the probability that a point a distance R away is also in the cluster. $G_c(R)$ is also a power-law

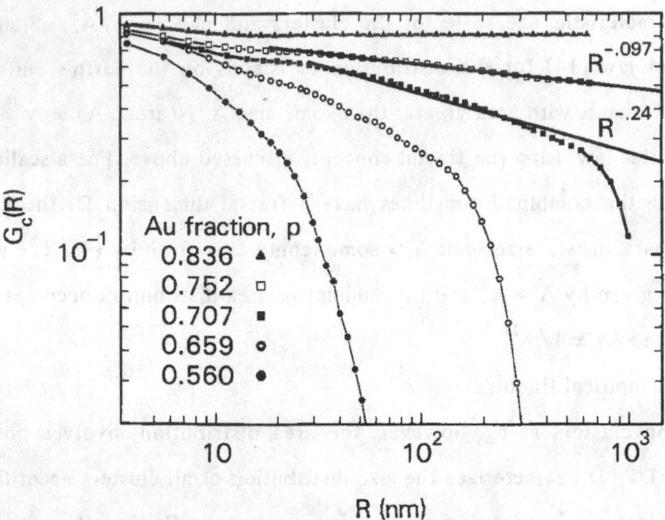

Figure 6. The single cluster correlation function $G_c(R)$ for the largest cluster in each the Au samples from Fig. 2. The solid lines show power-law fits near p_c.

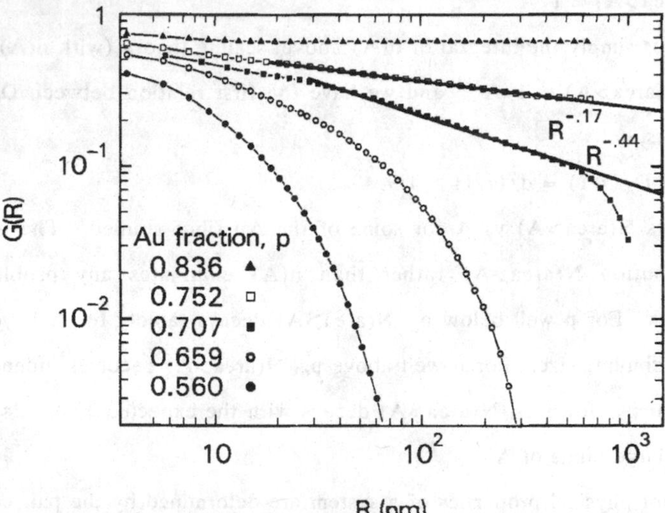

Figure 7. The pair correlation function G(R) vs R for the same samples. G(R) is the probability that two points separated by a distance R are both in the same cluster, averaged over all clusters. The solid lines show power-law fits near p_c to give the exponent η .

with exponent η_c, $G_c(R) \propto 1/R^{\eta_c}$, up to the cluster size λ . For $R >> \lambda$, $G_c(R) \to 0$. $G_c(R)$ is directly related to the mass distribution given in (3) by

$$M(R) \propto 1/R^{D_c} \propto \int G_c(R) R dR \propto 1/R^{2-\eta_c}$$

and $D_c = 2 - \eta_c$. Figure 6 shows $G_c(R)$ for the largest cluster in some of our Au samples. For $p < p_c$, $G_c(R)$ falls off rapidly with increasing R. Above p_c, $G_c(R)$ becomes a constant at large R. Near p_c, however, $G_c(R)$ does show large ranges of power-law behavior. The solid lines in Fig. 6 show least squares fits to these portions where $0.097 < \eta_c < 0.24$.

G(R), however, involves both $G_c(R)$ and the distribution over all clusters sizes λ . G(R) is proportional to $G_c(R)$ times the probability that the origin is in a specific cluster with size $\lambda > R$. Thus,

$$G(R) \propto G_c(R) \times \int_R^\infty area(\lambda) n(\lambda) d\lambda,$$

where $n(\lambda) \propto 1/\lambda^{D-1}$ is the derivative of the cumulative distribution $N(\lambda > L)$ and $area(\lambda) \propto \lambda^{D_c}$. This leads to

$$G(R) \propto 1/R^{\eta_c} \times R^{D_c} \times 1/R^{D} \propto 1/R^{2\eta_c} \propto 1/R^{\eta}$$

and gives another relation between D_c and the universal exponents

$$D_c = d - \eta_c = d - \eta/2 \approx 1.9. \qquad (10)$$

Figure 7 shows the measured G(R) at different p. As in Fig. 6, for $p<p_c$, G(R) decays rapidly at large R while for $p>p_c$ G(R) approaches a constant. At $p=0.707$ G(R) shows a large power-law section with $\eta = 0.44\pm0.07$ but decreases more rapidly for $R > 500$ nm. At $p=0.752$, on the other hand, just above p_c, G(R) remains very close to a power-law with $\eta = 0.17\pm0.02$. The measured G(R) is thus in good agreement with the accepted $\eta \approx 0.2$ at p_c and consistent with the single cluster $G_c(R)$ in Fig. 6.

Figure 8 shows a direct estimate of D_c for the largest cluster in each sample through the variations in cluster shape with the minimum resolved length λ in the image. If the entire cluster image of size L by L is divided into $(L/\lambda)^2$ squares of side λ and $N_s(\lambda)$ represents the number of these squares intersected by the largest cluster boundary, one expects

$$N_s(\lambda) \propto [L/\lambda]^{D_c} \qquad (11)$$

from (2). Intuitively, a zero-dimensional boundary looks like a point, always intersects one square, and has $N_s(\lambda) \propto 1/\lambda^0 = 1$. Similarly, a 1D line intersects L/λ squares, and has $N_s(\lambda) \propto 1/\lambda^1$. In Fig. 8 we show $f_s(\lambda) = N_s(\lambda)/(L/\lambda)^2 \propto \lambda^{2-D_c}$, the fraction of squares occupied by the boundary as λ is varied. At small scales, $f_s(\lambda) \propto \lambda^1$ at all p and the local correlations lead to a boundary $D_c \approx 1$. At large scales, for $p < p_c$, $f_s(\lambda) \propto \lambda^2$, $D_c \approx 0$, and all finite clusters look point-like. Near p_c, however, $f_s(\lambda)$ does scale as λ^{2-D_c} with $D_c \approx 1.78$ at p=0.707 below p_c and $D_c \approx 1.92$ at p=0.752 just above p_c. This measured D_c agrees with that estimated from the measured η_c and η in Figs. 6 and 7.

This direct estimate of D_c from (11) can also be related to other universal exponents. A finite size L by L sample at p_c centered on the largest cluster is indistinguishable from any L by L sample at $p > p_c$ when $L \approx \xi$. (The typical L by L sample at p_c would miss the largest cluster entirely). Thus, $L \approx \xi \propto 1/|p-p_c|^{\nu}$ or $|p-p_c| \propto 1/L^{1/\nu}$. At this $|p-p_c|$, the number of sites in the L by L sample that are part of the largest cluster,

$$N_s(\lambda) \propto L^d |p-p_c|^{\beta} \propto L^{d-\beta/\nu}. \tag{12}$$

Comparing (11) and (12) we get another connection between D_c and the exponents.

$$D_c = d - \beta/\nu \approx 1.9. \tag{13}$$

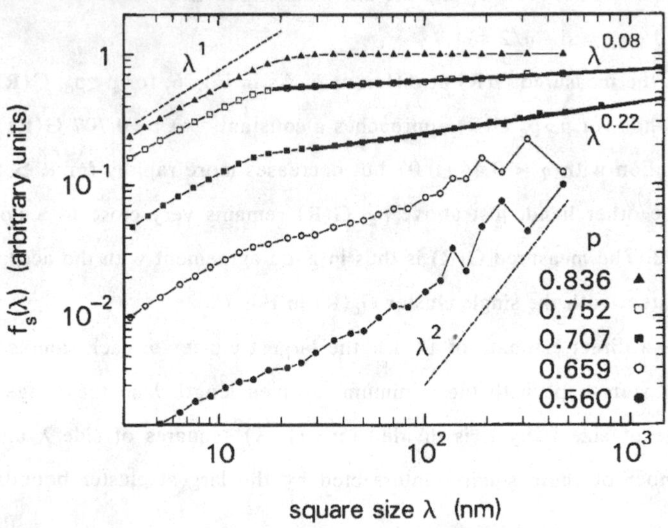

Figure 8. $f_s(\lambda)$, the fraction of squares of size λ in an image of the largest cluster boundary as a function of λ at different p. The solid lines show power-law fits near p_c to give the exponent $d-D_c$.

It is also possible to relate the fractal dimensions to the exponent γ which describes the divergence of the average area $<A> \propto 1/|p-p_c|^\gamma \propto \xi^{\gamma/\nu}$. $<A>$, like G(R), involves both the distribution fractal dimension D and the single cluster D_c. For clusters with size $\lambda < \xi$, $\text{area}(\lambda) \propto \lambda^{D_c}$. Thus, $<A>$ is given by area(λ) times the probability of getting a cluster of size λ, area(λ)n(λ), integrated over all $\lambda < \xi$,

$$<A> \; \propto \; \int_0^\xi area(\lambda) \times area(\lambda) n(\lambda) d\lambda \; \propto \; \xi^{2D_c - D}.$$

This gives yet another relation between D_c and the exponents

$$D_c = (d + \gamma/\nu)/2 \approx 1.9. \tag{14}$$

6. Conclusion

In conclusion, we have used computer analysis of TEM micrographs to show that at large scales actual Au clusters near p_c have the scaling properties one expects from both phase transition arguments and lattice percolation simulations and, hence, belong to the same universality class. Local Au-Au and Au-substrate correlations, however, set a lower cutoff to the scaling behavior of about 10 nm and raise p_c to about 0.74. The cluster boundaries (topologically 1D) are characterized by two fractal dimensions D and D_c. Taken together, the collection of all boundaries at large scales are characterized by D=d=2. At p_c (and for length scales up to ξ away from p_c) individual cluster boundaries are characterized by $D_c \approx 1.9$. We have demonstrated the equivalence between analytic scaling theory [2] with its universal exponents, and the geometric interpretation based on fractals [14]. The connection between D_c and D and the usual exponents τ, η, β, ν, and γ are given in Eqs. (9), (10), (13), and (14). Together, these relations provide an alternate geometric basis for the usual analytic scaling relations between the exponents.

ACKNOWLEDGEMENT

We are grateful to C. R. Guarnieri for expert sample preparation and to B. B. Mandelbrot, Y. Gefen, S. Kirkpatrick and Y. Imry for illuminating discussions and helpful suggestions.

References

1. See for example B. A. Abeles in *Applied Solid State Science,* edited by R. Wolfe (Academic, New York, 1976) Vol. 6, p. 1; B. A. Abeles, H. L. Pinsh, and J. I. Gittleman, *Phys. Rev. Lett. 35,* 247 (1976); or C. J. Lobb, M. Tinkham, and W. J. Skocpol, *Solid State Comm. 27,* 1253 (1978).

2. See the excellent review by D. Stauffer, *Phys. Reports 54,* 1, (1979) and references therein.

3. L. P. Kadanoff et al., *Rev. Mod. Phys. 39,* 395 (1967).

4. S. Kirkpatrick, *A.I.P Conf. Proc. 40,* 99 (1977) and *A.I.P Conf. Proc. 58,* 79 (1979).

5. H. E. Stanley *J. Phys. A 10,* L211 (1977).

6. P. L. Leath, *Phys. Rev. B14,* 5046 (1976).

7. R. J. Harrison, G. H. Bishop, and G. D. Quinn, *J. Stat. Phys. 19,* 53 (1978).

8. J. W. Halley and T. Mai, *Phys. Rev. Lett. 43,* 740 (1979).

9. R. B. Laibowitz, E. I. Allessandrini, and G. Deutscher, *Phys. Rev. B25,* 2965 (1982).

10. R. F. Voss, R. B. Laibowitz, and E. I. Alessandrini, *Phys. Rev. Lett. 49,* 1441 (1982).

11. A. Kapiltunik and G. Deutscher, *Phys. Rev. Lett. 49,* 1444 (1982).

12. R. B. Laibowitz and A. N. Broers, in *Treatise on Materials Science and Technology,* (Academic Press, New York, 1982), Vol. 24 p. 237.

13. The same metal-insulator assymetry is seen in the Pb films on Ge substrates [11].

14. For a general discussion of fractals see B. B. Mandelbrot, *The Fractal Geometry of Nature* (Freeman, San Francisco 1982) and references therein. Chapter 13 deals specifically with percolation.

15. Y. Gefen, A. Aharony, B. B. Mandelbrot, and S. Kirkpatrick, *Phys. Rev. Lett. 47,* 1771 (1981).

16. S. Lovejoy, *Science 216,* 185 (1982).

17. The multiple possibilities for "defining" a dimension are extensively discussed in both refs. 2 and 14.

18. H. Kunz and B. Souillard, *J. Stat. Phys. 19,* 77 (1978) and A. Coniglio and L. Russo, *J. Phys. A 12,* 545 (1979).

PERCOLATION AND CAPILLARY FLUID DISPLACEMENT

Joel Koplik, David Wilkinson, and Jorge F. Willemsen

Schlumberger-Doll Research
P.O. Box 307
Ridgefield, Connecticut 06877

ABSTRACT

We consider capillary displacement of immiscible fluids in porous media in the limit of vanishing flow rate. The motion is represented as a stepwise Monte Carlo process on a finite lattice, where at each step the fluid interface moves through the lattice element of least resistance. The displacement process exhibits considerable fingering and trapping of displaced phase at all length scales, with non-trivial associated fractal dimensions, and with some interesting qualitative differences between two and three dimensions. We interpret our results in terms of percolation theory concepts, and argue that capillary displacement corresponds to a modified percolation process with its own universality class.

1. Introduction

The study of fluid transport in disordered porous media draws upon a broad variety of physics and mathematics lore: hydrodynamics, percolation, critical phenomena, statistical geometry and stereology. The difficulty of the problem is twofold: the forces that control fluid motion, viscosity and capillarity, are determined in part by the random local geometry of the medium, and the path followed by a moving fluid is controlled both by the presence of other fluids and by the topology of the pore space (gravity effects can be neglected for lab-size experimental samples). If we focus on the boundary surface between two immiscible fluids, we face a difficult randomly-forced moving interface problem which has eluded solution despite years of effort [1].

We have obtained some quantitative understanding of a limiting case of this general problem - capillary displacement - in which one fluid immiscibly displaces another at flow rates so slow that the motion is driven by a discrete vestige of capillary force alone [2,3,4]. Not surprisingly, the theory of percolation [5,6] underlies our understanding of the phenomenon, but the relevant model is a variant that we refer to as invasion percolation. The crux of the distinction is that classical percolation refers to the properties of single configurations of bonds or sites on a lattice, whereas fluid displacement is necessarily a dynamic process in which the spatial distribution of fluids at any time determines the subsequent motion.

In the next section, we describe and motivate the discrete dynamics we use to model capillary displacement, and then in Section 3 discuss the conditions under which this simplification should apply. Section 4 presents our Monte Carlo results for two-dimensional (2-d) lattices, and interprets them in terms of percolation concepts. This provides the basis for understanding the 3-d results given in Section 5. Some concluding remarks appear in Section 6.

2. Model of Capillary Displacement

Consider a random disordered porous medium containing two viscous, immiscible, incompressible, Newtonian fluids in contact. A particular application of the ideas presented here is to the motion of oil and water in petroleum reservoirs, and for convenience we shall refer to the displaced fluid as oil, the displacing fluid as water, and the porous medium as rock. We suppose that the rock is "water-wet", i.e., that the contact angle between oil, water and rock is such as to draw water spontaneously into the rock. This case is also referred to as imbibition, in contrast to drainage in "oil-wet" rocks, in which case water would displace oil only if forced in under pressure.

In static equilibrium, the fluids arrange themselves so as to have a constant pressure p_o in the oil phase, a different constant pressure p_w in the water phase, with the difference given by the capillary pressure

$$p_o - p_w = p_c = \frac{2\gamma}{R} . \tag{1}$$

In Eq. (1), γ is the coefficient of surface tension and R the mean radius of curvature of the meniscus separating the fluids. In equilibrium each meniscus on the oil-water interface has the same value of R (see Fig. 1a). Now suppose that water is very slowly injected on the left, raising the water pressure slightly; the fluids will attempt to come to equilibrium, which in view of (1) requires each meniscus to flatten out slightly. Since the static contact angle does not change, a meniscus must move around to find a permissible position of decreased curvature. If a new position does not exist locally, the configuration is unstable and the meniscus must jump across one or more pores to a new equilibrium position. The motion of the oil-water interface then consists of quasi-static small advances of menisci interspersed with rapid jumps in which a pore or more of fluid is displaced. We model this process by retaining only the jumps and assuming that they occur at essentially random points along the oil-water interface.

Another, more heuristic, motivation for this random jump rule is based on modeling the pore space as a lattice of variable-sized channels in Fig. 1b. In any channel the magnitude of capillary pressure varies inversely with the channel radius, and we might suppose that in imbibition the fluid moves fastest where the capillary pressure is greatest. To obtain a discrete version of this process, neglect all motion but that of the most rapidly moving meniscus, so that the motion is modelled as a sequence of discrete steps in which water displaces oil from the smallest channel available on the interface. (Similarily, in drainage, the motion would be most rapid where the channel is widest and the resisting capillary force least.) In this model, the likelihood of motion is controlled by the <u>bonds</u> of the lattice representing the pore space.

If we choose a more realistic version of the second geometry, we can obtain a <u>site</u>-controlled problem: model the pore space as a lattice of random-sized pores connected by random-sized throats (Fig. 1c). In imbibition, the motion is rapid in the narrow throats but delayed by the slow filling of the larger pores. The advance of the interface is then controlled by the time required to fill the pores, and the smallest pore on the interface would be the position of discretized motion.

The upshot of this discussion is the following discrete rule of slow

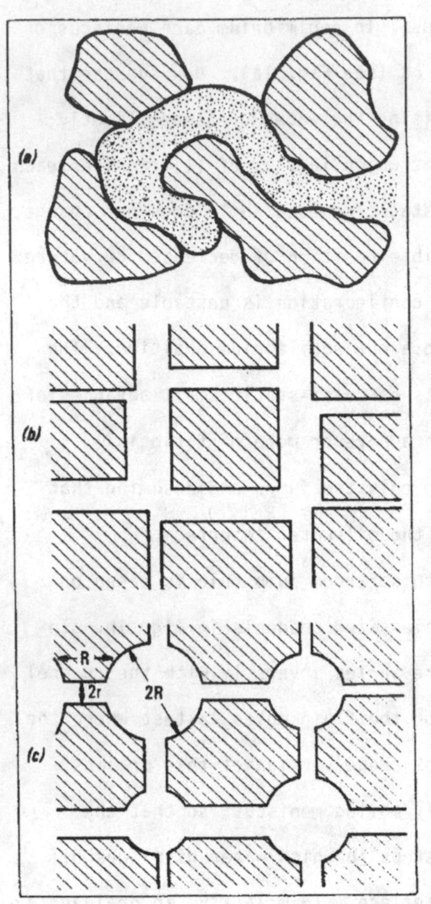

Figure 1

Illustration of capillary displacement
(a) semi-realistically
(b) in a network of random-sized channels and
(c) in a pore and throat model

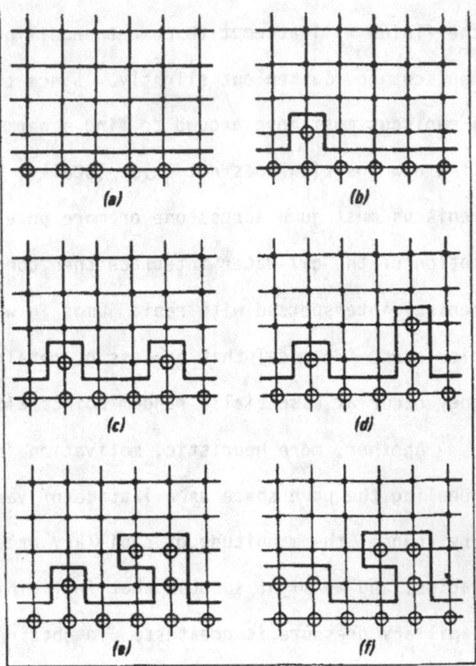

Figure 2

A possible sequence of fluid motions: solid dots are oil-filled pores, open circles are water-filled pores, light lines are throats, and the heavy line is the oil water interface.

capillary displacement. Model the porous medium as a lattice and assign "ranks" to the bonds or the sites of the lattice. Assume that the fluid motion occurs in a discrete sequence of time steps, where in each step the motion occurs in the lattice element of lowest rank available on the interface. Note that because the smallest of a set of numbers is chosen, the shape of the distribution of ranks is irrelevant and may be taken as uniform on the unit interval [0,1]. A possible sequence of displacements on a square lattice is shown in Fig. 2.

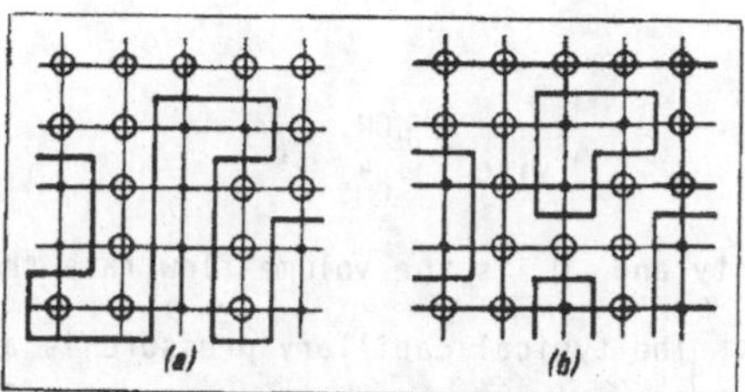

Figure 3. An example of oil trapping, before and after.

Oil can be surrounded by water as the interface moves through the lattice, as shown in Fig. 3. We regard surrounded blobs of oil as trapped or immobilized because they are incompressible and can only move if two steps occur at once. (Physically speaking, trapped blobs are held in place by capillary force.) The fraction of trapped oil left in the lattice after displacement is known as "residual oil saturation"; needless to say, residual oil is of great commercial interest.

Capillary displacement should be distinguished from the related process of "porosimetry" [1,7-9], wherein one forces non-wetting fluid into a porous medium under pressure. The first difference is that the second fluid is vacuum or (compressible) air, so that the trapping mechanism is absent. Furthermore, in porosimetry the imposed pressure is increased in finite steps and any amount of fluid is allowed into the porous medium at a step. Roughly speaking, porosimetry corresponds to constant pressure displacement, while in this paper we are concerned with constant (but very low) flow rate displacement.

3. The Capillary Length Criterion

To understand quantitatively the circumstances under which our discrete modeling of fluid motion might apply, we must estimate the ratio of viscous to capillary forces. As in Fig. 1c, consider the pore space to be made up of pores of typical radius R (of order a grain size) connected by throats of typical radius $r \ll R$ and length of order R. Let the flow be from left to right and suppose there are N such pore/throat combinations per unit area transverse to the flow. The viscous pressure drop across a pore-throat unit is dominated by the throat radius:

$$\Delta P_{visc} \sim \frac{\mu QR}{r^4} ,$$

where μ is the viscosity and Q is the volume flow rate through a single pore/throat combination. The typical capillary pressure is also dominated by the throat radius

$$\Delta P_{cap} \sim \frac{\gamma}{r} .$$

In order to elimate the flow rate Q we observe that the Darcy velocity v (flow per unit area) and porosity ϕ are given by

$$v \sim NQ , \phi \sim NR^2 .$$

Expressing Q and N in terms of v and ϕ, we obtain the ratio of viscous to capillary pressures

$$\frac{\Delta P_{visc}}{\Delta P_{cap}} \sim \frac{\mu v}{\gamma} \left\{\frac{R}{r}\right\}^3 \frac{1}{\phi} = \frac{Ca}{K} \quad (2)$$

where Ca is the conventional dimensionless "capillary number"

$$Ca = \frac{\mu v}{\gamma} , \quad (3)$$

and where the geometrical constant K is given by

$$K = \phi \left\{\frac{r}{R}\right\}^3 . \quad (4)$$

The value of K can be quite small, of order 10^{-4} if $r/R \sim 0.1$ and $\phi \sim 0.1$. In a more realistic representation of the medium the expression for K will be different, but we expect that it will always be a small number.

An important quantity which can be obtained from (2) is the capillary length, which is that distance over which viscous pressure drops are comparable to capillary pressures. Since Eq. (2) is computed from the viscous pressure across a distance of order R, we see that this capillary length is given by

$$L_{cap} = \frac{RK}{Ca} . \quad (5)$$

Using this length, we may make more precise the notion of capillary dominance. We define weak capillary dominance to be the situation where the capillary length is much greater than the pore size R, and from the definition (5), we see that this inequality holds when the capillary number is much less than the constant K. For $K \sim 10^{-4}$, this gives capillary numbers below say 10^{-6}. The latter number is typical of commercial oil-field practice where one tries to flush oil out of a reservoir by flooding with water. Strong capillary dominance on the other hand corresponds to the situation where the capillary length is much greater than the sample size. Clearly strong capillary dominance cannot hold over a macroscopic oil reservoir, but in a laboratory experiment it can easily be achieved since the sample size (centimeters) measured in units of the pore size R (say 100 microns) is only of the order 100 to 1000. The strong capillary regime then corresponds to capillary numbers below say 10^{-8}. The process considered in this paper corresponds to the strong capillary regime.

4. Two-Dimensional Displacement

Figure 4 shows a typical fluid displacement process on the central part of a periodic 100 × 150 rectangular lattice, showing the pores occupied by oil at intervals of 2,500 steps. In this case, we have assigned ranks to the lattice bonds. The black dots are oil-filled pores, the throats have not been drawn, and the white region has been invaded by water. The key qualitative features to note

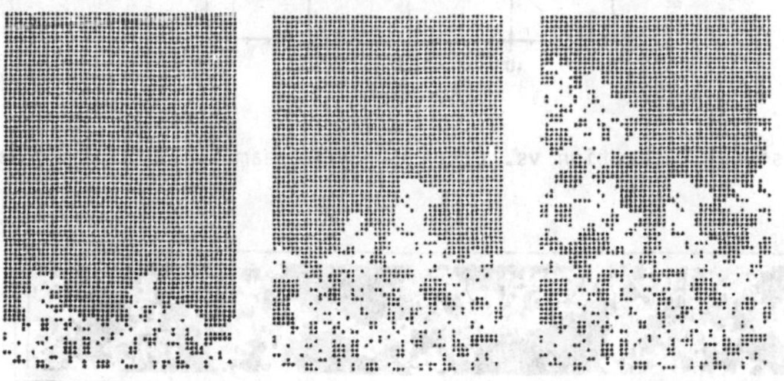

Figure 4. Sequence of fluid displacements on a subset of a 100 × 150 lattice at intervals of 2500 steps (water enters at left).

are the irregularity of the interface, and the relatively large regions of trapped oil. If one examines the process step-by-step, there is an alternation of two kinds of motion: an almost-smooth advance of the interface through a pore or two in different locations, and a "fingering" process in which the interface comes to a sequence of high-ranking throats and runs through a long tortuous path of random orientation. Large trapped oil blobs result when a region is pinched off by long fingers.

The first quantitative question to ask is how the amount of residual oil depends on the length L, width W, and coordination number of the lattice. In Fig. 5, we plot the fraction of trapped oil vs. row number (i.e., distance downstream) for a square lattice of width 50 and length 375, averaged over many runs. If the regions at the ends are disregarded as unrepresentative of the bulk of the lattice, then we find a "central plateau" whose height shows no significant length dependence. The origin of this behavior may be seen in Fig. 6, which shows the result of one flow on this (50 × 375) lattice: there is no evident systematic variation in the trapped oil blobs along the length. We infer that once the length of the lattice approximately exceeds the width, there is no longer any

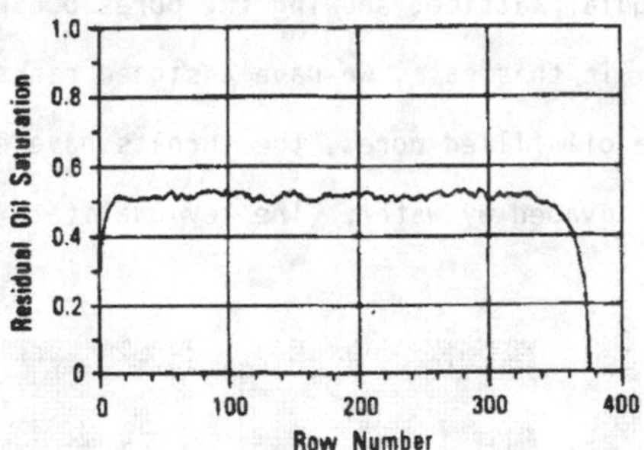

Figure 5. Residual saturation vs. distance downstream for a 50 × 375 lattice.

Figure 6. Example of a final state on a 50 × 375 lattice (water enters at top).

variation of average saturation with length. This has been verified by repeating the simulations at fixed W but varying L. The physical origin of this behavior is that, since the bond selection rule has no preferred direction, the interface frequently moves sideways and is unlikely to go further than O(W) pores downstream before closing itself off to the side.

If we examine the dependence of the water fraction S (the fraction of pores occupied by water) on the width W of the lattice (measured in number of pores) then all of our data, for sample lengths exceeding W, is nicely fitted by

$$S(W) = \text{const. } W^{-\alpha} \ . \tag{6}$$

With the bond selection rule, all 2-d planar lattices have $\alpha = 0.17$. If we instead assign ranks to the sites, we again find (6) but now with $\alpha = 0.18, 0.19$, and 0.12 for the honeycomb, square and triangular lattices respectively. Presumably the difference between the first two cases arises from numerical uncertainty but, for reasons discussed in [3], the triangular lattice may be special.

The data has the a priori surprising feature that $S \rightarrow 0$ as $W \rightarrow \infty$, originating from the fact that water fingers through the oil and surrounds regions that increase in size with the width of the lattice, while the fingers themselves (which contribute to water saturation) can remain at a narrow finite width. In view of our remarks at the end of Section 3, we do not regard this last result as directly relevant to very wide lattices, or, for that matter, to macroscopic oil reservoirs, which have the additional complicating feature of statistical inhomogeneity.

If we plot a histogram of the number of times a rank x was selected in a given run, as a function of x, we find in all cases a distinct threshold as illustrated in Fig. 7 for the square lattice using bond rankings. The observed threshold invariably agrees with the percolation threshold for the lattice in question. This fact, and the "universal critical behavior" of the saturation equation (6) suggests that capillary displacement is closely related to classical percolation. To understand the correspondence, recall that the probability that a given bond has a rank in the interval $[x, x + dx]$ is just dx, since the ranks are assigned randomly in [0,1]. Suppose we consider just those bonds with ranks in the interval $[0,p_c]$; these comprise a fraction

$$\int_0^{p_c} dx = p_c$$

of the bonds in the lattice, and the fundamental theorem of percolation ensures that this subset of bonds will form a percolating cluster. At any step in the simulation, the interface between oil and water regions must cross the percolating cluster (because the latter provides a path of bonds traversing the lattice in any direction). Therefore, at the next step of the simulation there is guaranteed to be a bond with rank $x < p_c$, and so the bond chosen for the advance of the interface must lie in the percolating cluster. The motion of the interface thus selects only bonds whose rank lies in the interval $[0,p_c]$. This argument accounts for the threshold in Fig. 7, but the difference between that curve and simple step function, the fact that not all of the bonds with $x < p_c$ have been filled with water, reflects the difference between invasion and pure percolation: not all of the percolating cluster need fill with water. Bonds with $x < p_c$ may fail to fill either by appearing on the interface but becoming surrounded by motion through other (lower-ranking) bonds, or because they are in the interior of a trapped region and hence inaccesible.

If not for the distinction between capillary displacement and percolation, we could have anticipated (6) from finite size scaling [10]. In bond percolation on a lattice of characteristic size W , the probability of a site belonging to the percolating cluster would vary as in (6) but with an exponent $\alpha = \beta/\nu = 0.11$ in 2d. The larger exponent we find means that the water cluster is smaller than the percolating cluster in accord with our remarks above. A more detailed discussion Of the distinction between the two processes appears in [2,3].

5. Three-Dimensional Displacement

In 3-d, we begin with the same rule as above, but it is now useful to focus attention on the overall saturations, both at the time of first arrival of water at the outlet face, and at the time when no further oil is recovered. In two

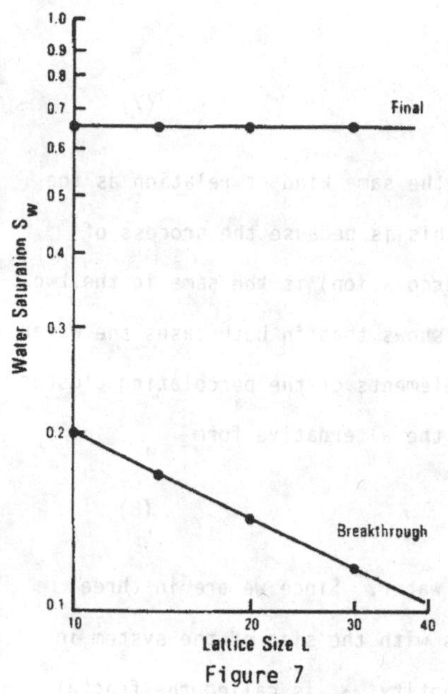

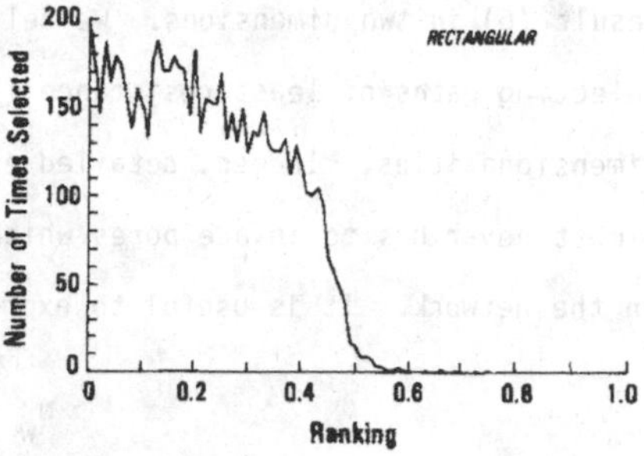

Figure 7

Figure 8

Histogram of bonds selected vs. ranking for a square lattice.

Water saturation as a function of sample size in 3d.

dimensions, these two points are essentially identical, since as soon as water spans the sample it disconnects the oil. In three dimensions, however, these two points are distinct, since it is topologically possible for both water and oil to span the sample simultaneously. We also investigate the manner in which the oil is distributed at the end of the simulation. The following results were obtained for a simple cubic lattice of coordination number 6, using the site-ranking version of the model.

Breakthrough

This is defined as the point at which water has just managed to make its way from the entrance face of the sample to the exit face. In Fig. 8, the water saturation at this point is plotted against the size L of the lattice. In each case, the lattice is of size $L \times L \times 2L$, with periodic boundary conditions on the sides. The saturation is measured over an $L \times L \times L$ region in the center to eliminate end effects. Water saturation at breakthrough is well fitted by the relation

$$S = AL^{-\alpha} \tag{7}$$

with $A = 0.60$ and $\alpha = 0.48$. Clearly this is the same kind of relation as the result (6) in two dimensions. We believe that this is because the process of selecting paths of least resistance (invasion percolation) is the same in the two dimensionalities. Indeed, detailed examination shows that in both cases the water almost never has to invade pores which are not elements of the percolating cluster on the network. It is useful to express (7) in the alternative form

$$N_W = AL^{\phi} \tag{8}$$

where N_W is the number of sites which contain water. Since we are in three dimensions, $\phi = 3 - \alpha = 2.52$. A set which grows with the size of the system in this manner is called a fractal set, and the quantity ϕ is called the fractal dimension. Intuitively, the fractal dimension is a measure of the "roughness" of the object, akin to the adjective "tortuosity", which suggests the twistiness of a single curve. Percolation clusters in three dimensions are highly "stringy", with many branches dangling off and dead-ending. The fractal dimension of an ordinary percolation cluster at percolation threshold is close to this value of 2.52.

Residual Oil Saturation

Water flow past breakthrough is modeled by allowing the breakthrough cluster to continue growing according to the same dynamical rule as before. Returning to Fig. 8, we see that by contrast with the breakthrough saturation, the water saturation at the end of the simulation has a weak dependence upon the size of the lattice. The results indicate a value of approximately 66% water in the limit of infinite sample size. This corresponds to a residual oil saturation $S_{or} = 0.34$.

In two-dimensional planar lattices, neglecting end effects, the water breakthrough point corresponds to the percolation threshold of the lattice. Topologically, this is also the point at which the oil in the interior breaks up into finite clusters which do not span end to end across the lattice. In 2-d nonplanar lattices and in 3-d lattices the above topological restriction does not hold. It is possible that specimen spanning clusters of both water and oil can exist, and, in fact, this is generally the case at water percolation threshold. The simulation continues until the oil no longer spans the sample, i.e. oil is at percolation threshold. This explains why $S_{or} = 0.34$ is close to though slightly larger than the known percolation threshold of 0.31 for a simple cubic lattice. We can also use this argument to predict S_{or} for simulations on lattices with different coordination numbers. The qualitative result here is that structures with better connectedness (high coordination number) have lower values for S_{or}. It is interesting to note that both the breakthrough point and the end of the simulation are percolation thresholds. In the former the water is at threshold but consists of a single cluster, which is a fractal set. In the latter, oil is at threshold but consists of many clusters which combine to give a finite oil saturation.

Residual Blob-size Distribution

The experimental evidence cited in [11,12] indicates clearly that the spatial connectivity of the oil influences its recoverability in a significant manner. It is important to understand how this works in detail. We believe a good first step is to understand the connectivity of the oil left behind after a capillary flood. It turns out that it is possible to make use of the fact that oil has just reached its percolation threshold to make a definite prediction as to how the residual oil is distributed.

Since in the capillary limit, capillary forces dominate viscous forces at all length scales, and we are at a percolation threshold, we expect to see oil blobs of all sizes, limited only by the finite size of the sample. The number of occurrences $n(s)$ of blobs containing s sites should scale with s according to the formula

$$n(s) = Bs^{-\tau} . \quad (9)$$

In the case of ordinary percolation, the exponent τ can be expressed as

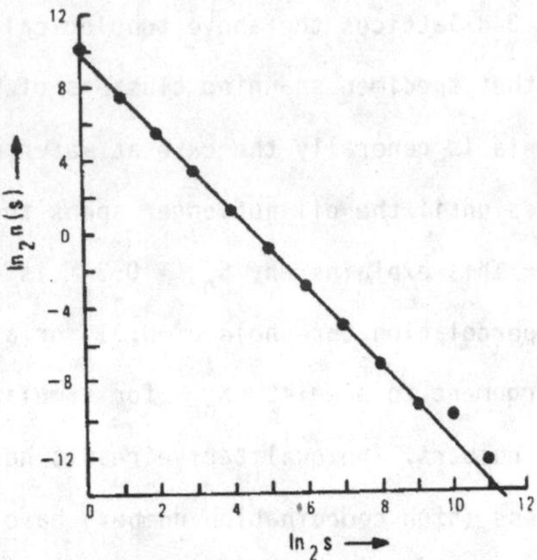

Figure 9. Blob size distribution in 3d.

$$\tau = \frac{d + \phi}{\phi} \tag{10}$$

where d is the space dimension and ϕ is the fractal dimension introduced earlier. Inserting the values $\phi = 2.52$ and $d = 3$ gives $\tau = 2.19$. This is in reasonable agreement with the results of simulations on a $30 \times 30 \times 60$ lattice, presented in Fig. 9. A fit to this data gives $\tau = 2.07$.

6. Conclusions

We have discussed the process of capillary displacement of fluids in porous media using a simplified discrete model of the motion. The model exhibits many of the familiar features of critical phenomena, and we have argued that it may be interpreted as a variant of pure percolation. From the point of view of mathematical physics, many fascinating problems remain: Can we think of the water fraction as a standard order parameter and construct field theories for it? Are there other observables in the system that are meaningful after ensemble averaging? Do other scaling relationships exist?

An equally important question not addressed here is the experimental validity of our ideas. The qualitative features and oil trapping in fluid didsplacement are known to occur in the laboratory and widely believed to occur in oil reservoirs [1,11,12]. Quantitative laboratory investigations of the scaling laws (6-8) are in progress. An independent theoretical confirmation of our assumptions can be made by numerically solving the appropriate limit of the equations of hydrodynamics in network models of porous media [13], where it is found that the fluid-fluid interface indeed moves a single random point at a time.

The outstanding problem in this subject is to obtain a quantitative macroscopic description of fluid flow in porous media at finite flow rates. This work treats the limit of vanishing flow rate but the general problem remains open.

REFERENCES

1. Scheidegger, A.E., The Physics of Flow in Porous Media (Toronto, 1974).
2. Chandler, R., Koplik, J., Lerman, K. and Willemsen, J.F., J. Fluid Mech. 119, 249 (1982).
3. Wilkinson, D. and Willemsen, J.F., "Invasion Percolation: A New Form of Percolation Theory", Schlumberger-Doll Research preprint (1983).
4. Lenormand, R., C.R. Acad. Sci.. 291B, 279(1980).
5. Broadbent, S.R. and Hammersley, J.M., Proc. Cambr. Phil. Soc. 53, 629(1957).
6. Kirkpatrick, S. (1979), in Ill-Condensed Matter: 1978 Les Houches Lectures, ed. R. Balian et al, p. 321, North Holland.
7. de Gennes, P.G. and Guyon, E., J. Mecan. 17, 403 (1978).
8. Chandler, R. and Willemsen, J.F., "Percolation and Porosimetry", Schlumberger-Doll Research preprint (1981).
9. Larson, R.G. and Morrow, N.R., Powder Tech. 30, 123 (1981).
10. Fisher, M.E., (1971), in Critical Phenomena, Proc. Intl. School of Physics E. Fermi, Course 51, ed. M.S. Green, Academic Press.
11. Payatakes, A.C., Ann. Rev. Fluid Mech. 14, 365 (1982).
12. Chatzis, I. and Morrow, N.R., SPE paper 10114 "Correlation of Capillary Number Relationships for Sandstones" (October, 1981).
13. Koplik, J., and Lasseter, T.J., SPE paper 11014, "Two-Phase Flow in Random Network Models of Porous Media" (September, 1982).

RECENT PROGRESS AND CURRENT PUZZLES IN PERCOLATION

Sidney Redner

Center for Polymer Studies and Department of Physics
Boston University
Boston, Massachusetts 02215

ABSTRACT

The basic physical phenomena of percolation are reviewed within the context of the modern theory of critical phenomena. The connection between percolation and the Potts model, a statistical mechanical model of ferromagnetism, is discussed. Recent advances in calculating critical exponents by position-space renormalization group methods are also described. Several open questions are also raised, including the nature of cluster structure and transport near the percolation threshold, and the anomalous geometrical properties of self-similar structures.

1. Introduction

The general problem of understanding the properties of disordered media is an area of considerable importance and interest. An important conceptual advance for describing such systems is the percolation model, first introduced by Broadbent and Hammersley in 1957 (for recent reviews and extensive references, see e.g., Stauffer 1979, Essam 1980). This is an idealized model which appears to capture the essential physical mechanism underlying many important features of random media, that of connectivity.

In order to gain an intuition for percolation, it is helpful to begin with a pictorial account. Consider bond percolation on the square lattice defined by randomly occupying each edge (or bond) of the lattice with a given probability p . Fig. 1a shows a 20×20 section of the square lattice whose edges are occupied with $p = 0.35$. From a geometrical point of view, there are two points that deserve emphasis: (i) Only small clusters up to a length scale denoted by

$\xi(p)$, the correlation length, occur. The number of clusters with radii $> \xi$ is exponentially small. (ii) The clusters are disconnected; it is not possible to find a continuous path that traverses the sample. This situation is analogous to the disordered phase of a ferromagnet.

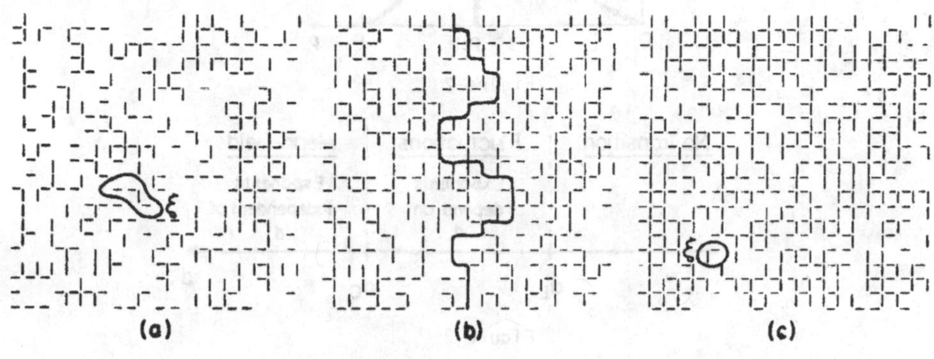

Figure 1

In Fig. 1c, a sample at $p = 0.65$ is shown. There now exists a single very large connected cluster which traverses across the lattice. This "state" of being connected is analogous to the ordered phase in the ferromagnetic transition. Notice also that the largest cluster has a strong propensity to absorb finite clusters as $p \to 1$. The correlation length, defined as the characteristic length scale of finite clusters, therefore vanishes in this limit.

Intermediate to the situations depicted in Figs. 1a and 1c is the percolation threshold, $p = p_c$. For the square lattice it is known by duality arguments that $p_c = 1/2$. An example of such a critical state is shown in Fig. 1b, on which a percolating path is indicated. In particular, notice that there exist clusters of all length scales (up to the linear dimension of the lattice) and ξ diverges. This singular behavior of ξ turns out to be crucial in understanding the physical properties of random media.

The connectivity transition at the percolation threshold is second order so that various geometrical properties exhibit power law singularities as $p \to p_c$. One example is the correlation length, $\xi(p)$, which diverges as $p \to p_c$:

$$\xi(p) \sim |p - p_c|^{-\nu} \qquad (1)$$

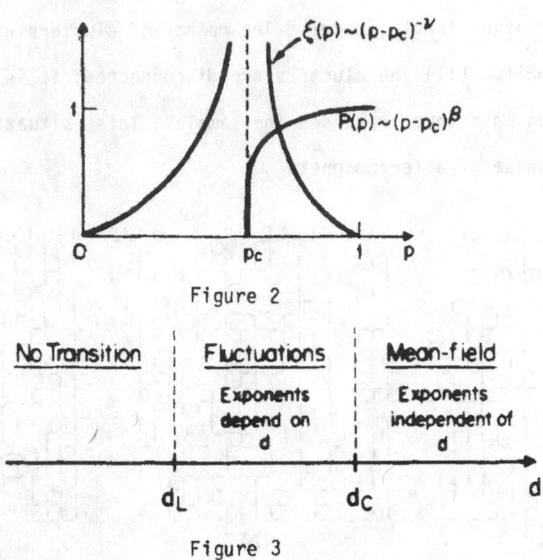

Figure 2

Figure 3

(see Fig. 2). A second important quantity is $P(p)$, the <u>percolation probability</u>, defined as the probability that a randomly chosen bond belongs to the infinite cluster. For $p < p_c$, there is no infinite cluster and $P = 0$. Above p_c, the percolation probability is finite and it vanishes as $p \to p_c^+$:

$$P(p) \sim (p - p_c)^{\beta} . \qquad (2)$$

A central tenet in the modern theory of critical phenomena is that critical exponents are generally <u>universal</u> (see e.g., Ma 1976, Pfeuty and Toulouse 1977). That is, since the transition involves phenomena on all length scales, local details of a model, such as lattice structure, should be irrelevant. However, the exponents do depend on a very small number of basic parameters characterizing a macroscopic system. For percolation, the single basic quantity is the spatial dimension d of the underlying lattice. The dependence of the exponents on d is the key to a qualitative understanding of the critical phenomena of a given model. As Fig. 3 indicates, there are three regimes of interest. First, above an upper critical dimension d_c, spatial fluctuations occurring in a system near a phase transition may be neglected. Simple analytic mean-field theories are accurate in this regime, and one finds exponents independent of d. Between d_c and d_L, the lower critical dimension, fluctuations are important and they cause exponents to become dependent on dimension. Considerable effort has been devoted to the calculation of critical exponents in this regime by a variety of techniques. Finally, below d_L , the lattice does not possess sufficient topological connections to propagate order or connectivity, and no phase transition is possible.

While percolation is an intriguing and appealing geometrical problem, the model also has many immediate applications to disordered physical systems whose properties are mediated by connectivity. Several examples are summarized in Table 1.

Table 1: Connection between percolation and physical problems.

bond	system described	calculable properties
spring	sol-gel transition	viscosity, bulk and shear moduli
resistor	random resistor network	conductivity of metal-insulator mixtures
hopping rate	transport in random media	diffusion coefficients

2. Recent Progress

(a) Potts Model Formulation

An important mapping between percolation and the Potts model of ferromagnetism was elucidated by Kasteleyn and Fortuin (1969). This was an important advance as it provided a firm basis for the description of percolation as a second order phase transition, and the Potts model served as a starting point for many renormalization group calculations of critical behavior.

To define the Potts model, consider a regular lattice with spins σ_i at each site, which can assume any one of s values 1,2,...s . Geometrically one may think of the spins as pointing from the center to one of the s vertices of a symmetric tetrahedron embedded in an (s-1) dimensional space. Nearest-neighbor pairs of spins have a ferromagnetic interaction which is proportional to the dot product between the two spin vectors. The Hamiltonian or energy of the entire lattice can be written as

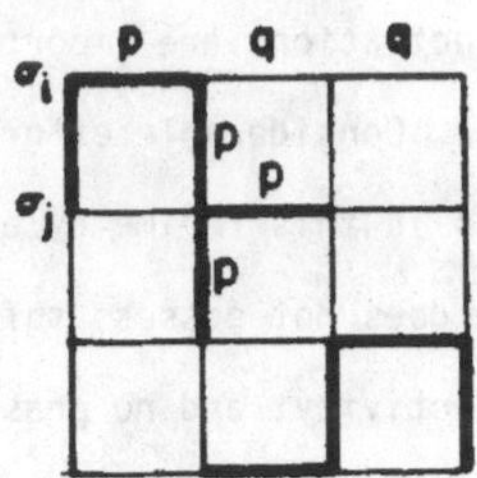

Figure 4

$$\mathcal{H} = -J \sum_{\langle i,j \rangle} (\delta_{\sigma_i \sigma_j} - 1) , \tag{3}$$

where the sum runs over all nearest-neighbor pairs of spins σ_i and σ_j. To make a correspondence with percolation, consider a high temperature expansion for the partition function Z. After several elementary manipulations one obtains,

$$Z = \mathrm{Tr}\, e^{-\beta \mathcal{H}} = \mathrm{Tr} \prod_{\langle i,j \rangle} e^{\beta J (\delta_{\sigma_i \sigma_j} - 1)}$$

$$= \mathrm{Tr} \prod_{\langle i,j \rangle} (e^{-\beta J} + (1 - e^{-\beta J}) \delta_{\sigma_i \sigma_j}) . \tag{4}$$

(Here $\beta = (kT)^{-1}$, with k Boltzmann's constant and T the absolute temperature.) Upon expanding out the product, the partition function is converted into a graphical expansion in which the factor $(1 - e^{-\beta J}) \delta_{\sigma_i \sigma_j}$ represents a bond between sites i and j. Fig. 4 shows a typical term in this expansion; each solid bond represents a factor $p = 1 - e^{-\beta J}$, and each site represents a spin state σ_i. When the trace over all spin states is performed, the product of delta functions forces the spin states in each connected cluster to be identical. There is one more sum than delta function constraints for every connected cluster, so that each such cluster is weighted by a factor of s. With these results, the partition function becomes

$$Z = \sum_{\text{bond configurations}} p^{N_b} q^{\mathcal{N} - N_b} s^{N_c} = \langle s^{N_c} \rangle , \tag{5a}$$

where $q = 1 - p$, $\mathcal{N}$ is the total number of bonds in the lattice, N_b is the number of occupied bonds, N_c is the number of clusters, and $<...>$ denotes a configurational average. Note that isolated sites are included as one-site clusters in the configurational sum of Eq. 5a.

From this partition function, we obtain the free energy per site as

$$f = \lim_{N \to \infty} \frac{1}{N} \ln Z \qquad (5b)$$

where N is the total number of lattice sites. Finally by differentiating f with respect to s and setting $s = 1$, it is straightforward to verify that $<N_c>$, the average number of clusters, is obtained. Various moments of this cluster generating function then yield all the physically important quantities of percolation theory (Wu 1982).

Another intriguing mapping arises by choosing the interaction strength J proportional to s and taking the limit $s \to 0$ (Fortuin and Kasteleyn 1972, Stephen 1976). In this limit, one obtains the generating function for all spanning trees, defined as connected subgraphs of the lattice which visit every site and which contain no closed loops. Spanning trees arise naturally in the solution to Kirchhoff's laws (see e.g., Wu 1982); consequently the zero-state Potts model may be used to describe random resistor networks.

In summary, the Potts model provides a concrete statistical mechanical model for percolation, and it may be used as a starting point for further developments. By making a correspondence between the Potts model and certain exactly soluble statistical mechanical models, several conjectures for the exponents of percolation in two dimensions have been proposed which appear to be exact (den Nijs 1979, Pearson 1980). The construction of mappings between apparently unrelated models is a very elegant tool for obtaining critical behavior.

(b) Position Space Renormalization Group Approach

Of the many methods developed to study phase transition phenomena, the position space renormalization group (PSRG) approach has proved to be of great utility. The primary advantage of the PSRG is that it is expressly constructed to treat the formidable problem of critical phenomena: all length scales contribute. By contrast, alternative methods such as series expansions are inherently perturbative, and therefore converge slowly near the critical point.

The basic idea underlying the PSRG is illustrated in Fig. 5 where schematic pictures of finite areas of three infinite random lattices at $p < p_c$, $p = p_c$, and $p > p_c$ are shown on the top row. Now imagine "zooming away" by a factor of 2 (rescaling all lengths by a factor of 2), and then turning down the contrast by a factor of 2 (summing out degrees of freedom at short distance scales) to arrive at the situation depicted in the bottom row of the figure. The change in cluster sizes under length rescaling may be represented by a "renormalization" of p to a new value p' . The functional relation $p' = R(p)$ is called a <u>renormalization transformation</u>. Under rescaling, a lattice at $p < p_c$ will renormalize to a new probability p' which is less than the initial probability p . The value of p is eventually renormalized to zero as the rescaling is repeated indefinitely. This represents a stable fixed point of the transformation. On the other hand, for $p > p_c$, the rescaling procedure iterated an infinite number of times leads

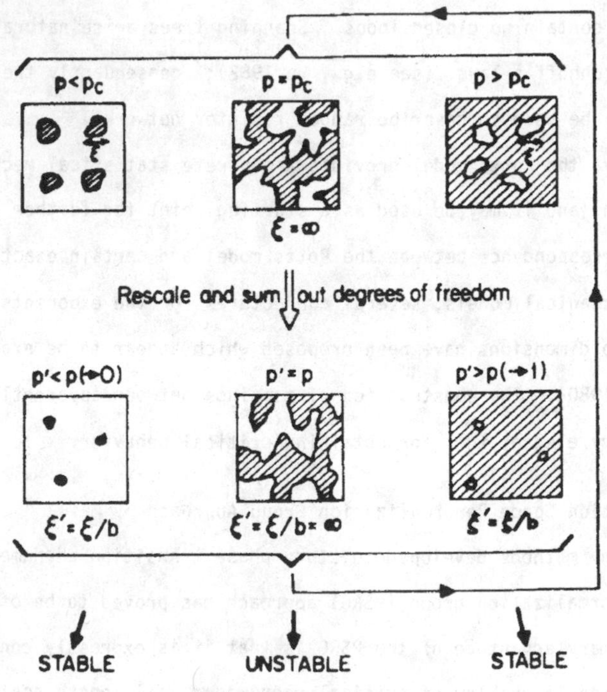

Figure 5

to a stable fixed point at $p = 1$ as finite "holes" eventually renormalize away.

The situation at $p = p_c$ is quite different. There is no longer any characteristic length scale, so that the system appears statistically identical both before and after rescaling. The critical point represents an unstable fixed point of the renormalization transformation. This "self-similarity" of a critical system under rescaling is the source of many interesting physical phenomena. The three fixed points just discussed represent the possible final states of repeated lattice rescalings: $p = 0$, $p = 1$, or $p = p_c$, the last being the critical system.

To obtain exponents through renormalization, note that if all lengths are rescaled by a factor of b then

$$\xi(\Delta p') = \xi(\Delta p)/b \ , \tag{6}$$

where $\Delta p = p - p_c$ is the deviation of the concentration from its critical value, and $\Delta p'$ is the renormalized value of the deviation. Furthermore, from Eq. (1) we have $\xi \sim (\Delta p)^{-\nu}$. Combining this with Eq. (6) then leads to

$$\nu = \ln b/\ln (dp'/dp)|_{p = p^*} \ , \tag{7}$$

where p^* is the unstable fixed point of the renormalization transformation, $R(p)$. This value of this fixed point represents approximation for the critical probability, p_c . At this fixed point, critical exponents may be computed from the linearization of $R(p)$ about p^* .

A particularly simple way of constructing a renormalization transformation was introduced by Reynolds et al. (1977). For bond percolation on the square lattice, the lattice is broken up into square $b \times b$ tessellating cells as indicated in Fig. 6. Upon rescaling, each such cell will map into a smaller $b' \times b'$ cell of the same topology. The simplest approximation is to choose $b = 2$ and $b' = 1$. Since the basic phenomenon of percolation resides in connectivity, a plausible definition for the renormalization transformation is to rescale all configurations which traverse the $b \times b$ cell to an occupied bond on the 1×1 cell. Hence p' is simply the total probability of spanning the relevant portion

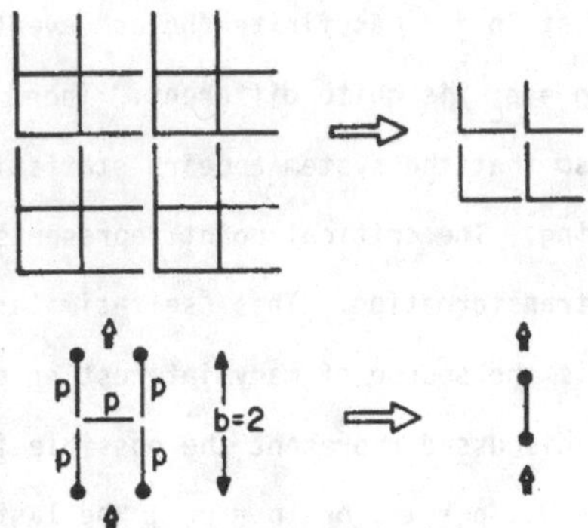

Figure 6

of the cell (bottom row of Fig. 6). For the $b = 2$ to $b' = 1$ rescaling, this is the probability of either or both of the bottom two sites of the cell being connected to either or both of the top two sites, i.e.

$$p' = R(p) = 2p^2 + 2p^3 - 5p^4 + 2p^5 . \qquad (8)$$

The unstable fixed point of Eq. (8) occurs at $p^* = 1/2$ (because of the self-duality of the cells chosen to construct the transformation), so that the exact value of p_c is predicted in this case. In addition, Eq. (7) gives $\nu \cong 1.428$. This is a reasonable approximation to the currently accepted value of $\nu = 4/3$ (den Nijs 1979), in view of the simplicity of the calculation.

Two attractive features of the cell renormalization methods are its simplicity and relative accuracy for a fixed amount of calculational effort. Furthermore, it is possible to reduce systematically the errors introduced in the approximation by considering larger cells (Reynolds et al. 1980). By this extension, the cell PSRG yields an accuracy which is comparable to other modern statistical mechanical methods. Finally, the method can be generalized to treat conductivity of random resistor networks (Bernasconi 1978), self-avoiding walks (de Queiroz and Chaves 1980, Redner and Reynolds 1981), lattice animals (Family 1980), directed percolation problems (Redner 1982, 1983), and many other models; a review of some of these methods is given in Stanley et al. (1982).

3. Current Puzzles

(a) Cluster Structure Near p_c

Although there has been great progress in understanding the critical phenomena of percolation, there is one important question that has not been adequately answered: "What is the qualitative picture of the infinite cluster just above p_c ?" The answer to this question would help one to understand the properties of transport in random media. In the recent past, two models of cluster structure have been introduced. One is the de Gennes-Skal-Shklovskii model (1975, 1976), in which the percolating cluster is viewed as a quasi-regular lattice of nodes -- points where there are at least three independent paths leading to infinity (Fig. 7(a)). The characteristic spacing between the nodes is assumed to be proportional to ξ . These considerations suggest a homogeneous lattice structure for the infinite cluster, which is scale similar to a regular lattice if all lengths are rescaled by a factor of ξ . It is then possible to calculate transport properties in terms of series and parallel combinations of the "macrolinks" which join neighboring nodes. An ingredient which is still unresolved in this picture is a model for these macrolinks.

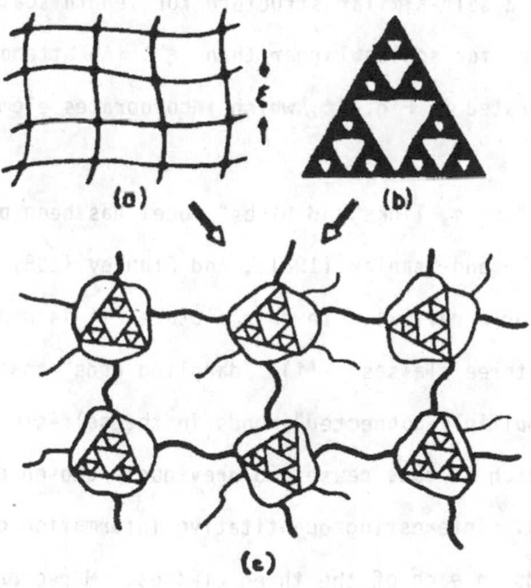

Figure 7

A more accurate picture of cluster structure would incorporate the fact that the infinite cluster is self-similar for length scales $< \xi$. The "fractal" model attempts to account for this self-similar aspect by representing the infinite cluster as a regular self-similar fractal object on which problems may be solved exactly (Gefen et al. 1981). The Sierpinski gasket, shown in Fig. 7(b), is a typical fractal that is obtained by removing a central triangle of linear dimension $\ell/2$ from a solid triangle of linear dimension ℓ . The construction is repeated indefinitely in all of the smaller solid triangles produced at each stage of construction. The resulting structure possesses no characteristic scale and has a non-integral Hausdorff or fractal dimensionality. Upon increasing the magnification by a factor of 2 , the area of the fractal increases by a factor of 3 (rather than a factor of 4 for a homogeneous two-dimensional object) leading to a <u>fractal dimension</u> $\overline{d} = \log 3/\log 2$.

The de Gennes-Skal-Shklovskii model works well in high dimensions where the formation of closed loops on many length scales does not dominate. In lower dimensions, loop formation becomes relatively more important and the fractal model works better. However, neither model is sufficient to give a complete description of cluster structure. The essential feature that has yet to be incorporated into a specific model is a self-similar structure for length scales less than ξ and a homogeneous structure for scales larger than ξ . An attempt to represent this situation is illustrated in Fig. 7c, which incorporates elements of both Figs. 7a and 7b.

The resulting "nodes, links and blobs" model has been proposed by Coniglio (1981a,b, 1982), Pike and Stanley (1981), and Stanley (1981) (see e.g., Stanley and Coniglio 1983 for a review). In this picture, it is possible to divide cluster bonds into three classes: (i) "dangling ends" that do not contribute to transport, (ii) "multiply-connected" bonds in the self-similar blobs, and (iii) "cutting" bonds, which if cut, cause two previously chosen connected sites to become disconnected. Interesting quantitative information can be obtained for the mean number of bonds in each of the three classes. Moreover, the cutting bonds play a dominant role in transport properties, and explicit predictions for experiments on inhomogeneous systems can be made.

(b) The conductivity problem.

An important variation of percolation is the random resistor network, obtained if each bond in a percolating sample is assigned a fixed resistance. Such an idealized model has been used to describe the conductivity of random metal-insulator composites (cf. Table 1). A basic feature of the random resistor network is that the conductivity σ vanishes as a power law as $p \to p_c^+$,

$$\sigma \sim (p - p_c)^t ,$$

thereby defining the conductivity exponent t. The calculation of this exponent is of fundamental theoretical significance in attempting to relate cluster structure and transport. Most of the theoretical attention has been confined to two-dimensional networks because of the computational difficulties associated with higher-dimensional lattices. Early work gave a wide spread in values for the conductivity exponent, but more recent estimates for t appeared to converge to a value of approximately 1.3 (Lobb and Frank 1979, Fogelholm 1980). It was thought that the coincidence of the value of t with $\nu = 4/3$ was not accidental, and that this fact might provide a bridge between cluster structure and transport. In the past year, however, two very accurate studies based on finite-size scaling gave $t = 1.28 \pm 0.02$ (Derrida and Vannimenus 1982), and $t = 1.30 \pm 0.02$ (Frank and Lobb 1983), apparently excluding the conjecture $t = \nu$ in two dimensions.

In attempting to resolve this situation, Alexander and Orbach (1982) made an important conceptual advance by studying a variety of physical problems on self-similar structures such as fractals. From scaling considerations, they were led to define a new fundamental dimension, the <u>fracton</u> dimension $\tilde{d}$, associated with the scaling behavior of the density of states on fractals, distinct from the fractal dimension of the structure. Roughly speaking, the fracton dimension may be thought of as a fractal dimension in reciprocal space. Each eigenmode of a linear problem, such as linear oscillations or diffusion on a fractal may be represented by a point in reciprocal space. The density of these points near the origin scales anomalously, leading to new predictions concerning the frequency dependence of the density of states (Rammal and Toulouse 1983). From existing numerical data, it appears that the fracton dimension of percolating clusters at p_c is 4/3, independent of the embedding spatial dimension. Assuming this result to be exact, and employing the connection between diffusion and conductivity, Alexander and Orbach conjectured that $t = 91/72 \cong 1.264$ in two dimensions, in reasonable agreement with the latest numerical work. It will be interesting to see whether their predictions will be ultimately borne out in two dimensions, and in three dimensions where there is presently a lack of accurate numerical data.

Since these new predictions are intimately connected with understanding diffusion on self-similar objects, it is worthwhile to consider this problem in some detail.

(c) Novel physics on self-similar structures: The diffusion problem.

Consider the problem of a discrete random walker on a percolating cluster (whimsically termed the "ant in the labyrinth" by de Gennes). If $p = 1$, then the medium is homogeneous and the mean-square displacement varies with time as,

$$\langle r^2 \rangle \cong Dt \,, \qquad t \to \infty \quad \text{or} \quad r \to \infty \,. \tag{9}$$

As the medium becomes random, the diffusion coefficient D decreases, and vanishes as $p \to p_c^+$. It is most convenient to write this critical behavior as function of ξ for what follows. Thus as $p \to p_c^+$, ξ diverges and D vanishes as

$$D \approx \xi^{-\theta} \,. \tag{10}$$

From the Einstein relation between the conductivity and the diffusion coefficient, the exponent θ is given by $\theta = (t - \beta)/\nu$ (Gefen et al. 1983).

It is important to emphasize that D has the behavior indicated in Eq. (10), only if the diffusing particle has travelled a distance r much greater than ξ. In this limit, the medium appears homogeneous and Fick's law, Eq. (9) applies. To obtain the behavior in the limit $r < \xi$, we use the following elegant scaling argument of Gefen et al. (1983):

The diffusion coefficient should be thought of as a function of the length scale r over which the diffusion takes place. Since r and ξ are the only two length scales in the problem, and since the behavior of $D(r)$ for $r \gg \xi$ is given by Eq. (10), we write for the general case

$$D(r) = \xi^{-\theta} f(r/\xi) \; . \tag{11}$$

In the limit $r \ll \xi$, the scaling function f is assumed to be a simple power law of its argument. Additionally, in this short time regime, ξ must drop out of the problem; the diffusing particle has not yet detected what the value of ξ should be. These considerations lead to

$$\begin{aligned} D(r) &\simeq \xi^{-\theta}(r/\xi)^{x} \\ &\rightarrow r^{-\theta} \quad \text{for} \quad r \ll \xi \; . \end{aligned} \tag{12}$$

As a result, we find <u>anomalous</u> diffusion in the short time regime

$$\begin{aligned} (d/dt)\langle r^2\rangle &\simeq D(r) \simeq r^{-\theta} \; , \\ \text{i.e.} \quad \langle r^2\rangle &\simeq t^{\{2/(2+\theta)\}} \; . \end{aligned} \tag{13a}$$

An alternative form of the latter equation is,

$$\langle r^2\rangle^{\bar{d}} \simeq t^{\tilde{d}} \tag{13b}$$

This indicates that Fick's law, $\langle r^2\rangle \simeq t$, is fundamentally modified on a fractal. Anomalous length rescaling properties enter through the fractal dimension $\bar{d}$, while anomalous time or frequency rescaling properties appear through the fracton dimension $\tilde{d}$ (Rammal and Toulouse 1983).

The transition from the short-time anomalous behavior to the asymptotic Fick's law behavior is shown in Fig. 8. The scaling argument predicts a crossover between the self-similar regime for small distance scales, to the homogeneous regime in the opposite limit at a "crossover" time, t_x, which scales as $\xi^{2+\theta}$. This is the time needed for a particle to diffuse a distance of the order of ξ. Presently, analytical and numerical work is in progress to check the predictions of anomalous diffusion.

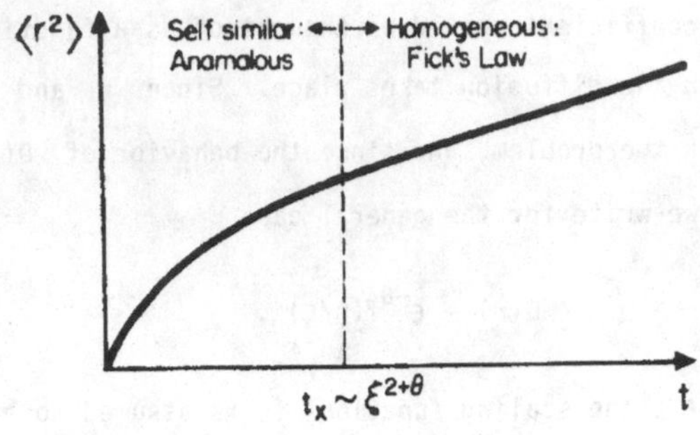

Figure 8

4. Concluding remarks

In summary, percolation is a very simple and appealing model for describing the physical properties of random media. Although extensive studies have provided much information about the model, there are still several open questions at the basic qualitative level. In particular, a simple picture of cluster structure near the percolation threshold is still not completely resolved. As a consequence, it is not clear if there exist direct relationships between cluster structure and transport in random media. Finally, a rapidly emerging area is the study of simple physical problems on self-similar structures. On these structures, translational symmetry is lost, but dilation symmetry holds. This feature appears to be the source of many new and intriguing physical effects which are currently under intensive study.

Acknowledgements

I wish to thank the Institute for Mathematics and its Applications of the University of Minnesota for inviting me to speak at this Symposium, and for the generous hospitality shown by the members of the Institute. I also wish to thank H.E. Stanley for a critical reading of the manuscript and helpful suggestions. This work was supported in part by grants from NSF, ONR and ARO to the Center for Polymer Studies, Boston University.

References

1. Alexander, S. and Orbach, R. 1982, J. de Physique 43, L625.

2. Bernasconi, J., 1978, Phys. Rev. B. 18, 2185.

3. Broadbent, S.R. and Hammersley, J.M., 1957, Proc. Camb. Philos. Soc. 53, 624.

4. Coniglio, A., 1981a, Phys. Rev. Lett. 46, 250.

5. Coniglio, A., 1981b, in Lecture Notes in Physics, Disordered Systems and Localization, Proc. Rome Conf., 1981, ed. C. Castellani, C. Di Castro and L. Peliti (Springer: Berlin).

6. Coniglio, A., 1982, J. Phys. A. 15, 3824.

7. Derrida, B. and Vannimenus, J., 1982, J. Phys. A. 15, L557.

8. Essam, J.W., 1980, Rep. Prog. Phys. 43, 833.

9. Family, F., 1980, J. Phys. A. 13, L325.

10. Fogelholm, R., 1980, J. Phys. C. 13, L571.

11. Fortuin, C.M. and Kasteleyn, P.W., 1972, Physica 57, 536.

12. Frank, D.J. and Lobb, C.J., 1983, unpublished.

13. Gefen, Y., Aharony, A., and Alexander, S., 1983, Phys. Rev. Lett 50, 77.

14. Gefen, Y., Aharony, A., Mandelbrot, B.B. and Kirkpatrick, 1981. Phys. Rev. Lett. 47, 1771.

15. de Gennes, P.G., 1976, La Recherche 7, 919.

16. Kasteleyn, P.W. and Fortuin, C.M. 1969, J. Phys. Soc. Japan (suppl) 26, 11.

17. Lobb, C.J. and Frank, D.J., 1979, J. Phys. C. 9, L827.

18. Ma, S.-K., 1976, Modern Theory of Critical Phenomena (Benjamin: Reading MA).

19. den Nijs, M.P.M., 1979, J. Phys. A. 12, 1857.

20. Pearson, R.B., 1980, Phys. Rev. B. 22, 2579.

21. Pfeuty, P. and Toulouse, G., 1977, Introduction to the Renormalization Group and to Critical Phenomena (Wiley: London).

22. Pike, R. and Stanley, H.E., 1981, J. Phys. A 14, L169.

23. Rammal, R. and Toulouse, G., 1983, J. de Physique 44, L13.

24. Redner, S., 1982. Phys. Rev. B 24, 3242.

25. Redner, S., 1983, in Percolation Structure and Processes, ed. G. Deutscher, R. Zallen, and J. Adler. Annals of the Israel Physical Society, vol. 5 (Hilger: Bristol).

26. Redner, S. and Reynolds, P.J., 1981. J. Phys. A 14, 2679.

27. Reynolds, P.J., Klein, W. and Stanley, H.E., 1977, J. Phys. C 10, L167.

28. Reynolds, P.J., Stanley, H.E., and Klein, W., 1980, Phys. Rev. B 21, 1223.

29. Skal, A.S. and Shklovskii, B.I., 1975, Sov. Phys. Semicond. 8, 1029.

30. Stanley, H.E., 1981, in Lecture Notes in Physics, Disordered Systems and Localization, Proc. Rome Conf. 1981, ed C. Castellani, C. Di Castro and L. Peliti (Springer: Berlin).

31. Stanley, H.E., and Coniglio, A., in Percolation Structure and Processes, ed. G. Deutscher, R. Zallen, and J. Adler. Annals of the Israel Physical Society, vol 5 (Hilger: Bristol).

32. Stanley, H.E., Reynolds, P.J., Redner, S., and Family, F., 1982 in "Real-Space Renormalization" Topics in Current Physics 30, chap 7 (Springer: Berlin).

33. Stauffer, D., 1979, Phys. Rept. 54, 1

34. Stephen, M.J., 1976, Phys. Lett. 56A, 149.

35. Wu, F.Y., 1982. Rev. Mod. Phys. 54, 235.

MARKOV FIELDS, CORRELATED PERCOLATION, AND THE ISING MODEL

By J. M. Hammersley and G. Mazzarino

Institute of Economics and Statistics, University of Oxford

ABSTRACT

We give an elementary introduction to the subject of correlated percolation. We illustrate it by diagrams and qualitative data for the Ising model on the square torus of order 100, which serves as a simple model for a binary alloy.

1. Disordered media.

The theory of a disordered medium consists in the construction of statistical models of its random microscopic structure, to explain and predict its macroscopic properties such as hardness or conductivity. The hardness of a tin-copper alloy is greatest when it contains 10% tin (this particular random mixture being known from historical times as *gun-metal*); yet the mere probabilistic statement that each individual atom has a 10% chance of being tin and a 90% chance of being copper is insufficient to explain the hardness of gun-metal. Interactions occur between adjacent atoms in the microstructure; so an adequate statistical model must also cater for the relative positions and spatial correlations of the two sorts of constituent atoms. For the sake of tractability, mathematical statisticians used to concentrate on correlations in a one-dimensional space (e.g. time), even though mathematical physicists had long known that the one-dimensional case is anomalous and quite atypical of what happens in more than one dimension. However, in the last decade or so, statisticians have made better progress towards multidimensional correlations and have caught up with the physicists. The aim of the present paper is to serve as a simple expository introduction to the recent literature, much of which has been highly technical. We shall avoid generality, and for clarity confine attention to *binary media*, namely materials with just two constituents conveniently labelled *black* and *white*. We shall eschew mathematical complications as far as possible; and we shall include some Monte Carlo simulations to illustrate a representative selection of two-dimensional patterns that arise from spatial correlation.

2. Graph theory.

We shall introduce the spatial topology of the situation in terms of a graph $\mathfrak{G} = (\mathfrak{V}, \mathfrak{E})$. Here $\mathfrak{V} = \{V_i\}_{i=1,2,\ldots}$ is a finite or countable collection of points (called *vertices* or *sites*) and $\mathfrak{E} = \{E_{ij} = E_{ji}\}$ is a collection of *edges* or *bonds* connecting certain specified pairs of vertices V_i and V_j. Two vertices connected by an edge are called *neighbours* of each other. We shall disallow any vertex being a neighbour itself, so no edge of the type E_{ii} can exist: apart from this restriction, the choice of which pairs of vertices are neighbours is simply a matter of initial postulation in each application. For convenience (and without any real loss of generality) we embed $\mathfrak{G}$ in d-dimensional Euclidean space $\mathfrak{R}^d$, and suppose that each vertex has integer coordinates in $\mathfrak{R}^d$. We shall write $\| V_i - V_j \|$ for the Euclidean distance between V_i and V_j. A few examples will illustrate some of the possibilities.

Example (i). The *simple cubic lattice* in $\mathfrak{R}^d$. Here all points with integer coordinates in $\mathfrak{R}^d$ are vertices, and V_i and V_j are neighbours if and only if $\| V_i - V_j \| = 1$.

Example (ii). The *body-centred cubic lattice* in $\mathfrak{R}^3$. The vertices are points with integer coordinates (x, y, z) such that $x \equiv y \equiv z$ (modulo 2), and V_i and V_j are neighbours if and only if $\| V_i - V_j \| = \sqrt{3}$.

Example (iii). The *face-centred cubic lattice* in $\mathfrak{R}^3$. The vertices are points with integer coordinates (x, y, z) such that $x + y + z \equiv 0$ (modulo 2), and V_i and V_j are neighbours if and only if $\| V_i - V_j \| = \sqrt{2}$.

Example (iv). The *Bethe tree* of valency 4. The vertices are all points with integer coordinates V in $\mathfrak{R}^1$, except that $V \neq 0$. The vertices V_i and V_j are neighbours if either $0 \leqslant \| V_i - 3V_j \| \leqslant 1$ or if $0 \leqslant \| 3V_i - V_j \| \leqslant 1$, and $(+1, -1)$ are also a pair of neighbours.

Example (v). The *square torus* of order N. The vertices are points with integer coordinates (x, y) in $\mathfrak{R}^2$ such that $1 \leqslant x \leqslant N$ and $1 \leqslant y \leqslant N$; and V_i and V_j are neighbours if $\| V_i - V_j \| = 1$. Additionally, $(1, y)$ and (N, y) are neighbours for each y; and $(x, 1)$ and (x, N) are neighbours for each x.

Examples (iv) and (v) illustrate the fact that the word *neighbours* does not necessarily imply closeness in terms of Euclidean distance: as already stated, the choice of which pairs are to be neighbours is simply a postulate. In example (v) the graph is finite; and in the other four examples it is countably infinite.

A subset $\mathfrak{h}$, composed of n vertices of $\mathfrak{G}$, is called an *n-clan* if each pair of distinct vertices in $\mathfrak{h}$ is a pair of neighbours. In this definition, the condition is trivially satisfied (in the void sense) if $\mathfrak{h}$ consists of only 1 vertex or if $\mathfrak{h}$ is empty. Hence each individual vertex of $\mathfrak{G}$ is automatically a 1-clan: and $\mathfrak{G}$ also has exactly one empty clan. The *valency* of any particular vertex V is the number of 2-clans that contain V: this is equivalent to saying that the valency of V equals the number of neighbours of V. We shall restrict ourselves to graphs in which every vertex has finite valency. In examples (i) to (v) respectively, the valencies of each vertex are $2d$, 8, 12, 4, and 4. In example (i), there are no n-clans for any $n \geqslant 3$; while, in example (iii), n-clans exist for each $n \leqslant 4$ but not for any $n \geqslant 5$.

Suppose that in $\mathfrak{R}^d$ we can find a set of d linearly independent vectors $\mathbf{e}_1, \mathbf{e}_2, \ldots, \mathbf{e}_d$ such that $\mathfrak{G}$ has the following property for each $k = 1, 2, \ldots, d$: if V_i and V_j are any pair of neighbours in $\mathfrak{G}$, then $V_i + \mathbf{e}_k$ and $V_j + \mathbf{e}_k$ are also a pair of neighbours in $\mathfrak{G}$, and $V_i - \mathbf{e}_k$ and $V_j - \mathbf{e}_k$ are also a pair of neighbours in $\mathfrak{G}$. Under these circumstances we say that $\mathfrak{G}$ is a *lattice*. Thus examples (i) to (iii) are lattices, while examples (iv) and (v) are not.

We define an *n-walk* on $\mathfrak{G}$ to be an ordered sequence of vertices $V_0, V_1, \ldots, V_n$ such that V_{i-1} and V_i are a pair of neighbours for each $i = 1, 2, \ldots, n$. This definition includes the trivial case of a 0-walk consisting just of a single vertex. If these $n + 1$ vertices are all distinct we say that the n-walk is an *n-saw* (acronym for *self-avoiding walk*)). The walk is said to *go from* its *first* vertex V_0 *to* its *last* vertex V_n, and to *visit* each of its vertices $V_0, V_1, \ldots, V_n$. A set of vertices $\mathfrak{h}$ is said to be *connected* if, given any two vertices V and V^* in $\mathfrak{h}$ there exists a finite saw from V to V^* which only visits vertices in $\mathfrak{h}$. We shall, for convenience, confine attention to graphs $\mathfrak{G}$ such that $\mathfrak{G}$ itself is connected. Given any two vertices V and V^* in $\mathfrak{G}$, we may accordingly define the *graphical distance* between V and V^* to be n if there exists an n-saw from V to V^* but no $(n-1)$-saw from V to V^*. Graphical distance, denoted by $\Delta(V,V^*)$, should not be confused with Euclidean distance $\| V-V^* \|$.

Given a set of vertices $\mathfrak{h}$, we define $\partial\mathfrak{h}$, the *boundary* of $\mathfrak{h}$, to be the set of vertices V which do not belong to $\mathfrak{h}$ but satisfy $\Delta(V,V^*)=1$ for some $V^*\in\mathfrak{h}$. In this definition, $\mathfrak{h}$ need not be a connected set. The *closure* of $\mathfrak{h}$, denoted by $\bar{\mathfrak{h}}$, is the union of $\mathfrak{h}$ and $\partial\mathfrak{h}$. We write $|\mathfrak{h}|$ for the number of vertices in $\mathfrak{h}$.

An *(n, b)-animal containing a vertex V* is defined to be a connected set $\mathfrak{h}$ of vertices such that $V\in\mathfrak{h}$ and $|\mathfrak{h}|=n$ and $|\partial\mathfrak{h}|=b$. Two animals, obtained from connected sets $\mathfrak{h}_1$ and $\mathfrak{h}_2$, are *distinct* if $\mathfrak{h}_1\neq\mathfrak{h}_2$. We write $a_{nb}(V)$ for the number of distinct (n,b)-animals containing V, and $c_n(V)$ for the number of distinct n-saws with first vertex V. These numbers, $a_{nb}(V)$ and $c_n(V)$, which are finite because of the assumption that all vertices have finite valency, play an important role in describing the topology of $\mathfrak{G}$.

THEOREM 2.1 (Leath and Reich (1978)). *For any vertex V in any graph* $\mathfrak{G}$,

$$0\leqslant a_{nb}(V)\leqslant\binom{n+b-1}{b}. \tag{2.1}$$

In this paper we shall not usually give proofs of theorems or assertions. However, Theorem 2.1 has an extremely simple proof, given below.

We can order the countable set $\mathfrak{V}$ of all vertices in $\mathfrak{G}$ in some fixed order; and then, given any subset of vertices, we can order that subset by means of this fixed order. We then say that the subset is in *standard order*. Let $\mathfrak{h}$ be a given (n,b)-animal containing V. We can order the vertices of $\mathfrak{h}$ in an ordered sequence $\mathfrak{h}_0, \mathfrak{h}_1, \mathfrak{h}_2,\ldots$, where $\mathfrak{h}_m$ denotes the set of vertices $V^*\in\bar{\mathfrak{h}}$ such that $\Delta(V,V^*)=m$ and $\mathfrak{h}_m$ is taken in standard order. To each vertex V^* in $\mathfrak{h}_0,\mathfrak{h}_1,\ldots$ we assign the symbol 0 if $V^*\in\partial\mathfrak{h}$ and the symbol 1 if $V^*\in\mathfrak{h}$. We write down these symbols in the order in which they occur in $\mathfrak{h}_0,\mathfrak{h}_1,\ldots$ to produce a coding of $\mathfrak{h}$ as an ordered sequence of b zeros and n ones. It is easy to see that two distinct (n,b)-animals will produce distinct codes. Moreover the first symbol in such a code is always 1 because $\mathfrak{h}_0=\{V\}$ and $V\in\mathfrak{h}$. The right-hand side of (2.1) is the number of ordered sequences of b zeros and n ones, with the first symbol equal to 1; and Theorem 2.1 follows at once.

Without further conditions on $\mathfrak{G}$, Theorem 2.1 is best possible. For example, if $\mathfrak{G}$ is the complete graph with exactly N vertices, in which every pair of distinct vertices is a pair of neighbours, then equality holds on the right-hand side of (2.1) whenever $n+b=N$ and $n>0$, while equality holds on the left-hand side of (2.1) whenever $n+b\neq N$.

As a further illustration, consider the Bethe tree of valency v (for which Example (iv) was the particular case $v=4$). The embedding in $\mathfrak{R}^1$ in Example (iv) somewhat obscures the structure of such a tree, and it is more enlightening to use an alternative characterization: namely each vertex shall have the same valency $v>1$ and no pair of vertices shall be connected by more than one saw. Hence, because $\mathfrak{G}$ is connected, there is exactly one saw between any given pair of vertices. Then

$$a_{nb}(V)=0 \text{ if } b\neq 2+(v-2)n, \tag{2.2}$$

while

$$a_{nb}(V)=\frac{v}{1+(v-1)n}\binom{n+b-1}{b} \text{ if } b=2+(v-2)n. \tag{2.3}$$

Indeed it is easy to see that $a_{nb}(V)$ is independent of V for all n, b; whereupon (2.2) follows by induction on n. When $b = 2+(v-2)n$, we notice that a_{nb} is the coefficient of x^{n-1} in the expansion of A^v, where $A = A(x)$ is the unique root of

$$A = 1 + xA^{v-1} \tag{2.4}$$

that tends to 1 as $x \to 0$. From Lagrange's inversion-expansion theorem (Whittaker and Watson (1940) p.133), we deduce

$$a_{nb} = \frac{1}{(n-1)!}\left[\frac{\mathrm{d}^{n-2}}{\mathrm{d}t^{n-2}}\left\{t^{(v-1)(n-1)}\frac{\mathrm{d}}{\mathrm{d}t}t^v\right\}\right]_{t=1} \tag{2.5}$$

which leads at once to (2.3).

In the principal physical applications, $\mathfrak{G}$ is a lattice; and we emphasize that we shall be using the term *lattice* in the particular mathematical sense already defined above. (Physicists, however, often use the term in a wider sense, and sometimes they call the Bethe tree a Bethe lattice; but to follow their terminology would nullify some of the important distinctions, which we shall encounter below in percolation theory, between trees and lattices). Except for trivial one-dimensional lattices, it is not possible to give any exact formula for the animal numbers a_{nb}. However, for macroscopic predictions, it is sufficient to know how $a_{nb}(V)$ behaves for large n and b; and here we have an asymptotic result involving the *entropy function* defined by

$$I(\rho) = -\rho\log\rho - (1-\rho)\log(1-\rho), \quad 0 \leqslant \rho \leqslant 1. \tag{2.6}$$

THEOREM 2.2. (Delyon (1980)). *Suppose $\mathfrak{G}$ is a lattice in $\mathfrak{R}^d$ with $d > 1$ and suppose p_H is the critical probability of $\mathfrak{G}$, as defined in (4.2) below. Then there exist numbers ρ_0 and ρ_1, satisfying $0 < \rho_0 < \rho_1 = p_H < 1$, and a function $\lambda(\rho)$, defined on $0 \leqslant \rho \leqslant 1$, such that*

$$\begin{aligned} \lambda(\rho) &= -\infty \quad \text{if} \quad 0 \leqslant \rho < \rho_0, \\ 0 \leqslant \lambda(\rho) &< I(\rho) \quad \text{if} \quad \rho_0 < \rho < \rho_1, \\ \lambda(\rho) &= I(\rho) \quad \text{if} \quad \rho_1 \leqslant \rho \leqslant 1, \end{aligned} \tag{2.7}$$

and

$$(n+b)^{-1}\log a_{nb}(V) \to \lambda(\rho) \tag{2.8}$$

when $(n+b) \to \infty$ and $n/(n+b) \to \rho \neq \rho_0$. The numbers ρ_0 and ρ_1 and the function $\lambda(\rho)$ depend upon $\mathfrak{G}$, but are independent of V. Also $\lambda(\rho)$ is concave and continuous for $\rho_0 < \rho \leqslant 1$.

The upper bound $I(\rho)$ follows at once from Stirling's approximation for the factorial function applied to the right-hand side of (2.1). What is remarkable about Theorem 2.2 is that this upper bound should be attained over a non-degenerate interval of values of ρ, irrespective of the detailed nature of the lattice $\mathfrak{G}$. This is in sharp contrast to the corresponding behaviour for the Bethe tree; for (2.3) shows that the upper bound can only be attained for the single value $\rho = \rho_1 = 1/(v-1)$ and that $\lambda(\rho) = -\infty$ for all $\rho \neq \rho_1$.

Stauffer (1978) discusses some more precise asymptotic expressions for a_{nb} in a lattice: his arguments are appealing and have considerable support on physical grounds, though they do not pretend to be mathematically rigorous.

The asymptotic behaviour of the number of n-saws depends upon a constant κ, which characterizes the structure of a lattice $\mathfrak{G}$, and is called the *connective constant* of $\mathfrak{G}$. (Note that the term *connective constant* is also used, especially by physicists, for e^{κ}; so care is needed in reading the literature to see which meaning is in use in any particular context).

THEOREM 2.3. (Hammersley (1957)). *If $\mathfrak{G}$ is a lattice with connective constant κ, then $n^{-1}\log c_n(V) \to \kappa$ as $n \to \infty$ independently of V.*

3. Markov fields.

We now impose a stochastic process on the graph $\mathfrak{G}$ by colouring the vertices of $\mathfrak{G}$, assigning at random a colour (either black or white) to each vertex of $\mathfrak{G}$. This will yield a pattern of colours θ on the whole of $\mathfrak{G}$; and the stochastic process will be specified once we specify the probability $P(\theta)$ that $\mathfrak{G}$ has any particular colouring θ. Certain mathematical technicalities arise because $\mathfrak{G}$ can have a countable infinity of vertices, and it would be distracting to enter here into these technicalities: suffice it to say that it is enough to specify the ratios $P(\theta)/P(\theta^*)$ for all pairs of colourings θ and θ^* that differ only at a finite set of vertices (depending on θ and θ^*). Let γ_i $(i=1,2,...)$ denote the clans of $\mathfrak{G}$. With the two available colours, an n-clan can be coloured in 2^n different ways. With each clan γ_i we associate a set of specified constants β_{ij}, where $j = 1, 2, ..., 2^n$ runs over the different possible colourings of γ_i. Given a colouring θ of $\mathfrak{G}$, we write $\chi_{ij}(\theta) = 1$ or 0 according as this colouring θ does or does not assign the jth possible colouring to the clan γ_i. Then we define the *Gibbsian potential*

$$H(\theta) = \textstyle\sum_{ij} \beta_{ij} \chi_{ij}(\theta). \tag{3.1}$$

The stochastic process is then defined by

$$P(\theta) = e^{-H(\theta)}. \tag{3.2}$$

If $\mathfrak{G}$ is a finite graph, the constant β associated with the empty clan is to be chosen so that the sum of $P(\theta)$ over all colourings θ is 1. If $\mathfrak{G}$ is countably infinite, the sum in (3.1), over all clans and all clan-colourings, is a formal sum; but, if θ and θ^* differ only at a finite number of vertices, the difference $H(\theta) - H(\theta^*)$ will be well-defined since it can only contain finitely many non-zero terms; and hence we have the required specification of the ratios $P(\theta)/P(\theta^*)$.

The foregoing stochastic process is called a *Gibbsian field*, and it possesses the following special property. Let $P(\theta_{\mathfrak{h}}|\theta_{\mathfrak{g}})$ denote the conditional probability that a subset of vertices $\mathfrak{h}$ has a colouring $\theta_{\mathfrak{h}}$ when it is known that another subset $\mathfrak{g}$ (disjoint from $\mathfrak{h}$) has a colouring $\theta_{\mathfrak{g}}$. Then, for all finite $\mathfrak{h} \subseteq \mathfrak{G}$,

$$P(\theta_{\mathfrak{h}}|\theta_{\mathfrak{G}-\mathfrak{h}}) = P(\theta_{\mathfrak{h}}|\theta_{\partial\mathfrak{h}}), \tag{3.3}$$

in which $\mathfrak{G} - \mathfrak{h}$ denotes the complement of $\mathfrak{h}$, and $\partial\mathfrak{h}$ is the boundary of $\mathfrak{h}$.

Equation (3.3) is the defining relation of a more general stochastic process called a *Markov field*. Every Gibbsian field is a Markov field, but not conversely: there do exist non-Gibbsian Markov fields, that is to say Markov fields in which $P(\theta)$ cannot be expressed in the simple form (3.1) and (3.2) (Moussouris (1974)). However, in this paper we shall confine ourselves to Gibbsian fields and Gibbsian potentials. The essential feature of a Markov field is the *contagious* nature of its statistical dependence: there may be a long-range dependence between the colours of two vertices V_1 and V_2 that are far apart; but this dependence has been transmitted by contagion between successive neighbours along the saws that connect V_1 and V_2. If V_1 lies in any set $\mathfrak{h}$, whose closure $\bar{\mathfrak{h}}$ does not contain V_2, then every saw from V_1 to V_2 must traverse the boundary $\partial\mathfrak{h}$; and now, if we limit the transfer of contagion in any way by prescribing only the colouring of $\partial\mathfrak{h}$, then by (3.3) we destroy the dependence of the colour of V_1 upon the colour of V_2. Since this holds for *every* boundary $\partial\mathfrak{h}$ surrounding V_1 (but excluding V_2), the transmission of contagion along saws must be from neighbour to neighbour.

The specified constants β_{ij} in (3.1) are arbitrary. However, in applications when $\mathfrak{G}$ is a lattice, it is natural to impose further restrictions upon the β_{ij}. In particular, we may require that these β_{ij} are invariant under the symmetries of the of the lattice: thus, if the clan γ_i is carried into the clan $\gamma_{i(k)} = \gamma_i + \mathbf{e}_k$ by a lattice translation along the basis vector $\mathbf{e}_k$, we impose the restrictions

$$\beta_{ij} = \beta_{i(k)j} \qquad \text{for all } i, j, k. \tag{3.4}$$

We then say that the Gibbsian field is *translation-invariant*. Similarly, we say that the Gibbsian field is *isotropic* if the β_{ij} are also invariant under the rotational symmetries of the lattice $\mathfrak{G}$.

The square torus of order N provides an illustration of an isotropic translation-invariant Gibbsian field. If the group of isotropies of the square torus includes improper rotations (i.e. reflections) as well as proper rotations, the coefficients associated with the 2-clans will not depend upon the directions and senses of these 2-clans, but only upon their unordered colourings; so the Gibbsian potential will be

$$H(\theta) = \beta_0 + \beta_B|B| + \beta_W|W| + \beta_{BB}|BB| + \beta_{BW}|BW| + \beta_{WW}|WW|, \tag{3.5}$$

in which $|B|$ denotes the total number of black vertices in the colouring θ of $\mathfrak{G}$, $|W|$ the total number of white vertices, $|BB|$ the total number of 2-clans with both vertices black, $|BW|$ the total number of 2-clans with one black and one white vertex, etc. The coefficient β_0 is the coefficient of the empty clan. Further compression of (3.5) results from the identities

$$|B| + |W| = N^2, \qquad |BW| = 4|B| - 2|BB| = 4|W| - 2|WW|, \tag{3.6}$$

in which N^2 is the total number of vertices in $\mathfrak{G}$. Thus

$$H(\theta) = \beta_0^* + \alpha|B| + \beta|BW|, \tag{3.7}$$

where β_0^*, α, β are fresh constants. We may also ignore β_0^* when we are only interested in probability ratios $P(\theta)/P(\theta^*)$. So finally the Gibbsian potential assumes a canonical form

$$H(\theta) = \alpha|B| + \beta|BW|. \tag{3.8}$$

As α and β vary over all real numbers, we shall get various probability structures for possible stochastic processes on the torus. In view of the minus sign in (3.2), increasingly negative values of α will favour an increasing preponderance of black vertices on the torus, positive values of α will favour white vertices. Neighbours will increasingly tend to be coloured alike or oppositely according as β is increasingly positive or negative. These effects can be seen in the diagrams reproduced in this paper, showing simulations of Gibbsian fields on a square torus with $N = 100$. The effect of reversing the sign of α is simply to interchange black and white in the stochastic process; so these diagrams concentrate on negative values of α, inasmuch as black diagrams on a white page are more readily assimilated by the eye — black is beautiful! Other symbols in the captions to the diagrams will be explained in the penultimate paragraph of §5.

4. Correlated percolation.

Given a particular random colouring θ of a graph $\mathfrak{G}$, obtained from a Gibbsian potential $H(\theta)$ in the manner described in §3, we can examine the properties of the pattern formed by this colouring. In particular, we define a *black cluster* to be a maximal connected set of black vertices, that is to say an (n, b)-animal with n black vertices and b white boundary vertices. Similarly, a *white cluster* is any (n, b)-animal with n white vertices having a boundary of b black vertices. A *grey cluster* is a maximal set of vertices such that each and every pair of neighbours in the set is a pair of oppositely coloured vertices. Percolation theory deals with questions about the shapes and sizes of clusters, and especially with the probability of getting large or infinite clusters.

The classical case of *Bernoulli percolation* arises when the coefficients β_{ij} in $H(\theta)$ in (3.1) are zero whenever the corresponding clan γ_i is an n-clan with $n > 1$. In that event, the vertices will be coloured black or white independently at random and we shall be dealing with *uncorrelated percolation.* Suppose further, for simplicity, that $\mathfrak{G}$ is a lattice and that each vertex has an independent probability p of being black and a probability $q = 1-p$ of being white. Then

$$P(V, p) = \sum_{nb} a_{nb}(V) p^n q^b \tag{4.1}$$

is the probability that a given vertex V belongs to a finite black cluster. The critical probability p_H, which is independent of V, is defined by

$$p_H = \sup\{p: P(V, p) = 1\}. \tag{4.2}$$

There is also another critical probability p_T, defined as the supremum of all p such that the expected size of the black cluster containing V is finite. It has long been conjectured that $p_T = p_H$ for Bernoulli percolation on any lattice; but this has only been proved in a few special cases.

We may now ask for corresponding results that apply in correlated percolation. Comparatively little seems to be known about this at present; and most of the remainder of this article will discuss Monte Carlo simulation of correlated percolation on the square torus, which is a finite approximation to the square lattice. This is the simplest typical case.

5. Monte Carlo simulation.

We have used the Metropolis method (Metropolis *et al.* (1953)) to obtain Monte Carlo samples. Hammersley and Handscomb (1964) give a detailed account of this Monte Carlo technique; so here it will suffice to sketch in the outline only. We work with a square torus of order 100, which consists of 10,000 vertices connected according to Example (v) in §2. Given any particular colouring θ of the torus, we choose a vertex uniformly at random and consider the colouring θ^* that has the opposite colour on this particular vertex but otherwise agrees with θ elsewhere. We calculate the difference

$$\delta H = H(\theta^*) - H(\theta) \tag{5.1}$$

from (3.8) for given α, β; and we draw a pseudorandom number ξ. If $\xi \leqslant e^{-\delta H}$, we adopt θ^* as the new colouring; otherwise we retain the previous colouring θ. This operation is called a *flip*. The process starts from a random colouring of $\mathfrak{G}$ and (for fixed α, β) proceeds through a sequence of flips until the system has settled down into an equilibrium state.

For most pairs (α, β), the system had settled down after a million flips. Thereafter we continued for a further million flips, with interruptions every 100,000 flips to collect appropriate statistics on the colouring at that stage. This gives, for each (α, β), a set of 10 samples of the equilibrium state of the system. The whole calculation is then repeated for other pairs (α, β) to assemble a representative story for the behaviour of percolation over the range $-1 \leqslant \alpha \leqslant 1$, $-1 \leqslant \beta \leqslant 1$. For a few values of α, β (discussed below) the convergence to equilibrium was slower; so that in these cases we continued the calculation to about 20 million flips.

At each interrupt, we counted the number of black clusters, and the values of n and b for each black cluster; and after the final interrupt we fed the colouring θ to the Lasercomp to print out a complete picture of the torus at that stage. This information is called the *tally*. At each interrupt we also compiled three other sets of data, called the *cotally*, the *antitally*, and the *anticotally*. The cotally is obtained by reversing the colour of every vertex: this is done in a separate part of the computer, so as not to disturb the sequential flipping in the generation of the tally. The cotally is then fed to the counting routines, which always count black clusters. The net effect, of course, is therefore to compile the corresponding information about the white clusters in the original process. Similarly, to form the antitally, we reverse the colours of vertices with coordinates (x, y) such that $x + y$ is even; and we feed the antitally to the counting routines. This time the result is to compile information about the grey clusters in the original output. Finally, in the anticotally, we reverse the colours when $x + y$ is odd. This gives information about the grey clusters of opposite phase to the grey clusters obtained through the antitally.

The programme also included an optional facility for fixing the colour of all sites with coordinates (x, y) such that either $x = N = 100$ or $y = 1$, and to prevent these sites being flipped during the generation of the samples. With this facility in operation, we could examine the effect of a fixed boundary upon the conditional probabilities of the Markov field within the boundary. A second optional facility enables examination of only the largest cluster on the torus, other clusters being suppressed in the diagram. By passing this largest cluster to and fro between the tally and the cotally, one can count the lengths of the external and internal boundaries of this cluster as well as the size and number of holes in the cluster. Our diagrams are labelled with two letters followed by the numerical values of α and β: the first letter is either T or A, indicating that the output comes from the tally or the antitally respectively, or L for largest cluster only; the second letter is either R or B or W. Here B or W indicate that the optional boundary facility is in operation with the boundary sites held all black or all white respectively; while R indicates that the boundary facility is not operation, so that all sites are under the control of the random process of flipping. Some diagrams are also labelled with a value of t: this indicates the number of flips (in units of 10^5) that have been used to generate the picture (e.g. $t = 30$ means that the picture is the one obtained after 3 million flips). If no value of t appears against a diagram, that diagram has the standard value $t = 20$ (i.e. 2 million flips).

The complete calculation was quite an extensive job, involving the generation of about 1500 million pseudorandom numbers. We used three computers, linked together by telephone line: a CDC 7600 (a fast but relatively small machine) in Manchester to carry out the flipping and cluster counting, an ICL 2988 (a larger but slower machine) in Oxford to collect and assemble the data and to drive the Lasercomp, and a VAX-11 780 (quite a small machine) in Oxford for programme development for the other two larger computers. The Lasercomp is a photocomposition machine, with an 80 megabyte disc drive for storing digitised founts and associated software, which we have used for typesetting the text of this article, including the diagrams. When producing diagrams, the Lasercomp functions as a high-resolution graphics device with a resolution of a million pixels per square inch.

6. The two-dimensional Ising model.

The main objective of our enquiry is to obtain qualitative, and to a lesser extent quantitative, information about percolation for the two-dimensional Ising model on the infinite square lattice, namely Example (i) in §2 with $d = 2$. The square torus of order 100 is a finite approximation to the square lattice; and we need to know how large a cluster on this torus represents the onset of percolation (i.e. an infinite cluster) on the square lattice. We can be guided to some extent by three known cases. In the special case of Bernoulli percolation, which occurs when $\beta = 0$, the critical probability is known to be $p_H = 0.59$ approximately. When $\beta = 0$, the relation between α and the proportion p of black vertices on the torus is

$$p = \tfrac{1}{2} - \tfrac{1}{2} \tanh \tfrac{1}{2}\alpha, \tag{6.1}$$

so $\alpha = -0.36$ when $p = p_H$. Hence the point $(\alpha, \beta) = (-0.36, 0)$ lies on the edge of the region of black percolation in the (α, β)-plane. Since changing the sign of α is equivalent to an interchange of the roles of black and white, the point $(+0.36, 0)$ lies on the edge of the region of white percolation in the (α, β)-plane. The third known case comes from the Onsager solution of the Ising model in the absence of an external magnetic field, namely when $\alpha = 0$ (Onsager (1944). The critical value of β is then

$$\beta_0 \;=\; \sinh^{-1} 1 \;=\; +0.88137... \tag{6.2}$$

[Note that the formulation used in (3.8) implies $\beta = 2K$ in terms of the constant K used by physicists. Some mathematical writers, e.g. Durrett (1981), who use a different notation $\beta = K$, will instead quote $\beta = K = +0.44069$ as the critical value of β]. Thus the point (0, +0.88) lies on the edge of the region of black percolation and also on the edge of the region of white percolation.

For selected values of α and β, Tables 1 to 12 summarize the information yielded by the Monte Carlo simulations. Most of these tables have three sections, respectively devoted to black, white, and grey clusters. The values of α in these tables are restricted to $\alpha \leqslant 0$; since the corresponding results for $\alpha \geqslant 0$ come at once from a interchange of the roles of black and white, while the data for grey clusters does not depend upon the sign of α. Thus according to Table 12, the mean number of black clusters is 73.2 when $\alpha = \beta = +1.0$, this being the mean number of white clusters when $\alpha = -1.0$ and $\beta = +1.0$. Actually, the Monte Carlo simulations were carried out both for positive and negative values of α, using independent sequences of pseudorandom numbers; and then the tabular entries followed by taking an appropriate average of the tally and cotally (or of the antitally and the anticotally) for the two computer outputs with opposite signs of α. The independence of these two outputs, for opposite signs of α, facilitated a check on the reliability and sampling errors in the data. Table 1, giving the estimated percentage of black vertices, does not quote the corresponding percentages for white vertices because white percentages follow immediately by subtracting the black percentages from 100%. Nor does Table 1 quote grey percentages because these are always 50% (or 100%, depending upon the definition used).

In some of the Tables there are two entries in each cell, one entry in larger type and the other in smaller type. In these cases, the larger type entry is the estimate of the quantity designated in the title of the Table, and the associated smaller type entry is the standard error of that estimate. As a rough working rule, the difference between the true value of the designated quantity and its estimate will not exceed twice the standard error with probability 0.95. For example in Table 1 with $\alpha = -1.0$ and $\beta = 0$, the estimated percentage of black vertices is quoted as 73.0 ± 0.2, indicating that there is a 95% chance that the true percentage lies between 72.6% and 73.4%. In the special case when $\beta = 0$, the true percentages can be found exactly from (6.1), which gives 73.11% when $\alpha = -1.0$; and this does indeed lie between 72.6% and 73.4%. In Table 1, other exact results for $\beta = 0$ and for $\alpha = 0$ are given in italic type. When $\alpha \to 0$ from above or below, the proportion $p(\alpha,\beta)$ of black vertices is always $\frac{1}{2}$ whenever $\beta \leqslant \beta_0$, the critical Onsager point (6.2). On the other hand, when $\beta > \beta_0$ there is a discontinuity in p at $\alpha = 0$, given by

$$p(0\pm,\beta) = \tfrac{1}{2} \mp \tfrac{1}{2}(1-\operatorname{cosech}^4\beta)^{1/8}. \tag{6.3}$$

This formula was originally stated without proof by Onsager (1949); and Durrett (1981) gives an account of the work leading to its eventual proof many years later. It is worth quoting a few values for this function, because they well illustrate the fierce character and sudden onset of the discontinuity between the adjacent black and white percolation regions on the half-line $\alpha = 0$, $\beta > \beta_0$, especially with β nearly equal to β_0.

$$\begin{array}{lll}
\beta & p(0-,\beta) & p(0+,\beta) \\
+0.88137359 = \beta_0 & 0.5000 & 0.5000 \\
+0.881375 & 0.6112 & 0.3888 \\
+0.8814 & 0.6604 & 0.3396 \\
+0.89 & 0.8298 & 0.1702 \\
+0.90 & 0.8620 & 0.1380 \\
+0.95 & 0.9199 & 0.0801 \\
+1.0 & 0.9436 & 0.0564 \\
+\infty & 1.0000 & 0.0000
\end{array} \tag{6.4}$$

The true percentage 94.36% at the top of the final column of Table 1 comes from (6.4).

TABLE 1. *Percentage black vertices.*

β	$\alpha=-1.0$		$\alpha=-0.8$		$\alpha=-0.6$		$\alpha=-0.4$		$\alpha=-0.2$		$\alpha=-0.1$		$\alpha=0.0$
+1.0	99.0	0.1	99.0	0.1	98.8	0.1	98.0	0.1	97.5	0.1	96.8	0.1	*94.36*
+0.8	98.0	0.1	97.4	0.1	96.5	0.1	95.1	0.1	91.8	0.1	88.1	0.2	*50.00*
+0.6	95.2	0.1	93.8	0.1	91.3	0.1	86.9	0.2	77.0	0.2	66.9	0.3	*50.00*
+0.4	89.7	0.1	86.3	0.1	81.5	0.2	74.2	0.1	63.2	0.2	57.0	0.3	*50.00*
+0.2	81.1	0.1	76.6	0.1	71.5	0.2	64.7	0.2	57.9	0.1	53.8	0.2	*50.00*
0.0	73.0	0.2	69.0	0.2	64.9	0.1	59.6	0.1	55.1	0.1	52.8	0.2	*50.00*
0.0	*73.11*		*69.00*		*64.57*		*59.87*		*54.98*		*52.50*		*50.00*
−0.2	67.0	0.1	63.6	0.1	60.3	0.1	56.8	0.1	53.3	0.1	51.9	0.1	*50.00*
−0.4	62.5	0.1	60.0	0.1	57.4	0.1	55.0	0.1	52.6	0.1	51.1	0.1	*50.00*
−0.6	59.1	0.1	57.1	0.1	55.5	0.1	53.6	0.1	51.7	0.1	51.0	0.1	*50.00*
−0.8	56.0	0.1	55.0	0.1	53.5	0.1	52.3	0.1	51.1	0.1	50.6	0.1	*50.00*
−1.0	53.0	0.1	52.1	0.1	51.9	0.1	51.0	0.1	50.6	0.1	50.0	0.1	*50.00*

TABLE 2. *Largest cluster size, $\hat{n}$.*

Black

β	$\alpha=-1.0$		$\alpha=-0.8$		$\alpha=-0.6$		$\alpha=-0.4$		$\alpha=-0.2$		$\alpha=-0.1$		$\alpha=0.0$	
+1.0	9917.6	2.7	9894.9	2.0	9862.6	3.7	9817.6	3.9	9748.7	4.4	9675.8	6.6	—	
+0.8	9802.6	3.7	9741.6	4.9	9645.3	6.1	9505.8	6.2	9175.6	10.7	8804.3	19.1	4110.5	86.5
+0.6	9527.4	7.2	9370.0	5.0	9126.6	7.4	8678.7	11.6	7669.9	18.4	6534.6	37.1	2183.3	268.3
+0.4	8970.5	9.9	8620.3	9.2	8132.6	12.0	7378.4	13.5	6079.1	29.4	4774.8	91.9	1198.2	161.4
+0.2	8105.1	7.2	7639.8	12.2	7075.4	19.2	6230.0	22.0	4577.3	142.8	1623.7	150.4	540.5	66.1
0.0	7228.6	16.3	6743.6	18.0	6125.6	18.1	4419.2	207.4	1077.9	164.6	444.6	31.3	248.5	35.4
−0.2	6394.9	20.7	5654.1	48.0	3075.8	246.2	690.6	66.0	268.4	20.8	179.4	9.3	139.6	10.8
−0.4	3492.7	194.8	1012.1	83.3	388.8	34.1	236.7	22.4	126.1	11.5	93.6	3.7	73.2	4.9
−0.6	363.9	32.0	171.0	11.6	126.5	8.4	91.0	7.1	60.6	3.9	54.8	3.0	49.4	3.9
−0.8	79.8	3.3	65.7	4.6	50.5	2.4	38.5	1.6	35.0	1.3	27.9	1.1	27.7	1.8
−1.0	37.7	1.8	27.3	1.0	24.1	1.1	19.9	0.7	18.6	0.7	18.5	1.0	17.0	0.9

White

β	$\alpha=-1.0$		$\alpha=-0.8$		$\alpha=-0.6$		$\alpha=-0.4$		$\alpha=-0.2$		$\alpha=-0.1$		$\alpha=0.0$	
+1.0	3.0	0.2	3.8	0.4	4.1	0.3	4.9	0.3	7.9	0.4	12.5	1.3	—	
+0.8	4.7	0.3	5.7	0.4	7.6	0.6	11.5	0.8	23.5	1.7	40.2	2.8	4110.5	86.5
+0.6	7.4	0.5	8.7	0.4	12.6	0.4	22.3	1.0	62.6	4.0	164.9	9.2	2183.3	268.3
+0.4	10.7	0.5	14.4	0.6	24.9	1.1	44.9	2.3	127.6	6.7	255.1	20.4	1198.2	161.4
+0.2	16.0	0.5	21.7	1.2	34.6	1.6	69.1	3.4	121.0	6.9	251.8	17.1	540.5	66.1
0.0	20.1	0.9	27.2	1.3	39.6	2.0	62.8	2.9	104.3	6.7	152.5	8.4	248.5	35.4
−0.2	22.6	0.8	29.1	1.6	38.1	2.6	58.1	2.7	85.3	4.1	99.9	5.9	139.6	10.8
−0.4	22.4	1.3	27.3	1.3	35.9	2.4	48.4	2.2	57.7	2.8	67.4	4.0	73.2	4.9
−0.6	19.6	0.9	22.7	1.0	26.1	1.3	28.0	0.8	36.4	1.9	43.1	1.9	49.4	3.9
−0.8	14.9	0.5	15.7	0.4	18.6	0.9	22.4	1.1	25.0	1.1	26.9	1.0	27.7	1.8
−1.0	11.2	0.4	12.0	0.4	12.8	0.6	15.1	0.5	15.8	0.6	18.3	1.1	17.0	0.9

Grey

β	$\alpha=-1.0$		$\alpha=-0.8$		$\alpha=-0.6$		$\alpha=-0.4$		$\alpha=-0.2$		$\alpha=-0.1$		$\alpha=0.0$	
+1.0	8.0	0.4	9.0	0.5	10.3	0.4	10.0	0.4	11.7	0.6	13.0	0.7	19.4	1.4
+0.8	10.1	0.3	11.6	0.6	14.0	0.6	15.9	0.6	19.0	0.9	24.0	2.0	26.1	1.2
+0.6	15.3	0.6	17.5	0.7	22.1	0.9	24.9	1.0	37.9	2.4	45.4	2.5	49.2	2.2
+0.4	25.3	1.1	31.6	1.4	39.9	1.9	56.8	2.8	78.2	5.5	88.1	5.9	82.6	4.8
+0.2	44.1	2.1	57.1	3.2	83.1	4.6	103.0	9.2	135.4	9.8	143.1	8.1	150.9	15.7
0.0	105.3	5.8	132.6	6.4	170.5	11.6	223.4	16.7	218.1	12.8	247.6	20.4	269.2	40.5
−0.2	196.5	9.6	278.0	20.6	433.8	54.9	428.0	31.9	473.1	53.7	570.7	35.8	560.1	87.9
−0.4	591.9	63.3	753.8	92.0	837.1	110.8	974.5	113.5	1078.6	109.2	1286.3	90.1	1300.4	187.0
−0.6	1946.3	204.1	1922.0	196.1	2273.7	221.3	2586.2	252.0	3096.1	165.2	2577.2	209.3	2596.6	249.0
−0.8	4294.2	153.9	4114.4	193.0	4330.6	257.7	3969.7	180.1	4136.9	199.0	4519.8	97.8	4930.1	96.9
−1.0	5286.5	43.6	5930.5	61.7	5867.1	56.3	5451.6	46.4	6171.4	33.4	6911.8	59.8	6638.0	36.8

TABLE 3. *Largest cluster boundary, $\hat{b}$.*

Black

β	$\alpha=-1.0$		$\alpha=-0.8$		$\alpha=-0.6$		$\alpha=-0.4$		$\alpha=-0.2$		$\alpha=-0.1$		$\alpha=0.0$	
+1.0	310.8	9.9	391.8	6.5	498.8	12.0	645.4	11.7	837.2	10.5	1027.5	15.9	1392.9	37.3
+0.8	720.2	12.0	916.0	14.8	1208.6	16.1	1587.6	17.3	2297.3	19.9	2873.7	26.2	2943.9	72.4
+0.6	1633.1	22.7	2090.3	15.2	2726.5	19.4	3660.0	21.4	5003.9	22.8	5600.8	21.9	2408.4	291.7
+0.4	3325.9	26.9	4168.4	21.3	5103.7	20.7	6170.2	15.6	6762.9	20.8	5900.4	106.7	1599.1	219.0
+0.2	5673.3	15.1	6513.5	24.0	7214.3	29.9	7602.5	36.8	6355.3	198.6	2351.2	216.9	818.2	99.7
0.0	7682.3	27.5	8093.7	18.5	8222.2	26.9	6502.3	298.0	1664.9	254.4	708.4	49.3	407.6	56.9
−0.2	8751.1	16.7	8359.7	71.7	4817.7	387.9	1128.2	106.3	451.1	35.0	305.0	16.2	243.6	18.9
−0.4	5550.3	304.3	1667.9	136.7	664.1	56.7	412.4	39.1	225.5	20.3	167.3	6.6	131.8	9.2
−0.6	623.1	55.7	302.9	19.5	228.9	14.8	164.2	12.7	111.0	7.1	102.4	6.1	91.9	7.6
−0.8	147.9	5.7	122.1	8.6	95.2	4.6	74.1	2.8	67.2	2.7	53.8	1.7	54.4	3.5
−1.0	71.8	3.2	54.2	2.0	47.7	2.2	40.1	1.3	38.0	1.2	37.4	2.0	34.6	1.9

White

β	$\alpha=-1.0$		$\alpha=-0.8$		$\alpha=-0.6$		$\alpha=-0.4$		$\alpha=-0.2$		$\alpha=-0.1$		$\alpha=0.0$	
+1.0	7.8	0.4	9.1	0.6	9.4	0.4	10.8	0.4	15.3	0.5	20.3	1.2	1392.9	37.3
+0.8	10.9	0.5	12.2	0.6	14.4	0.6	19.6	0.9	34.2	2.0	51.4	3.1	2943.9	72.4
+0.6	14.7	0.8	17.3	0.5	22.2	0.6	34.2	1.9	91.4	5.4	203.3	12.3	2408.4	291.7
+0.4	20.7	0.9	27.1	1.0	41.0	1.9	71.9	3.4	188.3	10.0	357.6	28.0	1599.1	219.0
+0.2	30.2	1.0	40.5	2.1	63.6	2.9	116.2	6.1	195.6	11.7	389.3	26.4	818.2	99.7
0.0	39.3	1.9	51.5	2.0	70.6	3.7	110.0	4.5	177.1	11.5	257.5	13.7	407.6	56.9
−0.2	44.3	1.5	55.5	3.2	71.5	4.7	104.7	5.2	150.7	7.1	174.0	10.6	243.6	18.9
−0.4	44.4	2.6	54.6	2.7	68.1	4.4	91.4	4.1	106.5	5.3	122.9	6.3	131.8	9.2
−0.6	38.6	1.7	46.2	2.0	50.4	2.6	54.4	1.6	70.3	3.7	81.3	3.7	91.9	7.6
−0.8	30.9	0.9	31.6	0.7	38.1	1.7	44.8	2.3	49.9	1.9	51.8	2.1	54.4	3.5
−1.0	23.8	0.7	25.2	0.8	26.8	1.2	31.0	0.9	32.7	1.3	37.6	2.1	34.6	1.9

Grey

β	$\alpha=-1.0$		$\alpha=-0.8$		$\alpha=-0.6$		$\alpha=-0.4$		$\alpha=-0.2$		$\alpha=-0.1$		$\alpha=0.0$	
+1.0	17.5	0.8	19.1	0.6	21.6	0.8	21.4	0.7	24.3	1.1	26.8	1.2	39.8	2.6
+0.8	21.6	0.7	24.0	1.2	28.6	1.2	32.1	1.1	38.7	1.8	46.6	3.7	51.4	3.9
+0.6	30.9	1.0	34.9	1.2	42.4	1.4	48.2	1.8	72.1	4.9	86.0	4.6	93.0	7.3
+0.4	47.1	2.2	59.9	2.7	72.1	3.4	103.6	5.1	141.7	9.6	160.4	10.6	150.4	8.6
+0.2	81.2	3.9	102.7	5.7	145.2	8.5	183.6	16.2	231.9	16.9	245.4	14.1	259.0	28.4
0.0	173.8	9.1	222.6	9.5	279.2	19.6	370.5	27.3	353.3	21.2	408.8	34.9	432.2	67.6
−0.2	306.4	15.5	432.3	32.5	659.4	82.3	646.7	50.5	712.5	79.5	853.6	54.2	821.0	129.4
−0.4	816.3	88.1	1043.6	124.1	1129.9	146.2	1306.3	151.7	1451.2	143.9	1709.5	121.6	1711.4	241.2
−0.6	2314.4	239.2	2180.4	220.4	2571.1	249.6	2871.3	276.1	3398.2	183.2	2848.5	229.7	2827.8	263.8
−0.8	3627.4	126.1	3267.4	144.6	3189.7	185.9	2976.9	129.9	2937.2	148.1	3162.4	76.0	3215.6	79.4
−1.0	1885.2	19.6	1805.8	25.1	1802.0	21.3	1647.9	24.8	1657.8	18.3	1734.4	21.3	1659.8	38.6

TABLE 4. *Mean cluster size, $\bar{n}$.*

Black

β	$\alpha=-1.0$		$\alpha=-0.8$		$\alpha=-0.6$		$\alpha=-0.4$		$\alpha=-0.2$		$\alpha=-0.1$		$\alpha=0.0$	
+1.0	9917.6	2.7	9894.9	2.0	9862.6	3.7	9817.6	3.9	9748.7	4.4	9675.8	6.6	48.5	8.9
+0.8	9802.6	3.7	9298.6	328.8	9645.3	6.1	8713.9	409.1	4020.3	730.5	1991.6	477.6	32.1	9.8
+0.6	8733.9	501.2	7604.7	624.0	5070.8	811.4	2390.4	641.2	390.3	160.8	102.9	31.2	22.9	6.6
+0.4	4529.9	681.2	2556.2	379.9	793.5	172.1	265.6	67.7	62.5	18.4	31.5	10.9	15.5	4.8
+0.2	818.5	208.6	353.1	75.4	146.0	32.7	60.7	18.3	25.7	7.6	16.1	6.2	11.1	4.7
0.0	146.7	34.0	81.1	25.2	45.6	12.3	23.3	6.7	12.4	4.0	9.8	4.1	7.7	2.5
−0.2	43.6	16.3	26.4	10.0	16.5	5.5	10.2	2.9	7.2	2.7	6.2	2.3	5.3	1.8
−0.4	14.3	4.0	10.0	4.6	7.4	2.0	5.8	1.8	4.6	1.2	4.1	0.9	3.7	1.1
−0.6	5.9	2.4	4.7	1.3	4.0	1.6	3.4	0.9	3.0	1.1	2.8	0.9	2.6	0.6
−0.8	2.9	0.8	2.5	0.8	2.2	0.6	2.1	0.8	1.9	0.5	1.8	0.5	1.7	0.4
−1.0	1.5	0.5	1.4	0.3	1.4	0.4	1.3	0.3	1.3	0.2	1.3	0.3	1.2	0.3

White

β	$\alpha=-1.0$		$\alpha=-0.8$		$\alpha=-0.6$		$\alpha=-0.4$		$\alpha=-0.2$		$\alpha=-0.1$		$\alpha=0.0$	
+1.0	1.1	0.1	1.2	0.1	1.2	0.2	1.3	0.1	1.5	0.2	1.6	0.4	48.5	8.9
+0.8	1.2	0.2	1.3	0.2	1.4	0.2	1.6	0.3	2.1	0.6	2.9	1.0	32.1	9.8
+0.6	1.4	0.2	1.5	0.2	1.7	0.4	2.2	0.7	3.9	1.2	7.1	3.1	22.9	6.6
+0.4	1.6	0.2	1.9	0.5	2.3	0.6	3.2	0.8	5.9	1.7	8.9	3.5	15.5	4.8
+0.2	1.9	0.3	2.2	0.4	2.8	0.8	3.9	1.0	5.8	1.4	7.8	2.2	11.1	4.7
0.0	2.1	0.5	2.4	0.6	2.9	0.8	3.8	0.9	5.1	1.5	6.2	2.0	7.7	2.5
−0.2	2.1	0.4	2.4	0.7	2.8	0.8	3.4	0.7	4.2	0.8	4.7	0.9	5.3	1.8
−0.4	2.0	0.4	2.2	0.5	2.4	0.7	2.8	0.8	3.2	0.9	3.5	1.1	3.7	1.1
−0.6	1.8	0.4	1.9	0.4	2.0	0.5	2.2	0.6	2.4	0.7	2.5	0.6	2.6	0.6
−0.8	1.5	0.3	1.5	0.4	1.5	0.3	1.6	0.4	1.7	0.4	1.7	0.3	1.7	0.4
−1.0	1.2	0.2	1.2	0.1	1.2	0.3	1.2	0.2	1.2	0.3	1.2	0.3	1.2	0.3

Grey

β	$\alpha=-1.0$		$\alpha=-0.8$		$\alpha=-0.6$		$\alpha=-0.4$		$\alpha=-0.2$		$\alpha=-0.1$		$\alpha=0.0$	
+1.0	1.0	0.1	1.0	0.1	1.1	0.1	1.1	0.1	1.1	0.1	1.1	0.1	1.2	0.2
+0.8	1.1	0.1	1.1	0.1	1.1	0.1	1.2	0.2	1.3	0.2	1.4	0.2	1.8	0.4
+0.6	1.2	0.2	1.3	0.2	1.4	0.2	1.6	0.2	2.0	0.4	2.4	0.5	2.6	0.6
+0.4	1.5	0.3	1.7	0.3	2.0	0.4	2.5	0.4	3.3	0.8	3.6	0.7	3.8	0.7
+0.2	2.2	0.3	2.6	0.6	3.2	0.9	4.1	0.8	4.9	1.1	5.2	1.2	5.3	1.2
0.0	3.5	0.9	4.2	1.0	5.2	1.2	6.3	1.5	7.2	1.5	7.4	1.9	7.6	1.4
−0.2	5.6	1.4	6.8	2.0	8.1	2.2	9.6	2.2	10.5	3.0	11.0	2.9	11.0	3.3
−0.4	9.1	2.5	11.0	3.4	12.7	3.9	14.1	3.7	15.3	3.7	15.5	4.4	15.8	5.5
−0.6	14.8	4.6	17.4	4.1	19.4	4.7	21.2	8.2	22.3	3.4	22.7	7.6	22.9	5.0
−0.8	23.8	6.3	26.0	8.4	27.8	13.1	29.7	9.3	30.7	7.0	31.3	6.3	32.5	7.4
−1.0	39.1	7.2	52.2	12.8	49.2	9.8	51.6	12.7	61.2	13.5	80.8	28.6	67.8	9.2

TABLE 5. *Cluster size standard deviation, σ_n, for all clusters.*

Black

β	$\alpha=-1.0$	$\alpha=-0.8$	$\alpha=-0.6$	$\alpha=-0.4$	$\alpha=-0.2$	$\alpha=-0.1$	$\alpha=0.0$
+1.0	—	—	—	—	—	—	465.8
+0.8	—	—	—	—	5402.9	4031.1	339.6
+0.6	—	6354.2	5608.3	4234.3	1719.7	810.7	170.7
+0.4	5556.0	4581.4	2527.5	1394.6	604.3	356.7	85.0
+0.2	2556.0	1637.6	1010.5	603.0	308.6	105.9	40.8
0.0	1024.7	731.0	514.1	284.4	72.0	33.5	20.3
−0.2	515.6	364.8	179.1	44.8	20.1	14.6	11.0
−0.4	176.6	57.3	26.1	15.9	9.6	7.6	6.3
−0.6	21.0	12.3	8.6	6.4	4.8	4.3	3.9
−0.8	5.8	4.5	3.6	3.0	2.5	2.3	2.1
−1.0	2.1	1.7	1.6	1.3	1.2	1.1	1.1

White

β	$\alpha=-1.0$	$\alpha=-0.8$	$\alpha=-0.6$	$\alpha=-0.4$	$\alpha=-0.2$	$\alpha=-0.1$	$\alpha=0.0$
+1.0	0.4	0.5	0.6	0.7	1.1	1.5	465.8
+0.8	0.6	0.7	0.9	1.3	2.5	4.4	339.6
+0.6	0.8	1.0	1.4	2.4	6.4	16.5	170.7
+0.4	1.2	1.6	2.5	4.5	11.9	23.4	85.0
+0.2	1.6	2.3	3.4	6.0	11.1	19.8	40.8
0.0	2.0	2.7	3.7	5.9	9.5	13.2	20.3
−0.2	2.1	2.7	3.6	5.0	7.1	8.5	11.0
−0.4	2.0	2.4	3.0	3.8	4.9	5.8	6.3
−0.6	1.7	1.9	2.2	2.6	3.2	3.5	3.9
−0.8	1.3	1.4	1.5	1.7	1.9	2.0	2.1
−1.0	0.7	0.8	0.8	0.9	1.0	1.0	1.1

Grey

β	$\alpha=-1.0$	$\alpha=-0.8$	$\alpha=-0.6$	$\alpha=-0.4$	$\alpha=-0.2$	$\alpha=-0.1$	$\alpha=0.0$
+1.0	0.4	0.4	0.5	0.5	0.6	0.7	1.1
+0.8	0.6	0.7	0.8	0.9	1.2	1.5	2.1
+0.6	1.0	1.1	1.4	1.9	2.7	3.4	3.9
+0.4	1.7	2.2	2.8	4.0	5.6	6.3	6.6
+0.2	3.4	4.4	6.0	8.1	10.2	10.9	11.2
0.0	7.1	9.2	12.0	15.8	17.9	19.0	20.2
−0.2	14.6	19.6	27.3	31.6	36.1	38.8	39.6
−0.4	35.1	48.8	59.5	68.3	78.6	84.6	86.6
−0.6	110.9	126.3	153.4	180.4	188.3	185.1	179.2
−0.8	263.3	277.3	298.4	297.3	309.5	321.8	341.3
−1.0	409.1	532.2	482.7	500.5	573.9	707.4	625.4

TABLE 6. *Cluster size standard deviation, σ_n, omitting largest cluster.*

Black

β	$\alpha=-1.0$	$\alpha=-0.8$	$\alpha=-0.6$	$\alpha=-0.4$	$\alpha=-0.2$	$\alpha=-0.1$	$\alpha=0.0$
+1.0	—	—	—	—	—	—	12.8
+0.8	—	—	—	—	—	—	23.4
+0.6	—	—	—	—	1.9	4.9	63.1
+0.4	—	0.3	0.5	1.3	4.8	22.6	45.8
+0.2	0.6	1.3	2.7	5.4	24.7	46.6	30.9
0.0	1.6	2.7	5.2	42.3	38.8	27.0	17.5
−0.2	2.9	7.8	57.2	32.1	17.4	13.2	10.0
−0.4	41.0	37.2	21.6	13.7	8.8	7.2	6.0
−0.6	17.1	11.2	7.9	5.9	4.6	4.1	3.8
−0.8	5.5	4.3	3.5	2.9	2.4	2.2	2.1
−1.0	2.0	1.7	1.5	1.3	1.1	1.1	1.0

White

β	$\alpha=-1.0$	$\alpha=-0.8$	$\alpha=-0.6$	$\alpha=-0.4$	$\alpha=-0.2$	$\alpha=-0.1$	$\alpha=0.0$
+1.0	0.3	0.3	0.4	0.6	0.9	1.2	12.8
+0.8	0.5	0.6	0.8	1.1	2.2	4.0	23.4
+0.6	0.8	1.0	1.3	2.3	5.9	14.7	63.1
+0.4	1.1	1.6	2.4	4.2	10.8	20.1	45.8
+0.2	1.6	2.2	3.2	5.6	10.2	16.8	30.9
0.0	1.9	2.6	3.5	5.6	8.8	12.0	17.5
−0.2	2.1	2.6	3.4	4.7	6.7	7.9	10.0
−0.4	1.9	2.3	2.9	3.6	4.7	5.5	6.0
−0.6	1.6	1.9	2.2	2.6	3.1	3.4	3.8
−0.8	1.2	1.3	1.5	1.7	1.9	2.0	2.1
−1.0	0.7	0.7	0.8	0.9	0.9	1.0	1.0

Grey

β	$\alpha=-1.0$	$\alpha=-0.8$	$\alpha=-0.6$	$\alpha=-0.4$	$\alpha=-0.2$	$\alpha=-0.1$	$\alpha=0.0$
+1.0	0.3	0.4	0.4	0.5	0.6	0.7	1.0
+0.8	0.5	0.6	0.8	0.9	1.2	1.4	2.1
+0.6	0.9	1.1	1.4	1.8	2.6	3.3	3.7
+0.4	1.7	2.1	2.7	3.8	5.3	5.9	6.2
+0.2	3.3	4.2	5.6	7.6	9.3	10.0	10.2
0.0	6.6	8.5	10.8	13.9	16.0	16.6	17.5
−0.2	13.1	17.0	21.3	25.7	28.7	30.1	30.1
−0.4	26.6	33.4	40.8	44.1	46.4	51.9	46.6
−0.6	49.8	53.6	62.5	53.8	59.7	57.7	64.2
−0.8	45.0	45.1	50.6	59.7	50.4	41.2	28.0
−1.0	—	—	—	—	—	—	—

TABLE 7. *Regression coefficient, c_0, for all clusters.*

Black

β	$\alpha=-1.0$	$\alpha=-0.8$	$\alpha=-0.6$	$\alpha=-0.4$	$\alpha=-0.2$	$\alpha=-0.1$	$\alpha=0.0$
+1.0	—	—	—	—	—	—	5.22
+0.8	—	—	—	—	—	—	5.35
+0.6	—	—	—	—	3.85	4.10	4.33
+0.4	—	3.71	3.64	3.72	3.73	3.84	3.60
+0.2	3.54	3.55	3.49	3.39	3.35	3.38	3.13
0.0	3.30	3.28	3.22	3.19	3.11	2.89	2.80
−0.2	3.05	3.01	2.91	2.83	2.72	2.60	2.53
−0.4	2.90	2.73	2.62	2.51	2.41	2.38	2.36
−0.6	2.53	2.41	2.36	2.31	2.26	2.23	2.21
−0.8	2.25	2.20	2.18	2.15	2.12	2.12	2.10
−1.0	2.10	2.09	2.07	2.06	2.05	2.05	2.05

White

β	$\alpha=-1.0$	$\alpha=-0.8$	$\alpha=-0.6$	$\alpha=-0.4$	$\alpha=-0.2$	$\alpha=-0.1$	$\alpha=0.0$
+1.0	2.07	2.08	2.17	2.18	2.34	2.59	5.22
+0.8	2.05	2.15	2.25	2.38	2.79	3.13	5.35
+0.6	2.17	2.20	2.33	2.60	3.09	3.64	4.33
+0.4	2.20	2.28	2.46	2.70	3.10	3.42	3.60
+0.2	2.21	2.30	2.43	2.61	2.80	2.99	3.13
0.0	2.20	2.26	2.36	2.49	2.61	2.68	2.80
−0.2	2.17	2.22	2.28	2.35	2.43	2.46	2.53
−0.4	2.13	2.14	2.20	2.23	2.29	2.34	2.36
−0.6	2.09	2.09	2.12	2.14	2.17	2.20	2.21
−0.8	2.04	2.05	2.06	2.07	2.08	2.10	2.10
−1.0	2.02	2.02	2.03	2.04	2.04	2.04	2.05

Grey

β	$\alpha=-1.0$	$\alpha=-0.8$	$\alpha=-0.6$	$\alpha=-0.4$	$\alpha=-0.2$	$\alpha=-0.1$	$\alpha=0.0$
+1.0	2.01	2.02	2.02	2.02	2.03	2.04	2.04
+0.8	2.03	2.03	2.05	2.06	2.07	2.08	2.10
+0.6	2.07	2.08	2.10	2.12	2.16	2.19	2.21
+0.4	2.14	2.16	2.21	2.25	2.32	2.34	2.36
+0.2	2.27	2.31	2.40	2.44	2.53	2.55	2.53
0.0	2.50	2.56	2.65	2.72	2.77	2.76	2.81
−0.2	2.81	2.91	3.00	3.10	3.14	3.19	3.23
−0.4	3.29	3.38	3.53	3.65	3.65	3.74	3.79
−0.6	3.93	4.16	4.23	4.44	4.35	4.32	4.47
−0.8	4.78	4.92	5.21	5.21	5.36	5.35	5.66
−1.0	5.14	4.85	4.88	5.18	4.90	5.06	5.13

TABLE 8. *Regression coefficient, c_1, for all clusters.*

Black

β	$\alpha=-1.0$	$\alpha=-0.8$	$\alpha=-0.6$	$\alpha=-0.4$	$\alpha=-0.2$	$\alpha=-0.1$	$\alpha=0.0$
+1.0	—	—	—	—	—	—	0.29
+0.8	—	—	—	—	—	—	0.72
+0.6	—	—	—	—	0.65	0.86	1.10
+0.4	—	0.48	0.63	0.84	1.11	1.23	1.34
+0.2	0.70	0.85	1.02	1.22	1.39	1.45	1.51
0.0	1.06	1.20	1.34	1.47	1.55	1.59	1.63
−0.2	1.37	1.48	1.57	1.63	1.68	1.71	1.73
−0.4	1.59	1.65	1.70	1.74	1.78	1.79	1.80
−0.6	1.73	1.77	1.80	1.82	1.84	1.86	1.87
−0.8	1.84	1.86	1.88	1.89	1.91	1.91	1.92
−1.0	1.92	1.93	1.94	1.95	1.95	1.96	1.96

White

β	$\alpha=-1.0$	$\alpha=-0.8$	$\alpha=-0.6$	$\alpha=-0.4$	$\alpha=-0.2$	$\alpha=-0.1$	$\alpha=0.0$
+1.0	1.93	1.93	1.85	1.84	1.73	1.56	0.29
+0.8	1.95	1.87	1.80	1.71	1.48	1.33	0.72
+0.6	1.86	1.84	1.75	1.62	1.43	1.29	1.10
+0.4	1.85	1.80	1.71	1.61	1.49	1.41	1.34
+0.2	1.85	1.80	1.74	1.68	1.61	1.56	1.51
0.0	1.87	1.83	1.79	1.74	1.69	1.67	1.63
−0.2	1.89	1.86	1.83	1.80	1.77	1.76	1.73
−0.4	1.91	1.90	1.87	1.86	1.83	1.81	1.80
−0.6	1.93	1.93	1.92	1.91	1.89	1.87	1.87
−0.8	1.96	1.96	1.95	1.94	1.94	1.92	1.92
−1.0	1.98	1.98	1.98	1.97	1.97	1.97	1.96

Grey

β	$\alpha=-1.0$	$\alpha=-0.8$	$\alpha=-0.6$	$\alpha=-0.4$	$\alpha=-0.2$	$\alpha=-0.1$	$\alpha=0.0$
+1.0	1.99	1.98	1.98	1.98	1.97	1.97	1.96
+0.8	1.97	1.97	1.96	1.95	1.94	1.93	1.93
+0.6	1.94	1.93	1.91	1.90	1.88	1.87	1.87
+0.4	1.88	1.87	1.85	1.83	1.81	1.80	1.80
+0.2	1.80	1.78	1.75	1.75	1.73	1.73	1.73
0.0	1.69	1.68	1.66	1.65	1.63	1.64	1.63
−0.2	1.57	1.55	1.53	1.51	1.51	1.51	1.50
−0.4	1.40	1.39	1.36	1.35	1.35	1.34	1.32
−0.6	1.19	1.15	1.13	1.11	1.11	1.10	1.10
−0.8	0.85	0.81	0.76	0.75	0.73	0.71	0.69
−1.0	0.38	0.35	0.35	0.32	0.32	0.32	0.30

TABLE 9. *Regression coefficient, c_0, omitting largest cluster.*

Black

β	$\alpha=-1.0$	$\alpha=-0.8$	$\alpha=-0.6$	$\alpha=-0.4$	$\alpha=-0.2$	$\alpha=-0.1$	$\alpha=0.0$
+1.0	—	—	—	—	—	—	3.56
+0.8	—	—	—	—	—	—	4.60
+0.6	—	—	—	—	2.80	3.25	4.31
+0.4	—	2.00	2.00	2.22	2.90	3.41	3.51
+0.2	2.08	2.34	2.61	2.90	3.25	3.27	3.13
0.0	2.50	2.65	2.71	3.10	3.02	2.85	2.78
−0.2	2.44	2.75	2.91	2.82	2.71	2.59	2.52
−0.4	2.81	2.65	2.61	2.51	2.41	2.37	2.35
−0.6	2.49	2.39	2.35	2.29	2.24	2.22	2.20
−0.8	2.25	2.20	2.17	2.15	2.12	2.11	2.10
−1.0	2.09	2.09	2.07	2.06	2.05	2.04	2.04

White

β	$\alpha=-1.0$	$\alpha=-0.8$	$\alpha=-0.6$	$\alpha=-0.4$	$\alpha=-0.2$	$\alpha=-0.1$	$\alpha=0.0$
+1.0	2.00	2.00	2.08	2.11	2.28	2.41	3.56
+0.8	2.02	2.06	2.15	2.27	2.70	3.04	4.60
+0.6	2.10	2.17	2.29	2.55	3.07	3.53	4.31
+0.4	2.18	2.26	2.41	2.67	3.06	3.36	3.51
+0.2	2.20	2.29	2.43	2.59	2.78	2.94	3.13
0.0	2.20	2.26	2.35	2.47	2.61	2.68	2.78
−0.2	2.16	2.21	2.28	2.33	2.42	2.43	2.52
−0.4	2.12	2.15	2.19	2.23	2.28	2.33	2.35
−0.6	2.08	2.09	2.11	2.14	2.17	2.20	2.20
−0.8	2.04	2.05	2.06	2.07	2.08	2.10	2.10
−1.0	2.02	2.02	2.02	2.03	2.04	2.04	2.04

Grey

β	$\alpha=-1.0$	$\alpha=-0.8$	$\alpha=-0.6$	$\alpha=-0.4$	$\alpha=-0.2$	$\alpha=-0.1$	$\alpha=0.0$
+1.0	2.01	2.01	2.02	2.02	2.03	2.03	2.04
+0.8	2.02	2.03	2.04	2.05	2.07	2.08	2.09
+0.6	2.06	2.07	2.09	2.11	2.16	2.18	2.21
+0.4	2.13	2.16	2.19	2.25	2.31	2.33	2.36
+0.2	2.27	2.31	2.38	2.44	2.52	2.53	2.52
0.0	2.48	2.54	2.63	2.70	2.75	2.77	2.76
−0.2	2.78	2.89	2.97	3.08	3.10	3.12	3.15
−0.4	3.18	3.30	3.44	3.56	3.55	3.70	3.70
−0.6	3.79	3.88	4.01	4.11	4.18	4.24	4.23
−0.8	4.44	4.60	4.79	4.99	4.75	4.89	4.49
−1.0	4.32	3.87	4.10	3.96	3.32	3.60	3.83

TABLE 10. *Regression coefficient, c_1, omitting largest cluster.*

Black

β	$\alpha=-1.0$	$\alpha=-0.8$	$\alpha=-0.6$	$\alpha=-0.4$	$\alpha=-0.2$	$\alpha=-0.1$	$\alpha=0.0$
+1.0	—	—	—	—	—	—	1.01
+0.8	—	—	—	—	—	—	0.87
+0.6	—	—	—	—	1.38	1.21	1.10
+0.4	—	2.00	2.00	1.81	1.44	1.32	1.35
+0.2	1.93	1.73	1.53	1.43	1.41	1.46	1.51
0.0	1.60	1.54	1.55	1.48	1.56	1.60	1.64
−0.2	1.68	1.57	1.57	1.63	1.68	1.71	1.73
−0.4	1.60	1.66	1.70	1.74	1.78	1.79	1.81
−0.6	1.74	1.78	1.80	1.83	1.85	1.86	1.87
−0.8	1.84	1.86	1.88	1.89	1.91	1.91	1.93
−1.0	1.92	1.93	1.94	1.95	1.95	1.96	1.96

White

β	$\alpha=-1.0$	$\alpha=-0.8$	$\alpha=-0.6$	$\alpha=-0.4$	$\alpha=-0.2$	$\alpha=-0.1$	$\alpha=0.0$
+1.0	2.00	2.00	1.93	1.90	1.77	1.68	1.01
+0.8	1.99	1.95	1.88	1.79	1.53	1.37	0.87
+0.6	1.91	1.86	1.79	1.64	1.43	1.31	1.10
+0.4	1.87	1.81	1.73	1.62	1.50	1.42	1.35
+0.2	1.86	1.81	1.74	1.68	1.61	1.57	1.51
0.0	1.87	1.83	1.79	1.74	1.69	1.67	1.64
−0.2	1.89	1.86	1.83	1.81	1.77	1.77	1.73
−0.4	1.91	1.90	1.88	1.86	1.83	1.81	1.81
−0.6	1.94	1.93	1.92	1.91	1.89	1.87	1.87
−0.8	1.97	1.96	1.95	1.94	1.94	1.93	1.93
−1.0	1.98	1.98	1.98	1.97	1.97	1.97	1.96

Grey

β	$\alpha=-1.0$	$\alpha=-0.8$	$\alpha=-0.6$	$\alpha=-0.4$	$\alpha=-0.2$	$\alpha=-0.1$	$\alpha=0.0$
+1.0	1.99	1.99	1.98	1.98	1.98	1.97	1.96
+0.8	1.98	1.97	1.96	1.95	1.94	1.93	1.93
+0.6	1.94	1.93	1.92	1.91	1.88	1.87	1.87
+0.4	1.89	1.87	1.85	1.83	1.81	1.81	1.80
+0.2	1.80	1.79	1.76	1.75	1.73	1.73	1.73
0.0	1.70	1.69	1.66	1.65	1.64	1.64	1.64
−0.2	1.58	1.55	1.54	1.52	1.51	1.51	1.50
−0.4	1.42	1.39	1.37	1.36	1.36	1.34	1.33
−0.6	1.20	1.18	1.15	1.14	1.13	1.11	1.12
−0.8	0.91	0.85	0.82	0.78	0.83	0.79	0.91
−1.0	0.60	0.82	0.72	0.80	1.13	1.02	0.89

TABLE 11. *Percentage residual variance.*

Black

β	$\alpha=-1.0$	$\alpha=-0.8$	$\alpha=-0.6$	$\alpha=-0.4$	$\alpha=-0.2$	$\alpha=-0.1$	$\alpha=0.0$
+1.0	—	—	—	—	—	—	0.93
+0.8	—	—	—	—	—	—	0.24
+0.6	—	—	—	—	0.07	0.07	0.17
+0.4	—	0.06	0.05	0.02	0.04	0.05	0.24
+0.2	0.02	0.05	0.08	0.03	0.02	0.11	0.45
0.0	0.06	0.04	0.02	0.02	0.16	0.46	0.85
−0.2	0.02	0.02	0.02	0.24	0.75	1.05	1.41
−0.4	0.02	0.15	0.47	0.85	1.51	1.90	2.31
−0.6	0.57	1.10	1.67	2.31	2.96	3.22	3.47
−0.8	2.67	3.44	4.22	4.80	5.33	5.65	5.81
−1.0	7.01	7.61	7.64	7.59	7.50	7.42	7.32

White

β	$\alpha=-1.0$	$\alpha=-0.8$	$\alpha=-0.6$	$\alpha=-0.4$	$\alpha=-0.2$	$\alpha=-0.1$	$\alpha=0.0$
+1.0	2.68	3.45	4.27	4.87	7.33	10.74	0.93
+0.8	3.77	4.82	6.26	7.83	10.48	9.94	0.24
+0.6	5.45	5.86	6.94	7.78	5.80	3.05	0.17
+0.4	5.76	6.05	6.14	5.17	2.51	1.47	0.24
+0.2	5.39	5.20	4.68	3.30	1.97	1.16	0.45
0.0	4.91	4.39	3.86	2.89	1.91	1.37	0.85
−0.2	4.65	4.24	3.69	2.91	2.18	1.76	1.41
−0.4	4.69	4.35	4.01	3.40	2.83	2.56	2.31
−0.6	5.17	4.98	4.79	4.38	4.01	3.87	3.47
−0.8	5.82	5.95	6.08	6.02	5.89	5.98	5.81
−1.0	5.43	5.84	6.20	6.65	6.94	7.13	7.32

Grey

β	$\alpha=-1.0$	$\alpha=-0.8$	$\alpha=-0.6$	$\alpha=-0.4$	$\alpha=-0.2$	$\alpha=-0.1$	$\alpha=0.0$
+1.0	2.58	3.15	3.78	4.48	5.32	6.01	7.26
+0.8	4.87	5.58	6.56	7.33	7.68	7.52	5.66
+0.6	7.47	7.82	7.91	7.05	5.31	4.12	3.55
+0.4	7.82	6.89	5.68	4.03	2.70	2.34	2.27
+0.2	5.32	4.06	3.06	2.06	1.60	1.45	1.40
0.0	2.90	2.15	1.57	1.16	1.02	0.93	0.88
−0.2	1.56	1.12	0.75	0.66	0.56	0.53	0.52
−0.4	0.76	0.48	0.43	0.37	0.28	0.27	0.26
−0.6	0.27	0.24	0.21	0.16	0.18	0.16	0.18
−0.8	0.34	0.24	0.27	0.31	0.23	0.32	0.35
−1.0	0.81	0.80	0.72	0.76	0.50	1.28	1.47

TABLE 12. *Mean number of clusters.*

Black

β	$\alpha=-1.0$		$\alpha=-0.8$		$\alpha=-0.6$		$\alpha=-0.4$		$\alpha=-0.2$		$\alpha=-0.1$		$\alpha=0.0$	
+1.0	1.0	0.0	1.0	0.0	1.0	0.0	1.0	0.0	1.0	0.0	1.0	0.0	103.3	2.4
+0.8	1.0	0.0	1.1	0.1	1.0	0.0	1.1	0.1	2.3	0.3	4.5	0.4	161.5	4.5
+0.6	1.1	0.1	1.3	0.1	1.8	0.2	3.7	0.4	19.8	1.4	65.0	2.2	220.0	5.2
+0.4	2.1	0.2	3.4	0.3	10.3	0.6	28.0	1.4	101.4	2.7	180.8	4.0	321.9	6.7
+0.2	10.2	0.8	21.8	0.9	49.0	1.5	106.7	2.9	225.2	4.4	335.9	6.3	452.8	11.0
0.0	49.8	1.5	85.1	2.7	141.9	3.1	256.4	4.3	444.5	6.2	539.7	8.6	653.2	10.3
−0.2	153.8	4.5	240.5	5.6	364.5	6.2	555.6	6.0	741.1	9.1	836.5	9.6	949.0	12.8
−0.4	436.3	5.6	597.4	10.4	775.9	7.1	954.4	8.6	1141.2	8.2	1241.1	7.2	1344.5	12.8
−0.6	1000.5	12.0	1206.5	9.1	1395.2	13.4	1584.6	10.6	1743.2	12.8	1826.6	12.5	1896.8	13.4
−0.8	1947.2	11.7	2189.2	13.7	2403.6	11.0	2542.6	17.6	2701.8	10.8	2787.1	12.8	2863.7	15.3
−1.0	3446.5	15.8	3656.7	10.6	3724.3	16.9	3833.8	14.4	3955.2	9.7	3993.4	15.1	4025.4	17.4

White

β	$\alpha=-1.0$		$\alpha=-0.8$		$\alpha=-0.6$		$\alpha=-0.4$		$\alpha=-0.2$		$\alpha=-0.1$		$\alpha=0.0$	
+1.0	73.2	2.3	91.0	1.3	112.8	2.4	141.8	2.3	172.5	1.7	202.0	2.7	103.3	2.4
+0.8	163.1	2.6	201.1	3.0	253.5	2.4	311.9	3.8	389.1	3.4	414.8	3.3	161.5	4.5
+0.6	347.6	4.5	423.7	3.5	513.6	3.1	593.1	5.1	585.4	4.9	471.0	6.6	220.0	5.2
+0.4	649.4	4.7	744.3	5.2	802.8	3.7	799.0	5.3	624.8	5.0	484.6	6.9	321.9	6.7
+0.2	1006.4	4.4	1045.6	4.0	1028.4	6.7	904.2	5.1	725.5	5.3	591.6	5.5	452.8	11.0
0.0	1307.6	6.0	1266.9	5.6	1200.8	7.5	1052.2	6.7	883.7	8.0	767.5	7.1	653.2	10.3
−0.2	1567.4	7.0	1508.7	9.0	1413.0	8.4	1284.8	6.0	1122.6	5.9	1033.2	6.0	949.0	12.8
−0.4	1878.8	7.6	1831.5	5.5	1738.7	9.7	1612.9	9.7	1474.1	9.5	1398.6	10.2	1344.5	12.8
−0.6	2305.8	9.7	2282.2	8.8	2221.4	10.0	2130.0	12.1	2021.7	9.5	1974.2	9.9	1896.8	13.4
−0.8	2975.4	11.9	3004.2	12.6	3005.6	10.2	2939.2	13.5	2896.0	10.7	2888.6	8.2	2863.7	15.3
−1.0	4017.5	12.7	4071.2	6.9	4045.9	13.8	4037.7	11.6	4059.6	11.5	4036.8	14.8	4025.4	17.4

Grey

β	$\alpha=-1.0$		$\alpha=-0.8$		$\alpha=-0.6$		$\alpha=-0.4$		$\alpha=-0.2$		$\alpha=-0.1$		$\alpha=0.0$	
+1.0	4845.1	3.8	4805.2	2.8	4752.1	4.5	4679.5	4.4	4584.8	4.0	4491.0	5.9	4007.2	13.8
+0.8	4642.6	4.6	4545.8	5.4	4403.3	6.1	4218.4	6.8	3869.8	7.4	3585.2	9.4	2823.9	14.0
+0.6	4197.3	8.3	3975.8	6.0	3670.7	6.7	3217.9	7.6	2546.6	9.4	2127.3	8.0	1910.5	10.3
+0.4	3394.0	9.8	2996.8	8.7	2552.0	7.9	2023.6	6.6	1527.9	7.7	1389.6	6.3	1333.7	9.3
+0.2	2324.1	5.9	1929.1	8.9	1573.6	10.9	1233.1	6.0	1015.7	5.8	954.5	5.9	942.0	8.5
0.0	1447.5	9.2	1193.1	7.4	970.3	6.1	800.0	5.9	696.2	4.6	674.7	5.7	655.6	5.7
−0.2	895.3	6.4	735.7	6.9	613.9	5.3	523.5	4.7	476.1	5.3	456.7	4.8	454.3	7.6
−0.4	548.2	5.4	453.9	5.5	394.3	5.2	354.8	4.2	326.6	3.8	323.6	4.3	317.0	7.3
−0.6	338.9	4.9	288.0	3.4	258.3	3.3	235.8	4.8	225.1	1.9	221.2	4.1	218.6	4.0
−0.8	216.2	3.2	195.0	3.3	184.9	4.6	168.7	3.1	165.3	2.2	163.5	1.9	165.9	2.8
−1.0	138.8	1.9	126.0	1.9	115.4	1.7	107.9	1.8	99.9	1.9	102.3	1.5	105.0	1.3

Captions are explained in the penultimate paragraph of §5.

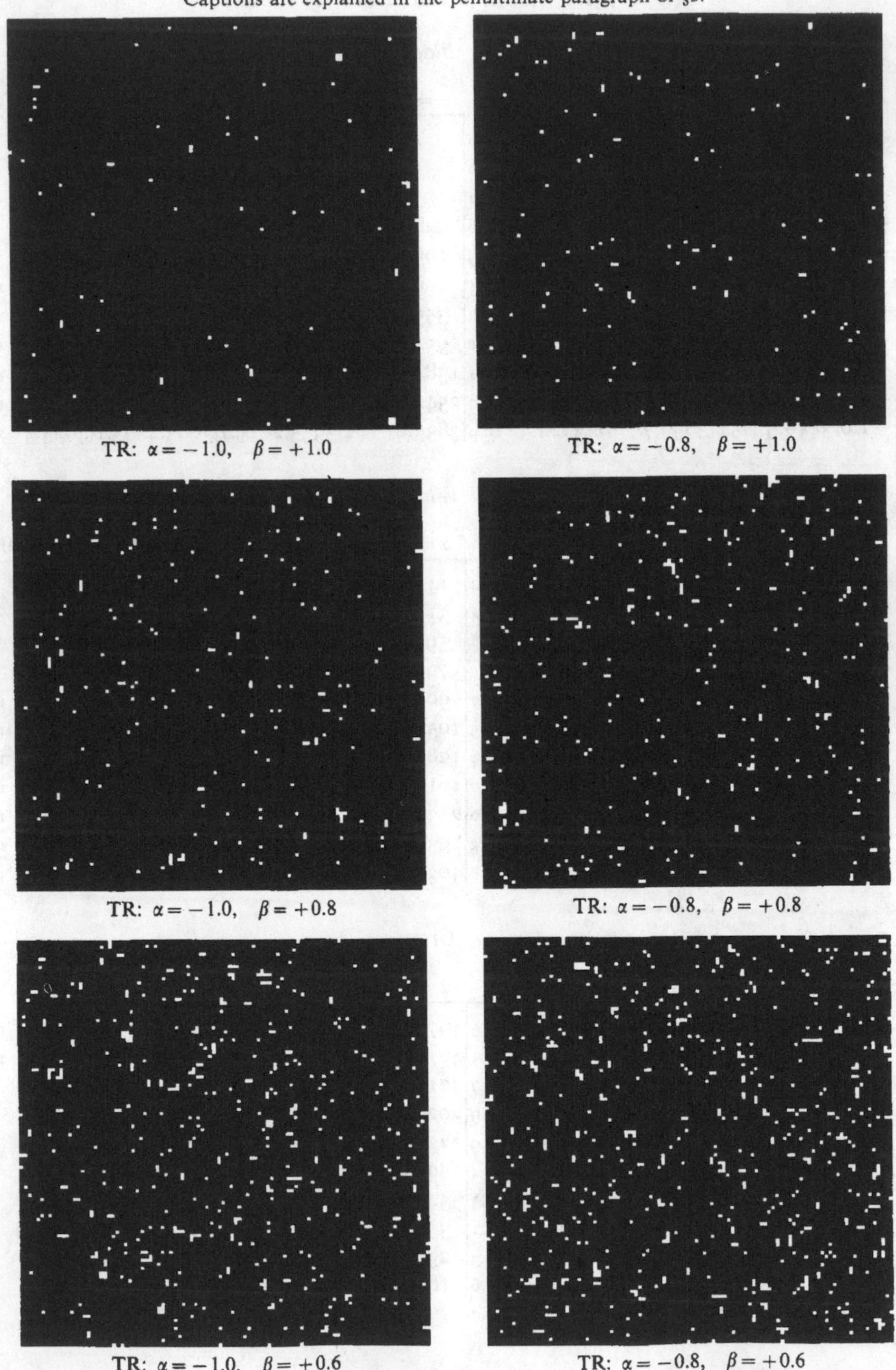

TR: $\alpha = -1.0, \quad \beta = +1.0$

TR: $\alpha = -0.8, \quad \beta = +1.0$

TR: $\alpha = -1.0, \quad \beta = +0.8$

TR: $\alpha = -0.8, \quad \beta = +0.8$

TR: $\alpha = -1.0, \quad \beta = +0.6$

TR: $\alpha = -0.8, \quad \beta = +0.6$

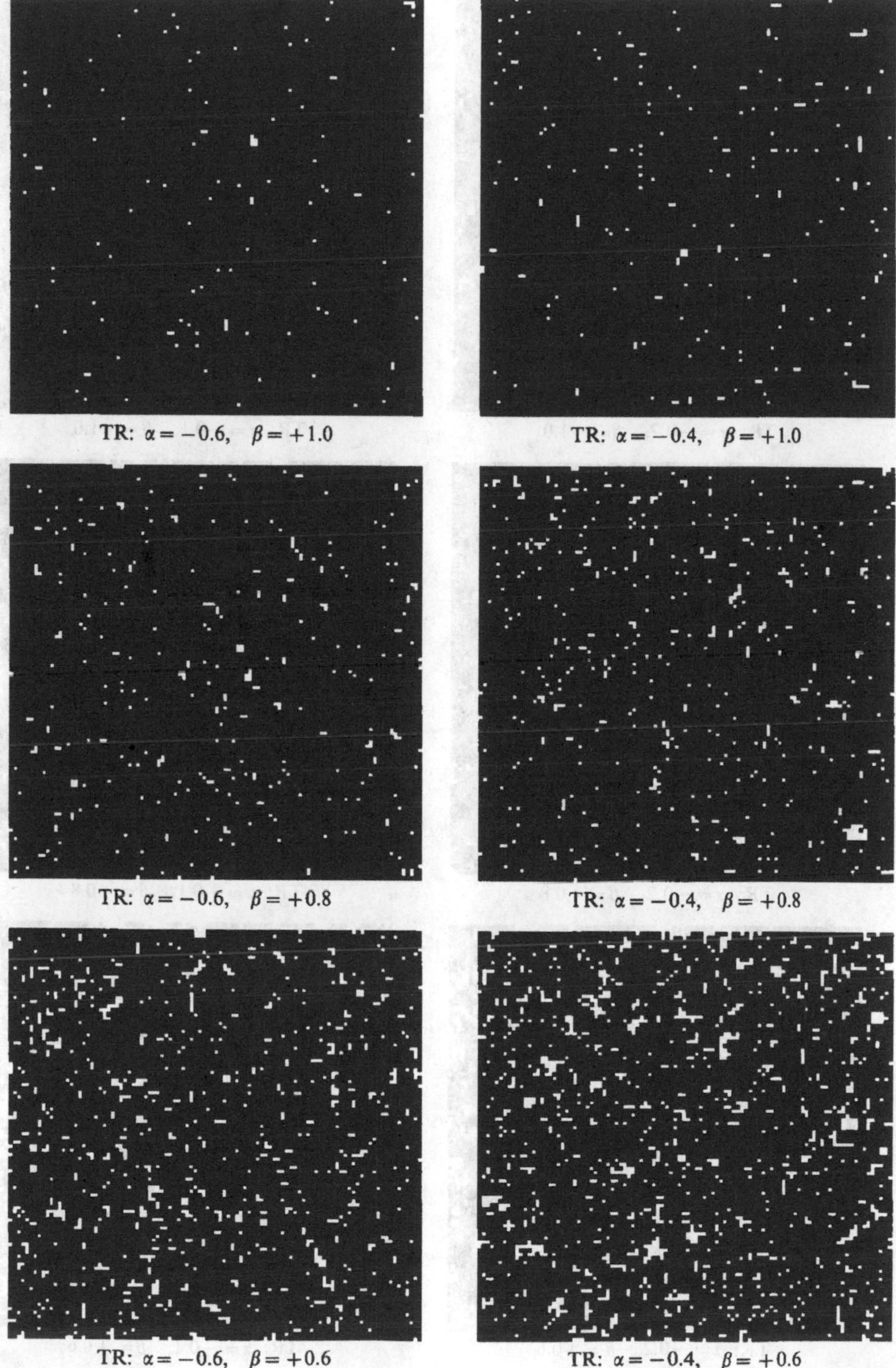

TR: $\alpha = -0.6$, $\beta = +1.0$

TR: $\alpha = -0.4$, $\beta = +1.0$

TR: $\alpha = -0.6$, $\beta = +0.8$

TR: $\alpha = -0.4$, $\beta = +0.8$

TR: $\alpha = -0.6$, $\beta = +0.6$

TR: $\alpha = -0.4$, $\beta = +0.6$

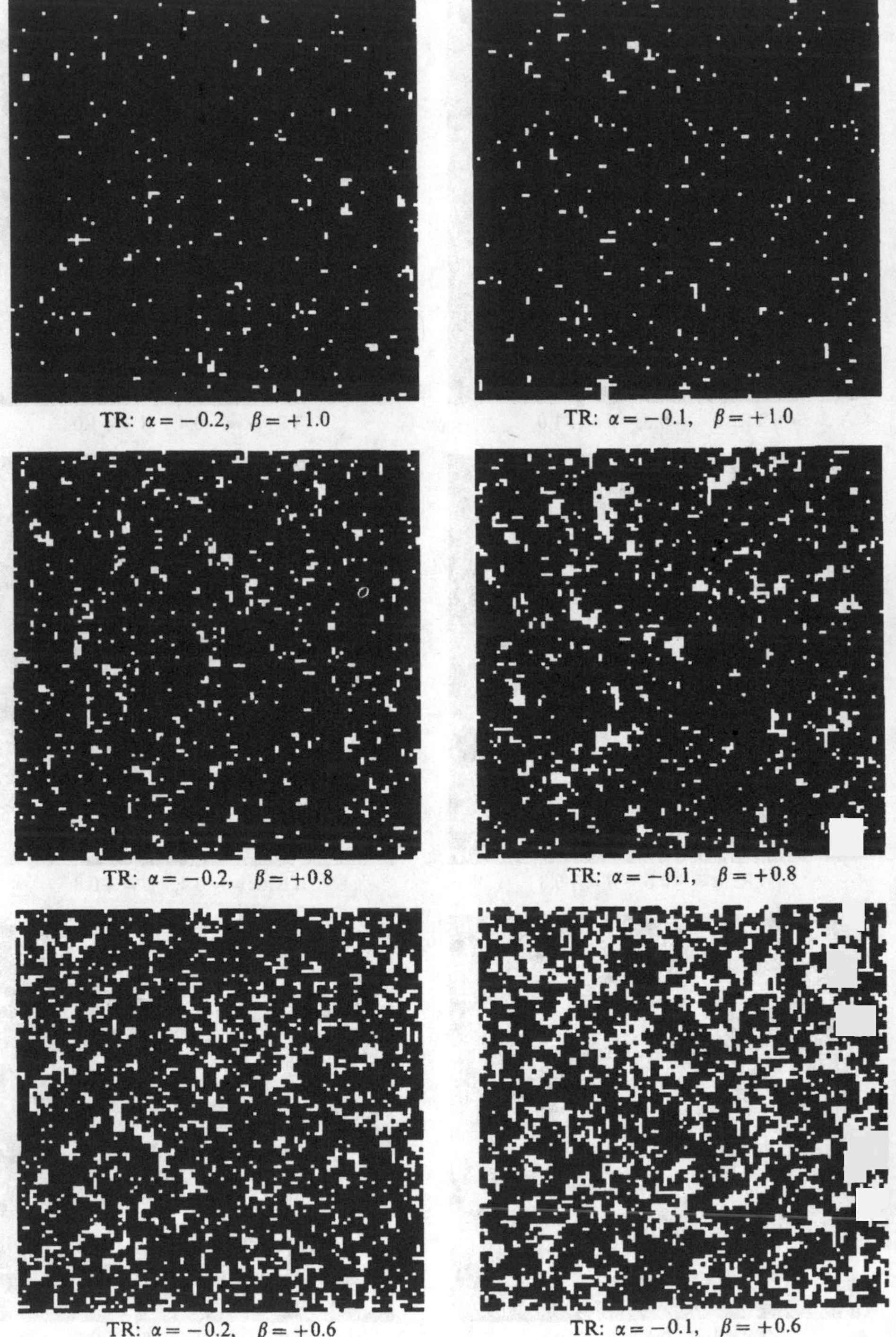

TR: $\alpha = -0.2, \quad \beta = +1.0$

TR: $\alpha = -0.1, \quad \beta = +1.0$

TR: $\alpha = -0.2, \quad \beta = +0.8$

TR: $\alpha = -0.1, \quad \beta = +0.8$

TR: $\alpha = -0.2, \quad \beta = +0.6$

TR: $\alpha = -0.1, \quad \beta = +0.6$

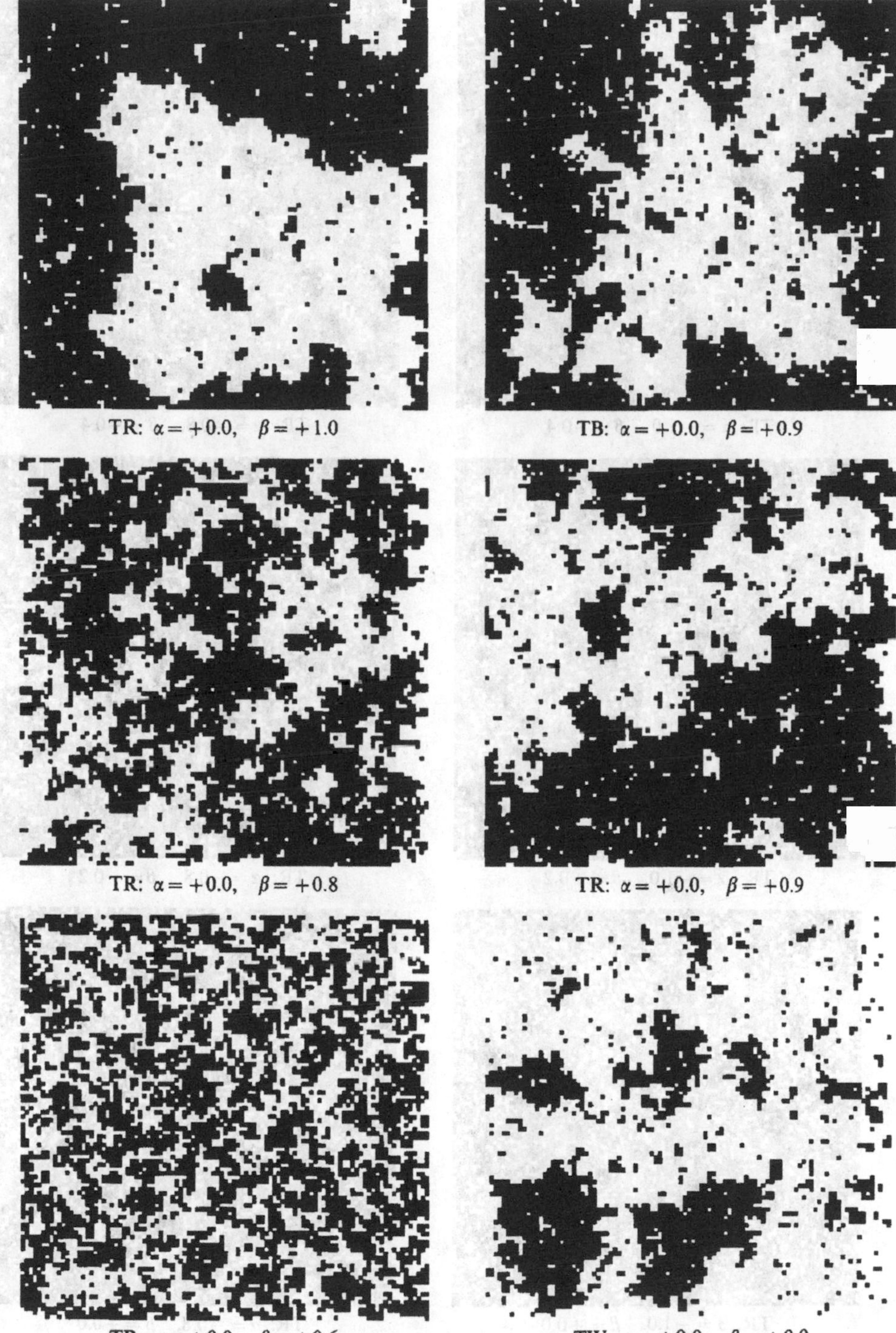

TR: $\alpha = +0.0$, $\beta = +1.0$

TB: $\alpha = +0.0$, $\beta = +0.9$

TR: $\alpha = +0.0$, $\beta = +0.8$

TR: $\alpha = +0.0$, $\beta = +0.9$

TR: $\alpha = +0.0$, $\beta = +0.6$

TW: $\alpha = +0.0$, $\beta = +0.9$

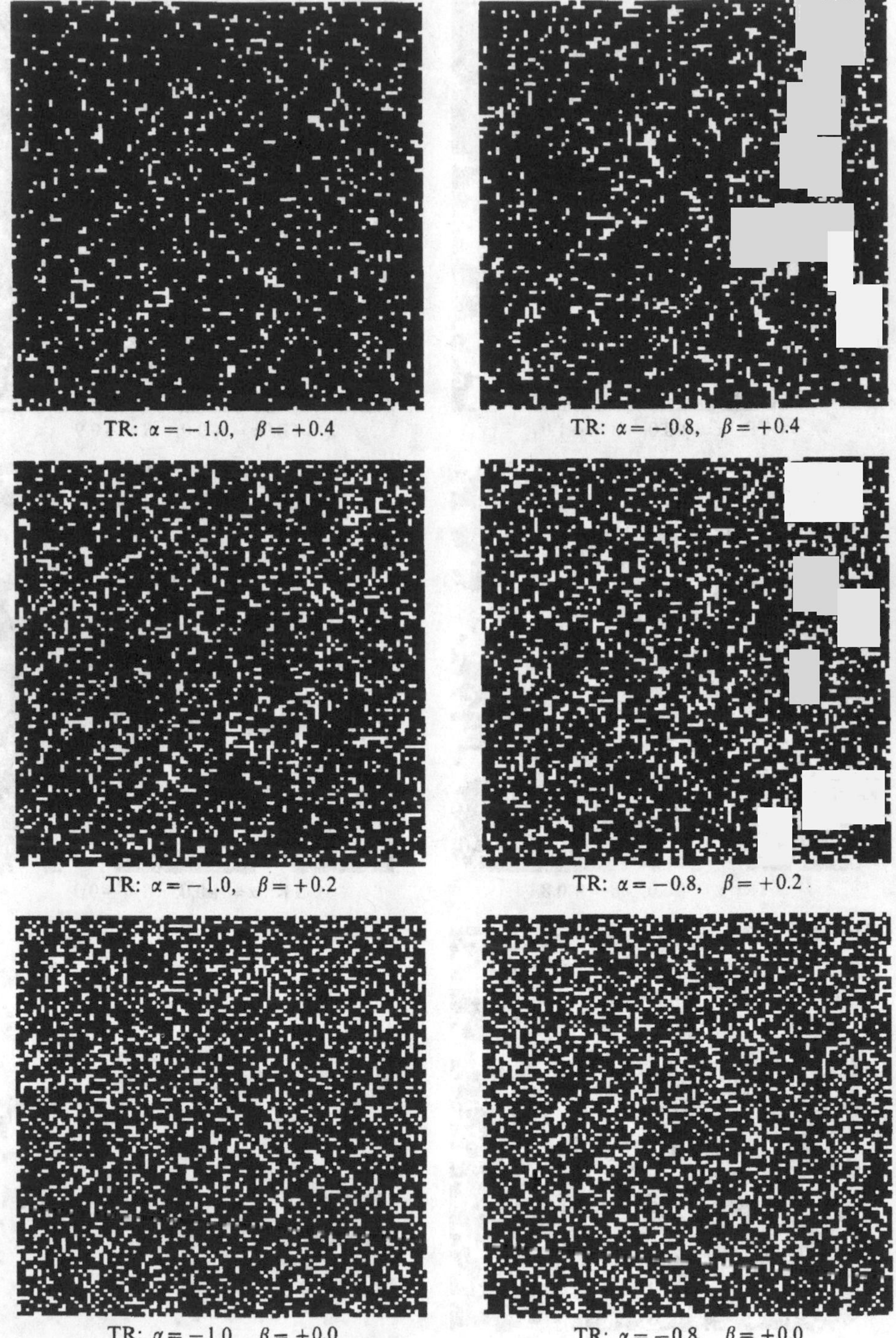

TR: $\alpha = -1.0$, $\beta = +0.4$

TR: $\alpha = -0.8$, $\beta = +0.4$

TR: $\alpha = -1.0$, $\beta = +0.2$

TR: $\alpha = -0.8$, $\beta = +0.2$

TR: $\alpha = -1.0$, $\beta = +0.0$

TR: $\alpha = -0.8$, $\beta = +0.0$

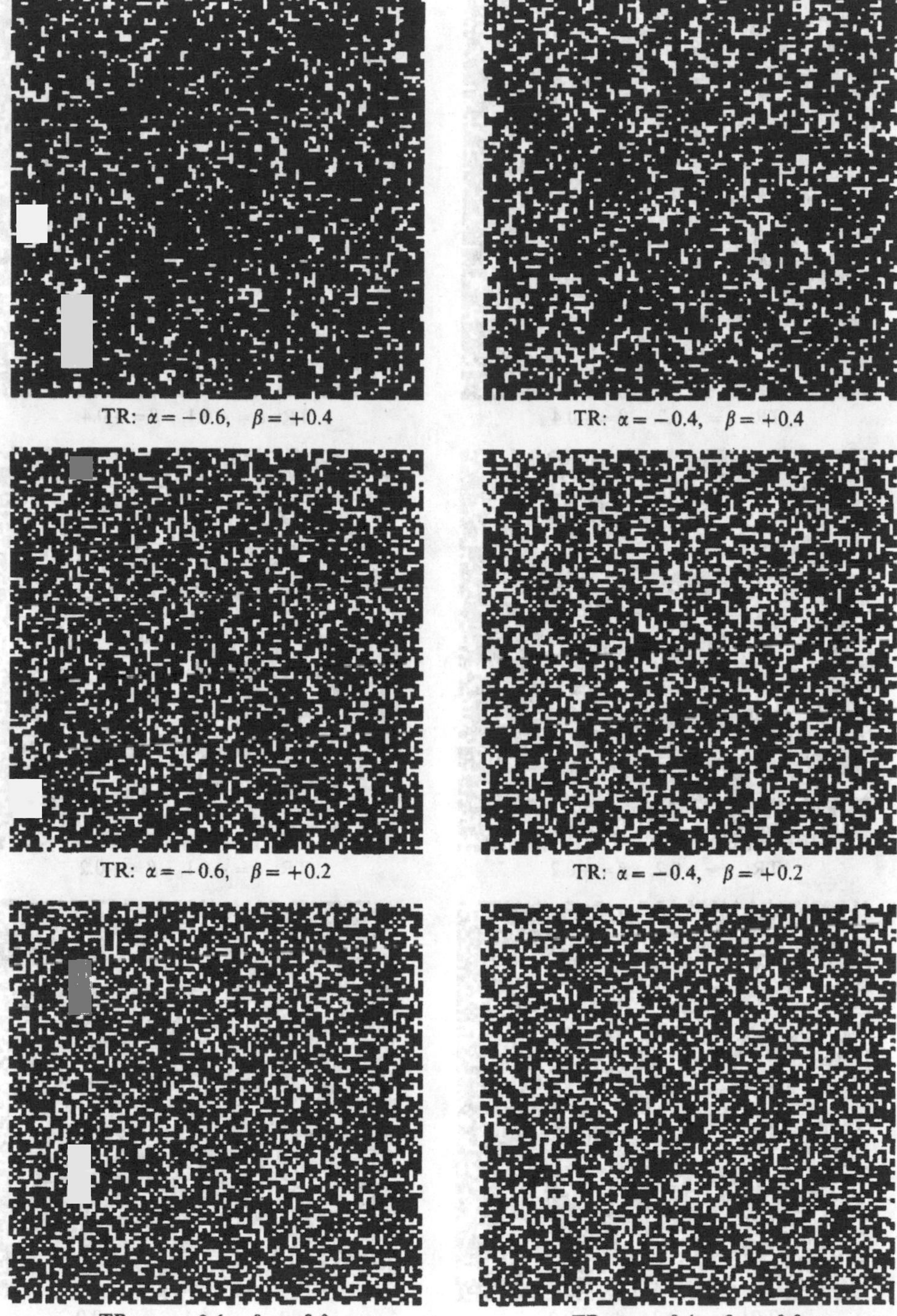

TR: $\alpha = -0.6$, $\beta = +0.4$

TR: $\alpha = -0.4$, $\beta = +0.4$

TR: $\alpha = -0.6$, $\beta = +0.2$

TR: $\alpha = -0.4$, $\beta = +0.2$

TR: $\alpha = -0.6$, $\beta = +0.0$

TR: $\alpha = -0.4$, $\beta = +0.0$

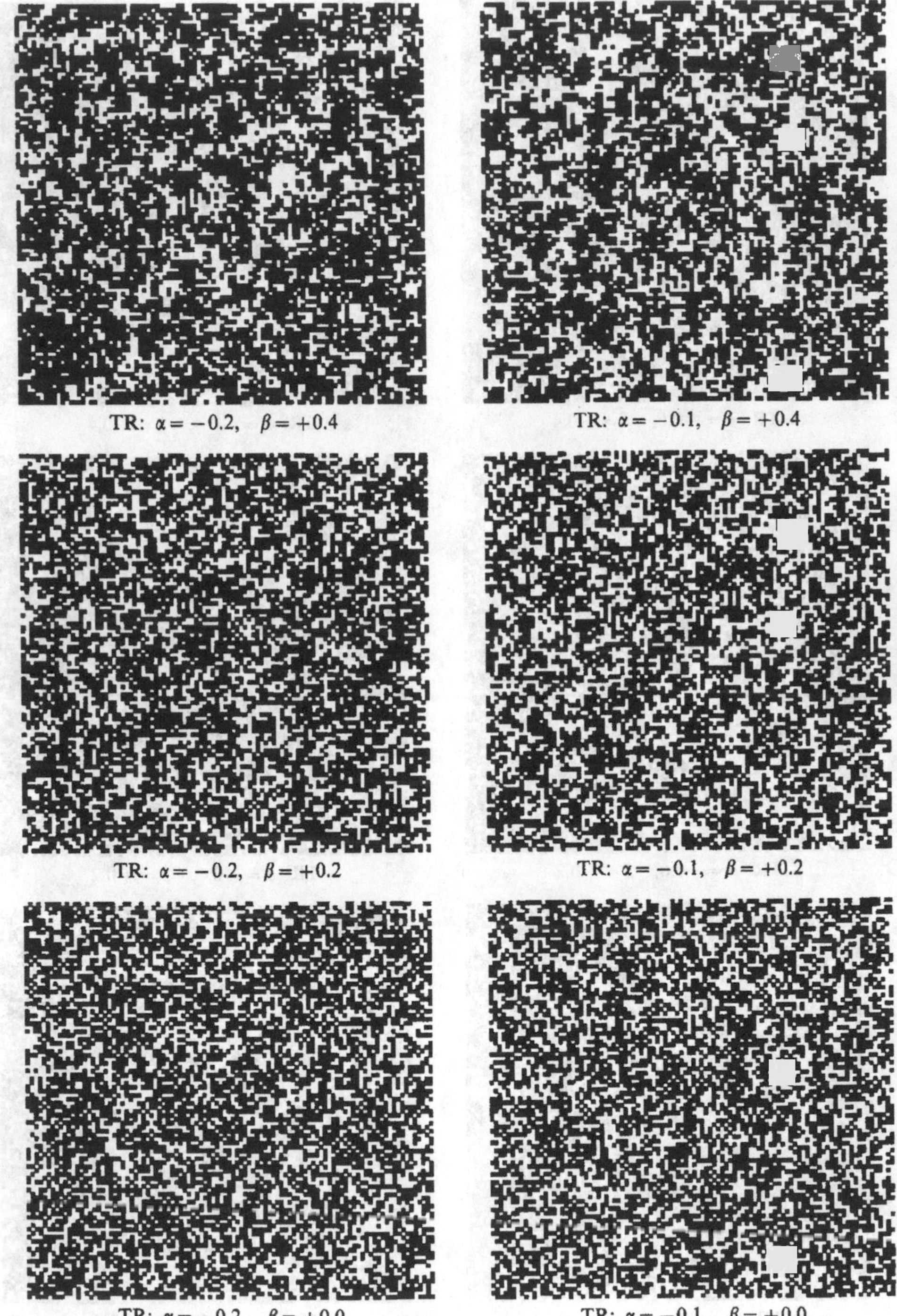

TR: $\alpha = -0.2, \quad \beta = +0.4$

TR: $\alpha = -0.1, \quad \beta = +0.4$

TR: $\alpha = -0.2, \quad \beta = +0.2$

TR: $\alpha = -0.1, \quad \beta = +0.2$

TR: $\alpha = -0.2, \quad \beta = +0.0$

TR: $\alpha = -0.1, \quad \beta = +0.0$

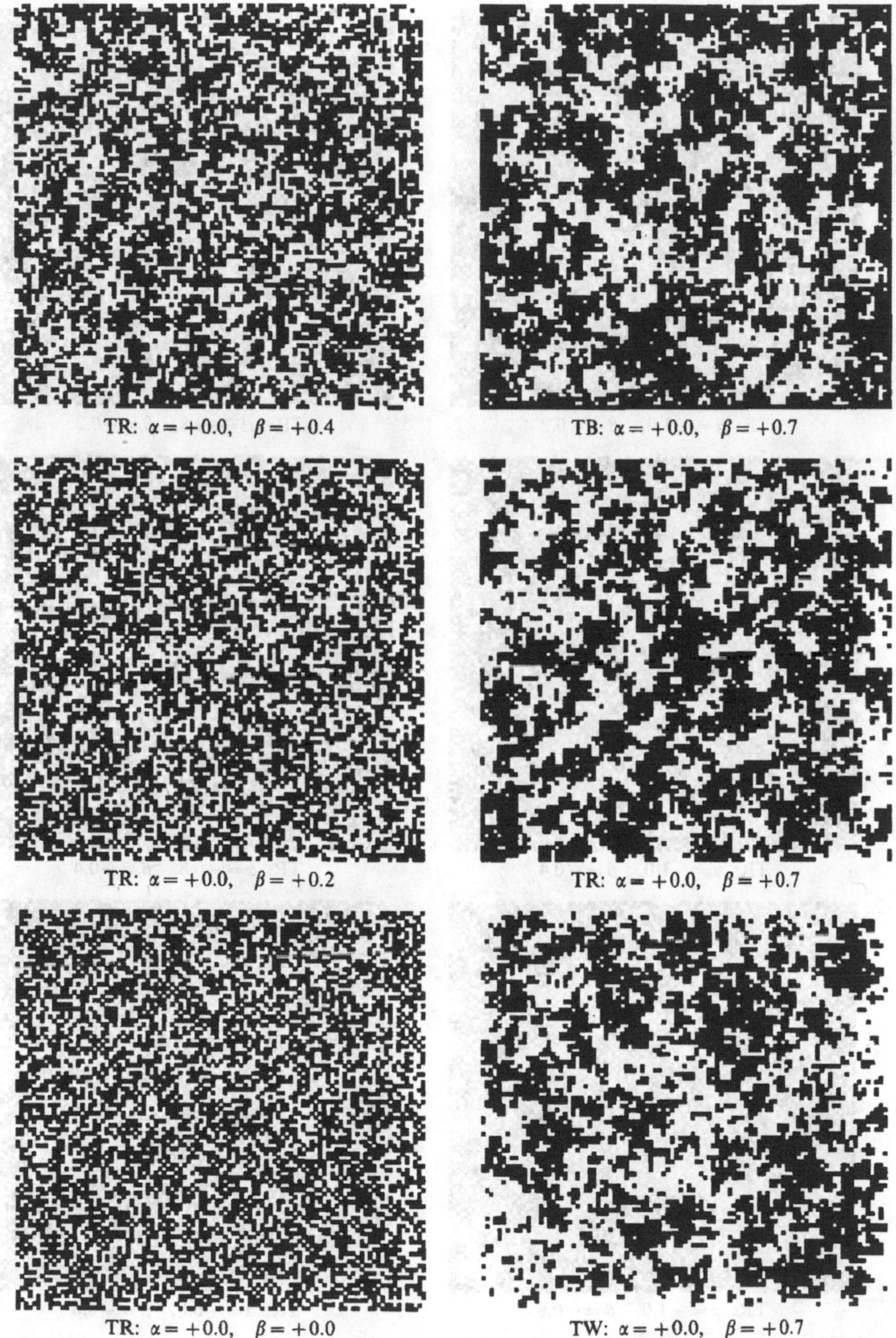

TR: $\alpha = +0.0$, $\beta = +0.4$

TB: $\alpha = +0.0$, $\beta = +0.7$

TR: $\alpha = +0.0$, $\beta = +0.2$

TR: $\alpha = +0.0$, $\beta = +0.7$

TR: $\alpha = +0.0$, $\beta = +0.0$

TW: $\alpha = +0.0$, $\beta = +0.7$

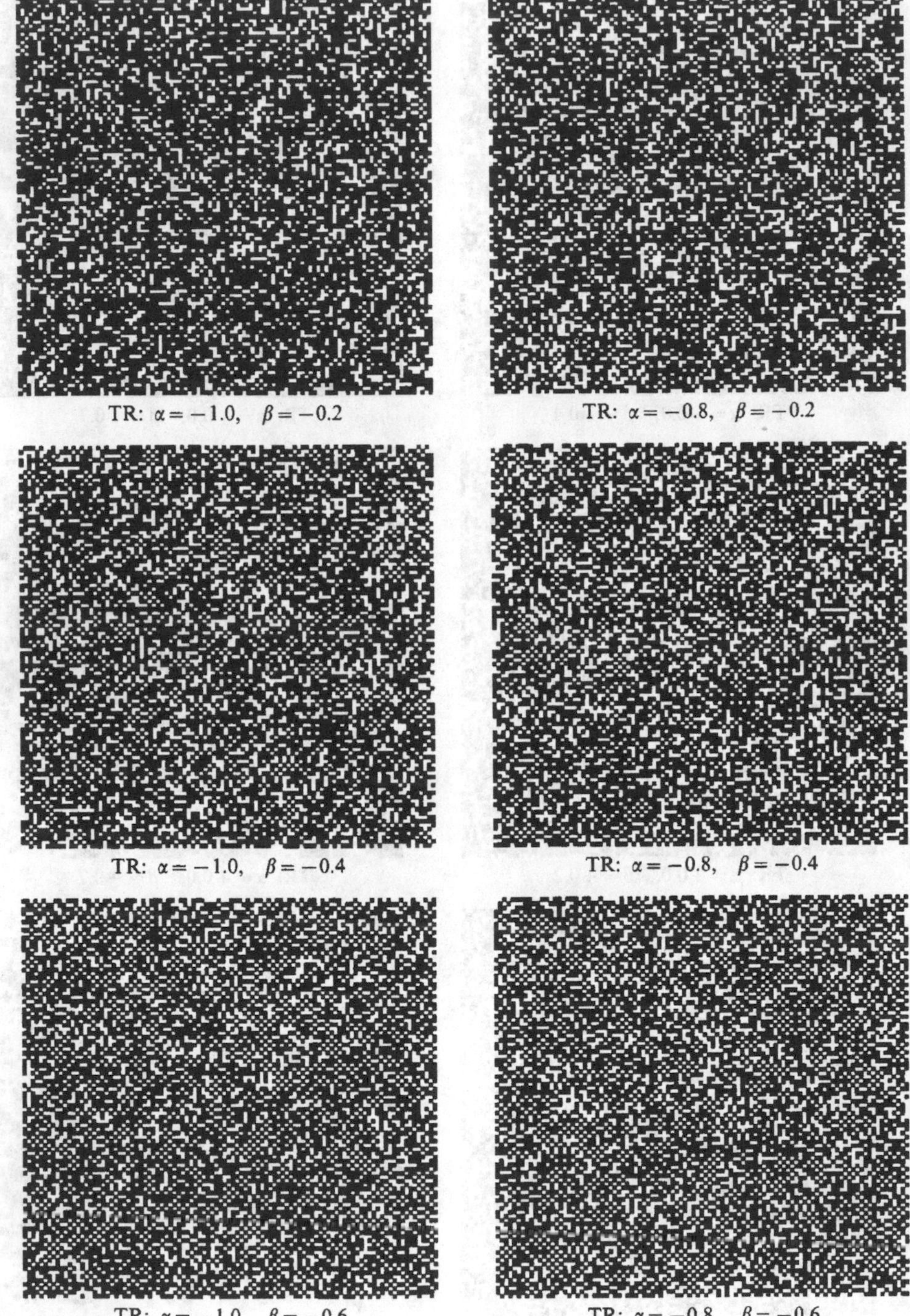

TR: $\alpha = -1.0$, $\beta = -0.2$

TR: $\alpha = -0.8$, $\beta = -0.2$

TR: $\alpha = -1.0$, $\beta = -0.4$

TR: $\alpha = -0.8$, $\beta = -0.4$

TR: $\alpha = -1.0$, $\beta = -0.6$

TR: $\alpha = -0.8$, $\beta = -0.6$

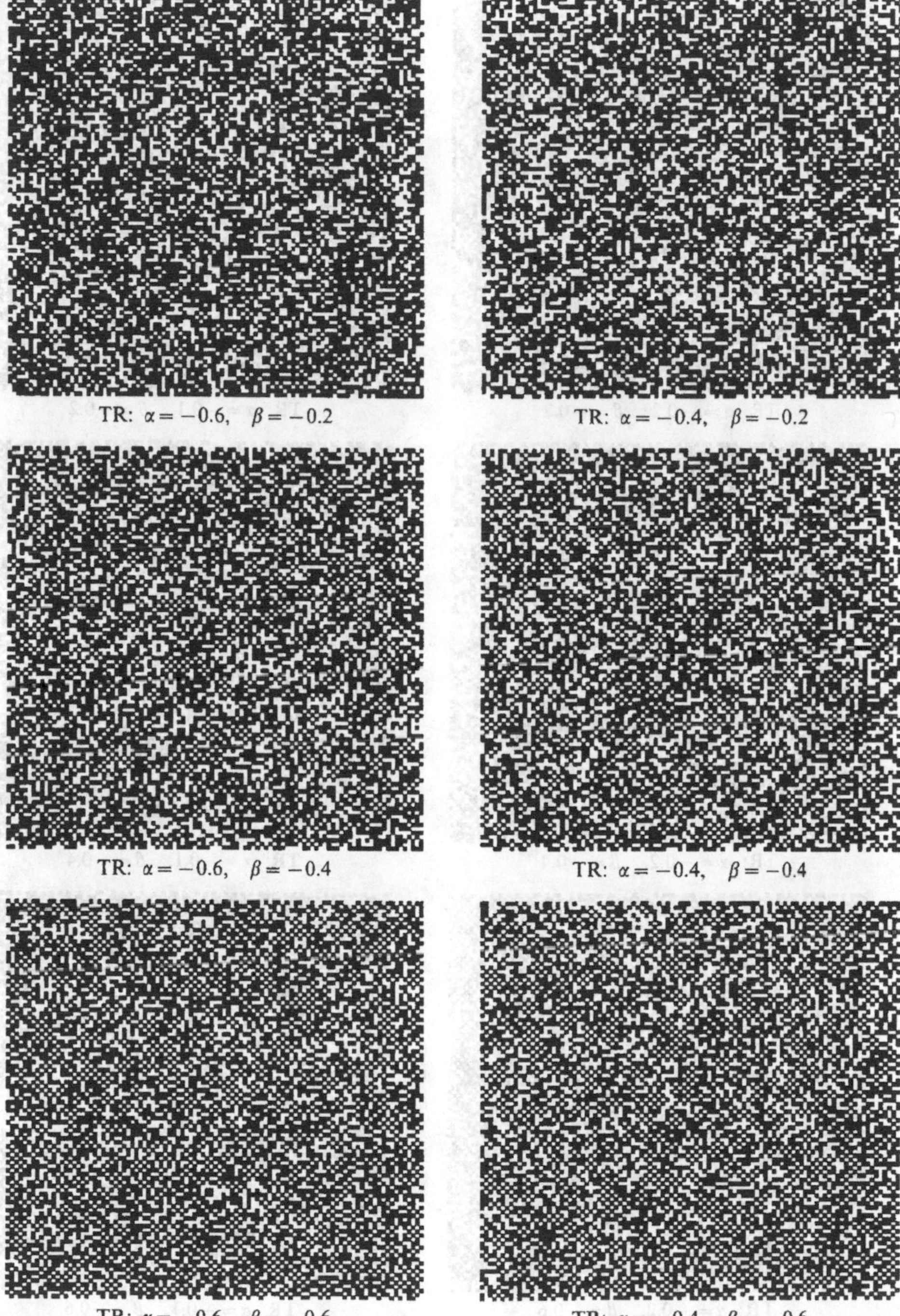

TR: $\alpha = -0.6$, $\beta = -0.2$

TR: $\alpha = -0.4$, $\beta = -0.2$

TR: $\alpha = -0.6$, $\beta = -0.4$

TR: $\alpha = -0.4$, $\beta = -0.4$

TR: $\alpha = -0.6$, $\beta = -0.6$

TR: $\alpha = -0.4$, $\beta = -0.6$

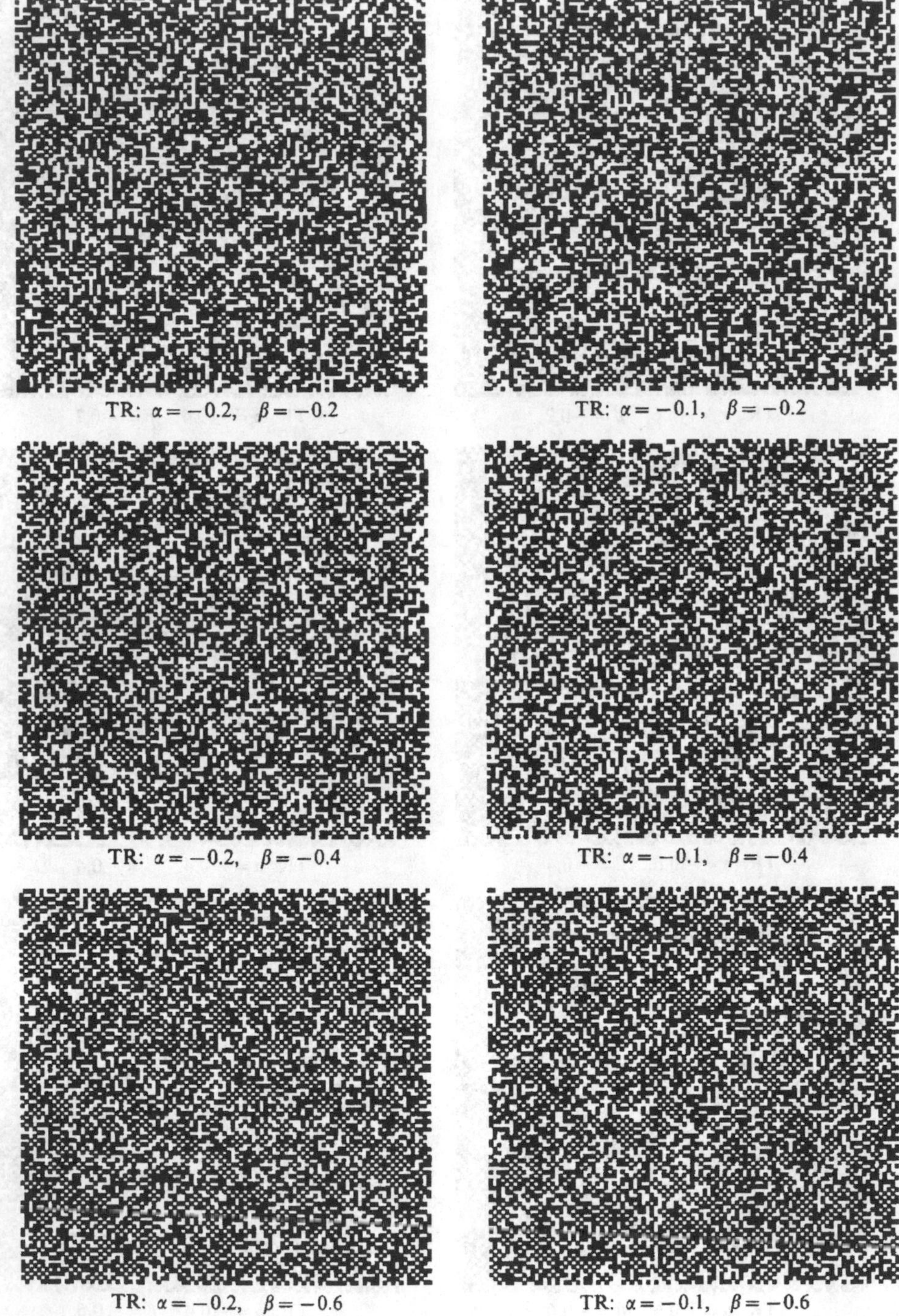

TR: $\alpha = -0.2, \quad \beta = -0.2$

TR: $\alpha = -0.1, \quad \beta = -0.2$

TR: $\alpha = -0.2, \quad \beta = -0.4$

TR: $\alpha = -0.1, \quad \beta = -0.4$

TR: $\alpha = -0.2, \quad \beta = -0.6$

TR: $\alpha = -0.1, \quad \beta = -0.6$

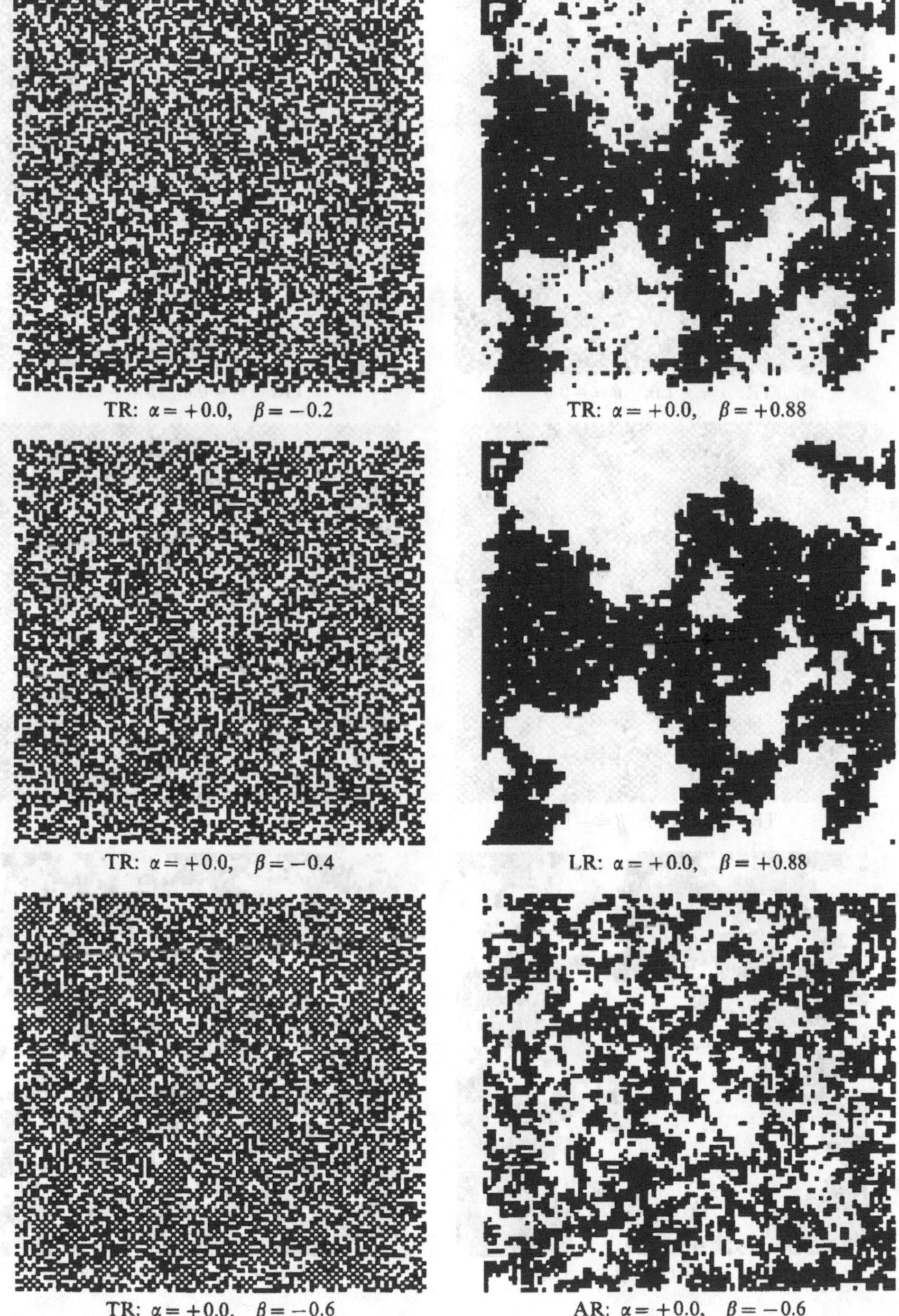

TR: $\alpha = +0.0, \quad \beta = -0.2$

TR: $\alpha = +0.0, \quad \beta = +0.88$

TR: $\alpha = +0.0, \quad \beta = -0.4$

LR: $\alpha = +0.0, \quad \beta = +0.88$

TR: $\alpha = +0.0, \quad \beta = -0.6$

AR: $\alpha = +0.0, \quad \beta = -0.6$

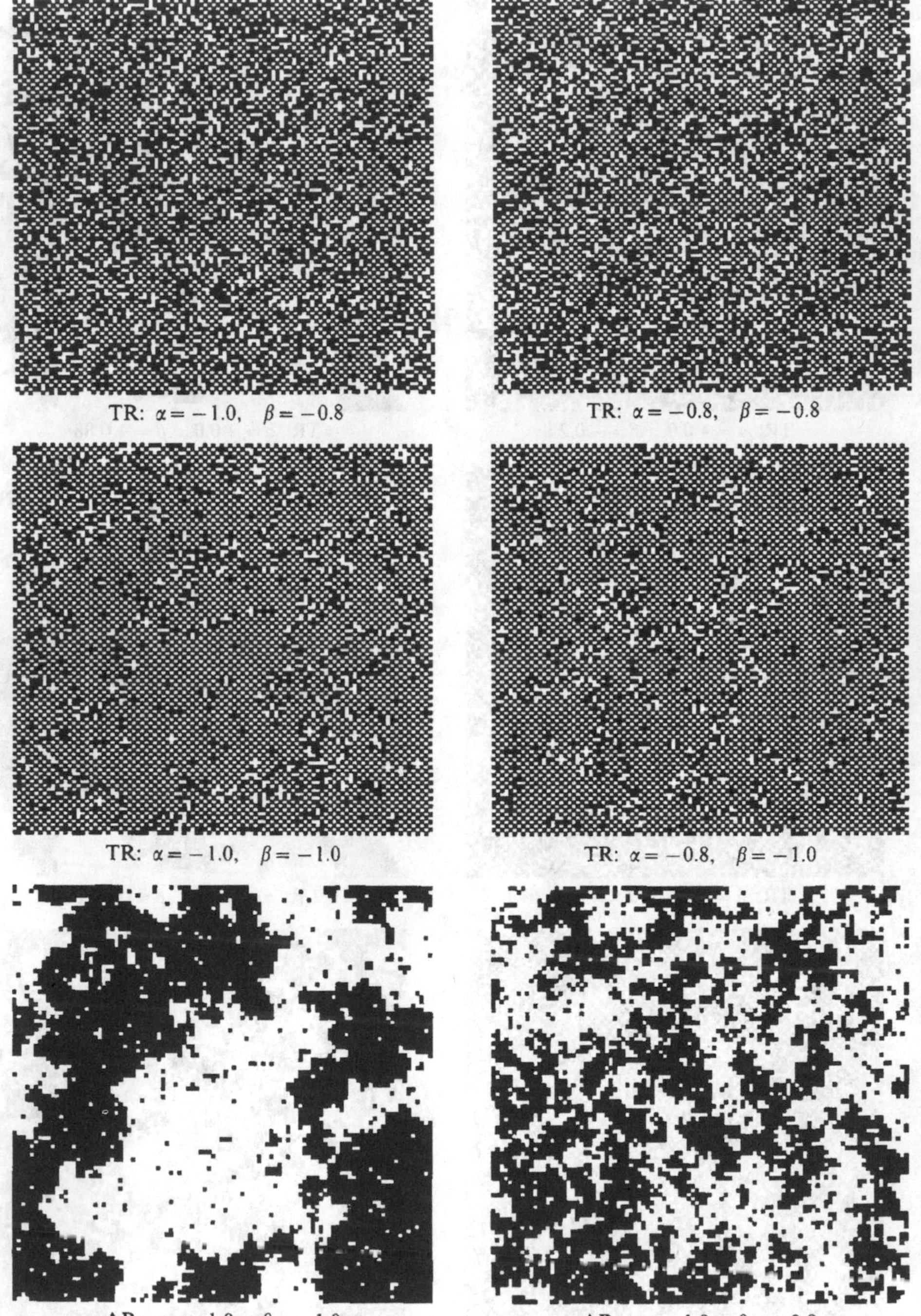

TR: $\alpha = -1.0, \quad \beta = -0.8$

TR: $\alpha = -0.8, \quad \beta = -0.8$

TR: $\alpha = -1.0, \quad \beta = -1.0$

TR: $\alpha = -0.8, \quad \beta = -1.0$

AR: $\alpha = -1.0, \quad \beta = -1.0$

AR: $\alpha = -1.0, \quad \beta = -0.8$

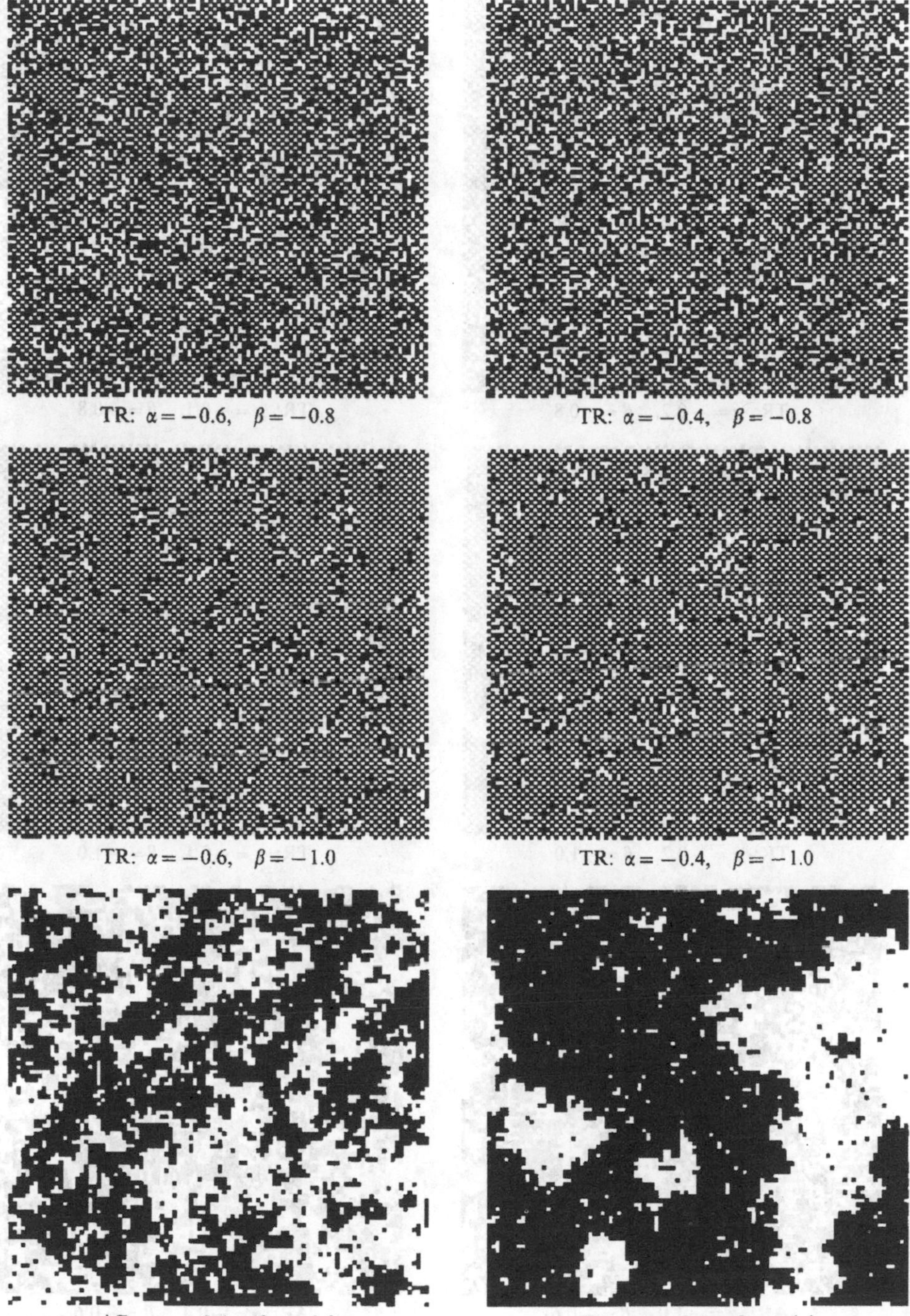

TR: $\alpha=-0.6,\quad \beta=-0.8$

TR: $\alpha=-0.4,\quad \beta=-0.8$

TR: $\alpha=-0.6,\quad \beta=-1.0$

TR: $\alpha=-0.4,\quad \beta=-1.0$

AR: $\alpha=-0.4,\quad \beta=-0.8$

AR: $\alpha=-0.4,\quad \beta=-1.0$

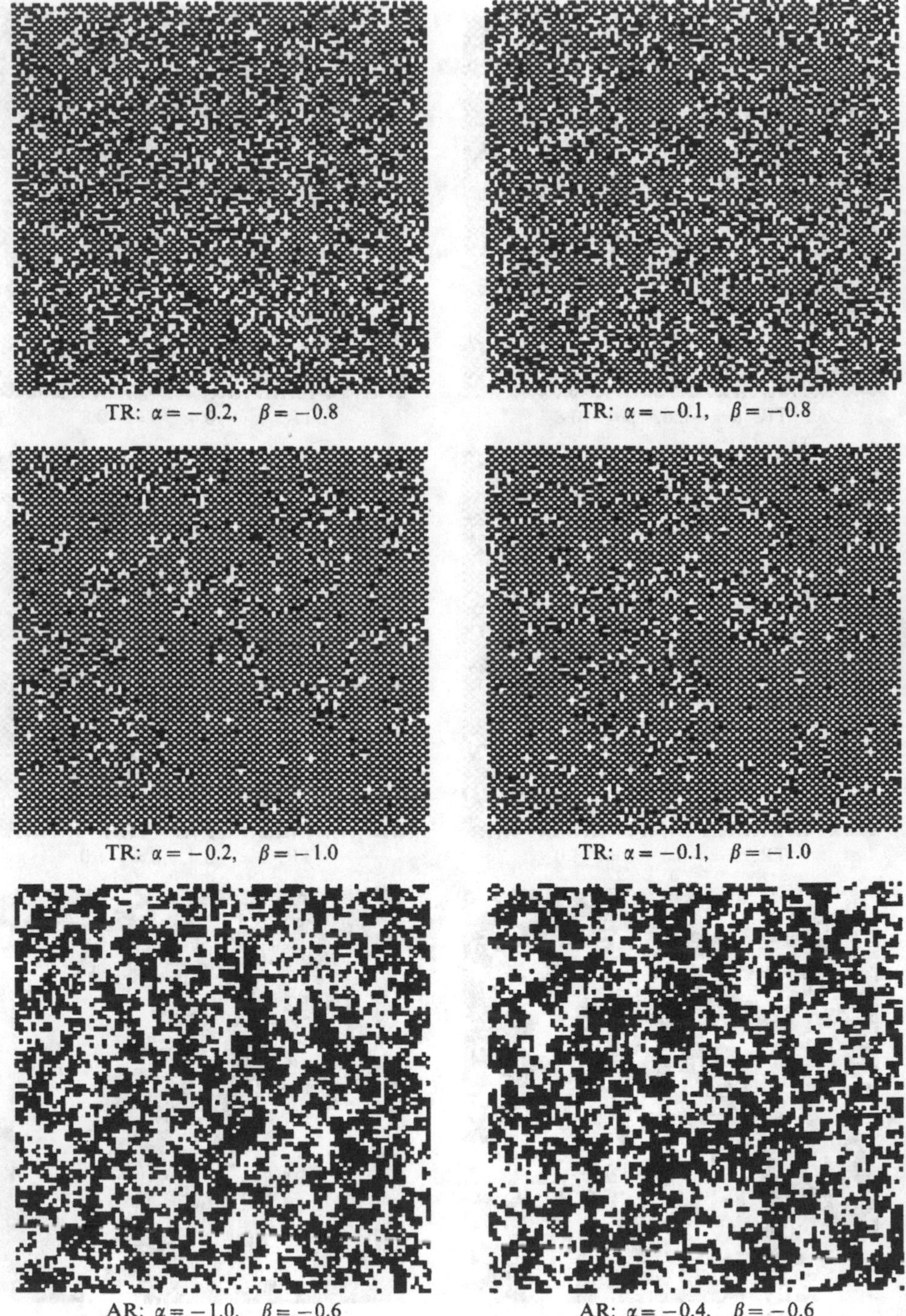

TR: $\alpha = -0.2, \quad \beta = -0.8$

TR: $\alpha = -0.1, \quad \beta = -0.8$

TR: $\alpha = -0.2, \quad \beta = -1.0$

TR: $\alpha = -0.1, \quad \beta = -1.0$

AR: $\alpha = -1.0, \quad \beta = -0.6$

AR: $\alpha = -0.4, \quad \beta = -0.6$

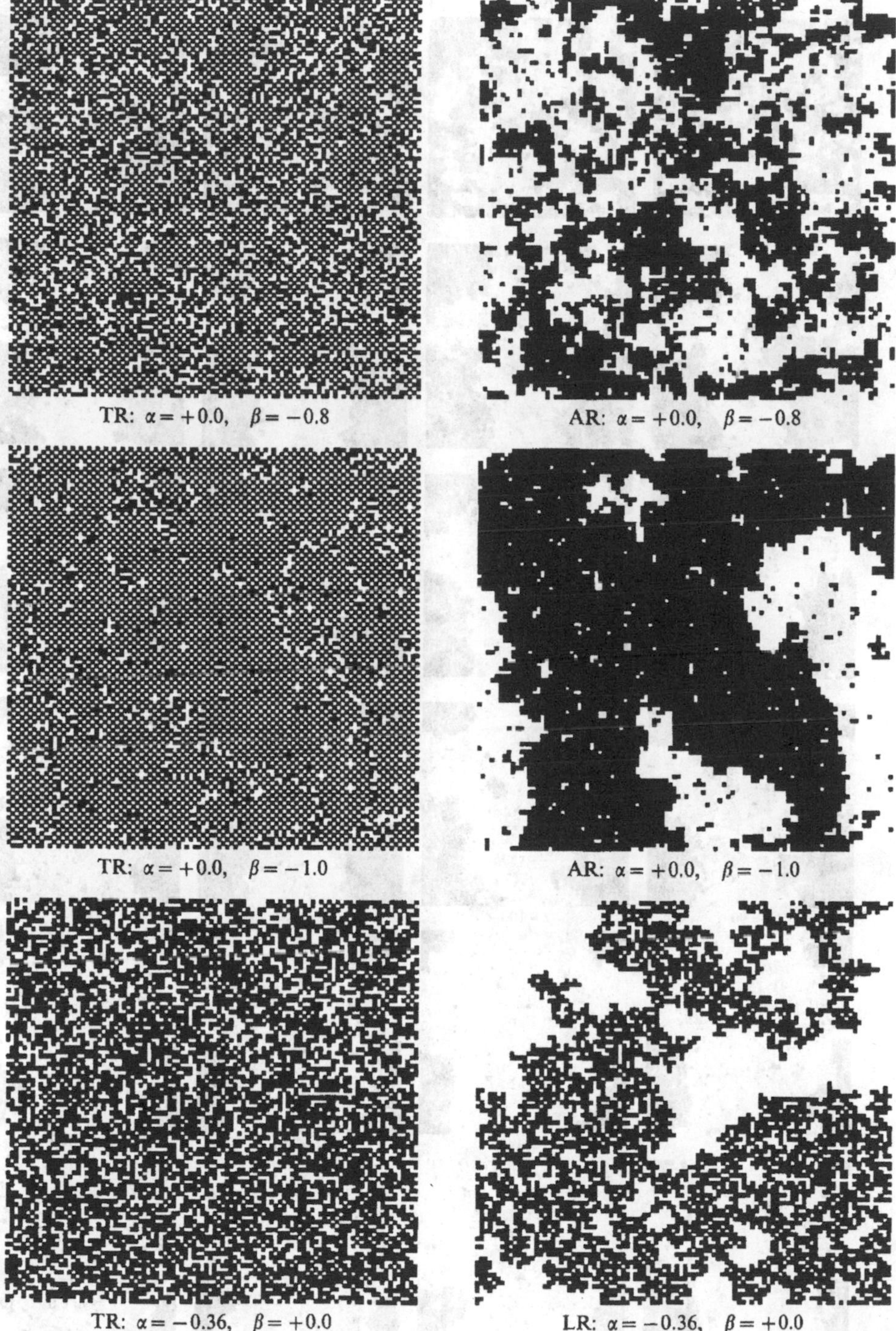

TR: $\alpha = +0.0$, $\beta = -0.8$

AR: $\alpha = +0.0$, $\beta = -0.8$

TR: $\alpha = +0.0$, $\beta = -1.0$

AR: $\alpha = +0.0$, $\beta = -1.0$

TR: $\alpha = -0.36$, $\beta = +0.0$

LR: $\alpha = -0.36$, $\beta = +0.0$

Time–lapse dynamics TR: $\alpha = +0.0$, $\beta = +1.0$

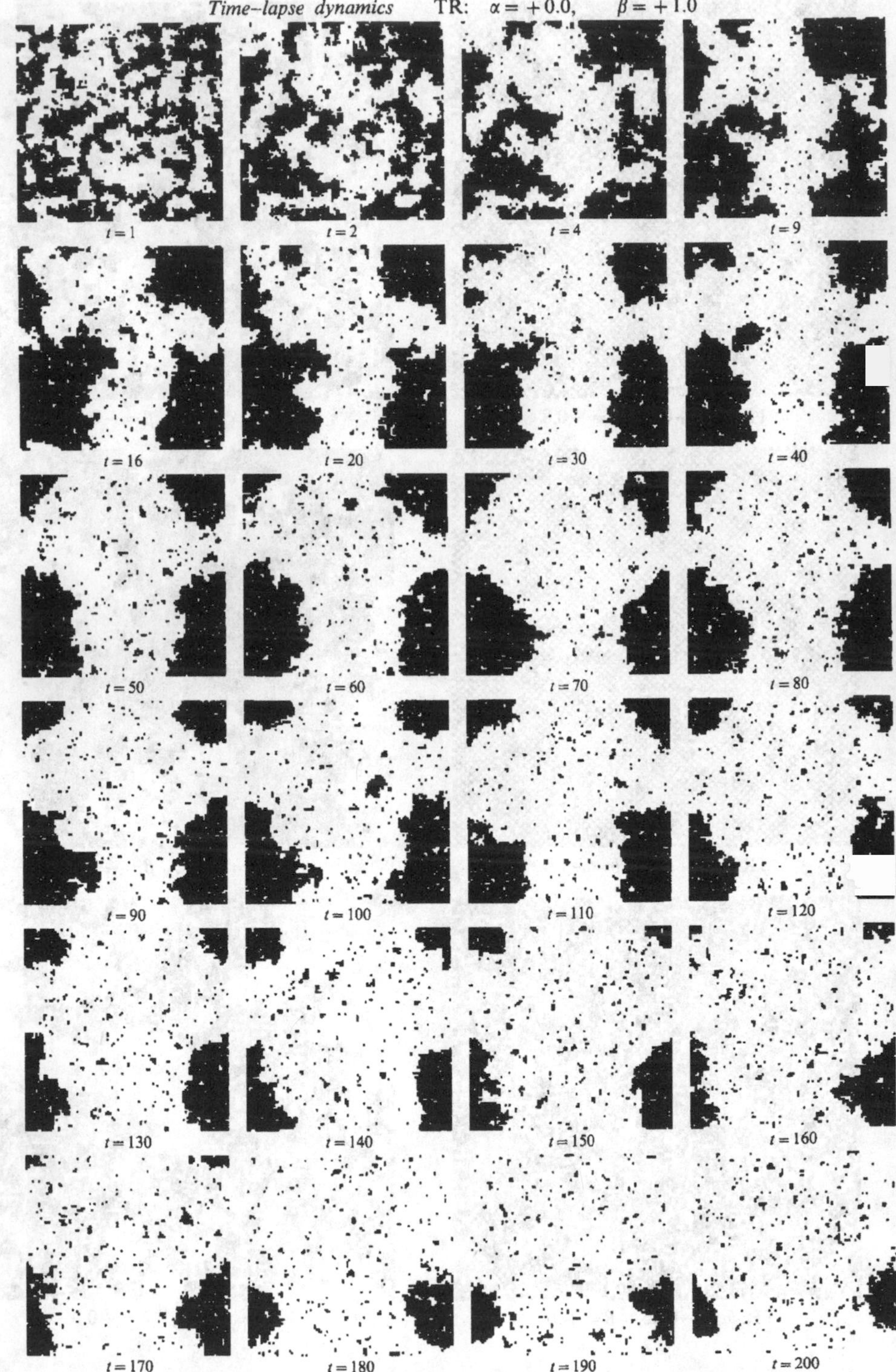

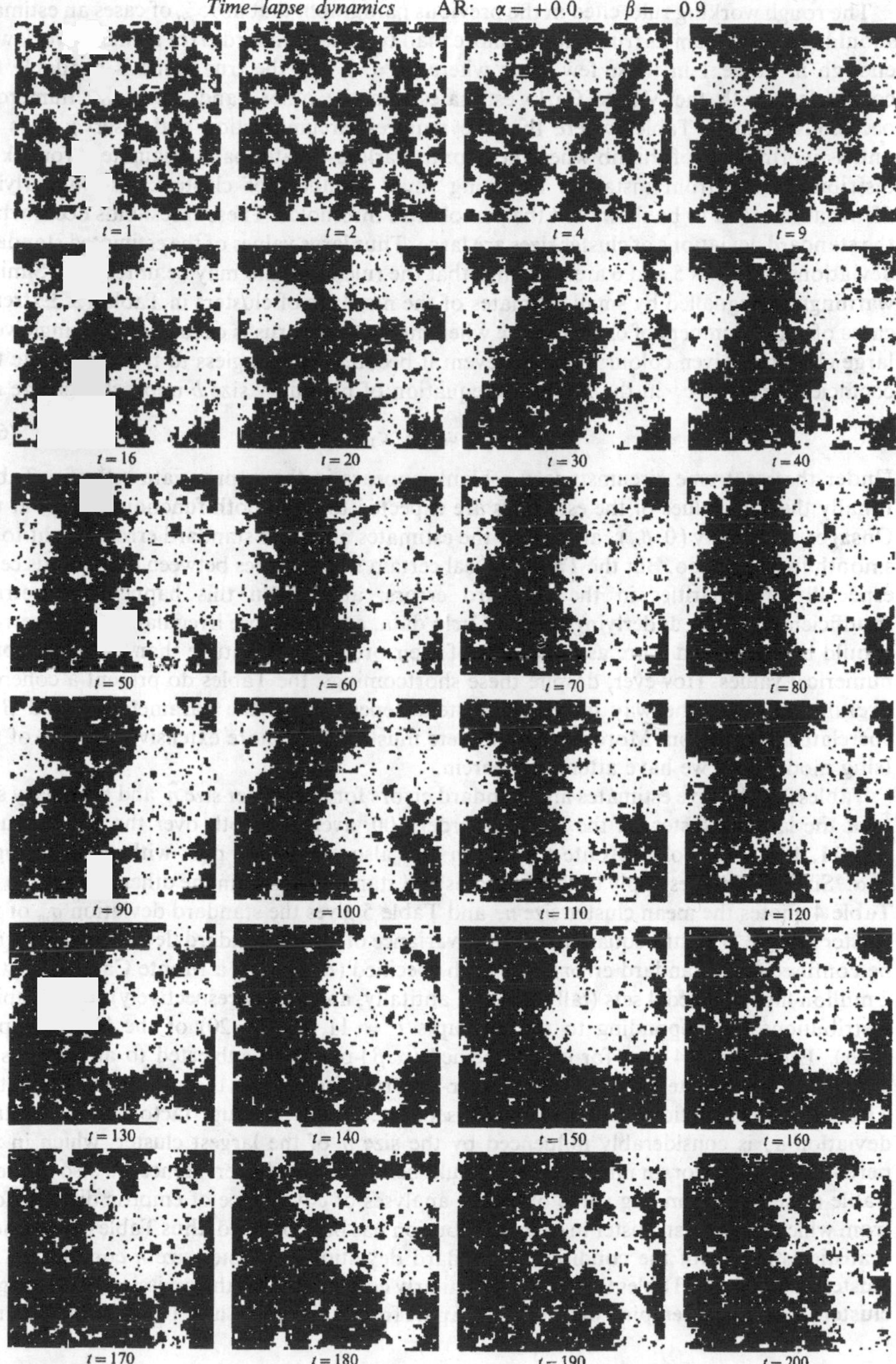

Time–lapse dynamics AR: $\alpha = +0.0$, $\beta = -0.9$

The rough working rule, cited in the previous paragraph, that in 95% of cases an estimate should not differ from its true value by more than twice its standard error seems to work well enough in Table 1 in those few cases where $\beta = 0$ and the true value is available for comparison with the Monte Carlo estimates. On the other hand, there are numerous occasions in these Tables where the rule requires circumspection and reserve. It is an untrustworthy rule of thumb when the Monte Carlo estimates have originated from skew distributions or from instances involving small numbers of clusters. The underlying distributions tend to be skew near the percolation thresholds. These thresholds occur when the standard deviations of cluster sizes are large. Thus large values of the estimated standard deviations in Tables 5 and 6 are warnings that the rule of thumb may be unreliable. Similar warnings are signalled by small estimates of the numbers of clusters in Table 12. Extreme cases of small numbers of clusters arise when most of the torus is occupied by a single very large cluster of given colour; in which event it becomes meaningless to try to estimate the coefficients c_0 and c_1 in the regression equation of boundary size b upon cluster size n:

$$b = c_0 + c_1 n. \tag{6.5}$$

Under these extreme circumstances a blank appears in the appropriate cell of a Table. Finally the true values of the estimates are expected to be smooth functions except at the Onsager singularity $(0, \beta_0)$; and hence the estimates and their standard errors ought to be smooth functions too. But the Tables reveal certain irregularities between contiguous cells, especially irregularities in the standard errors; and, when this happens, it betrays insufficiencies in the underlying Monte Carlo data, and therefore irregular standard errors should be considered more as indicators of their order of magnitude than of their quoted numerical values. However, despite these shortcomings, the Tables do present a coherent overall picture of the way in which clusters depend upon the parameters α and β in correlated percolation. More precise answers must await a more extensive analysis of the Ising model than we have attempted herein.

Tables 2 and 3 give estimates and standard errors for the cluster size $\hat{n}$ and boundary size $\hat{b}$ of the largest cluster on the torus of order 100, averaged both over the 10 interrupts $(t = 11, 12, ..., 20)$ of a Monte Carlo run and also over the 2 runs with opposite signs of α. Similar averages apply to the estimates and standard errors in the other Tables 4 to 12. Table 4 quotes the mean cluster size $\bar{n}$, and Table 5 gives the standard deviation σ_n of the cluster size (i.e. σ_n is the square root of the variance of n; and standard deviation should not be confused with standard error). For each specified (α, β)-pair, a Monte Carlo run (with 2 million flips) yielded 4 sets (tally, cotally, antitally, anticotally respectively) of 10 empiric distributions (corresponding to the interupts $t = 11, 12, ..., 20$) of the bivariate pair (n, b). Each of the 4 sets, for each specified (α, β)-pair, was subjected to an analysis of covariance to separate the effects of the inter-interrupt factor from the intra-interrupt factor. The standard deviation σ_n relates to the separated intra-interrupt factor. This standard deviation σ_n is considerably influenced by the size $\hat{n}$ of the largest cluster, which in the percolative region for an infinite lattice would be an infinite cluster. Hence we also quote in Table 6 the corresponding value of σ_n for analyses of covariance of empiric distributions from which the largest cluster in each distribution has been omitted: thus Table 6 gives more realistic estimates of the conditional standard deviation of cluster size n conditioned on clusters being finite. Tables 9 and 10 are similarly conditioned by the omission of the largest cluster size n in each empiric distribution. Apart from Tables 6, 9, and 10, all Tables refer to the unconditioned empiric distributions. In particular, Table 12 gives the mean sample size of these unconditioned empiric distributions.

Tables 7 to 10 provide estimates of the regression coefficients c_0 and c_1 in (6.5). We emphasize that these coefficients relate to the empiric distribution of clusters on a torus of order 100, each cluster in such a distribution carrying equal weight irrespective of its size. This is in sharp distinction to the regression which would have been obtained had we instead been dealing with the ergodic distribution of (n, b)-pairs for clusters containing a *fixed* vertex on the torus, for these clusters in the ergodic distribution would have been weighted according to their size n: in explanation of this remark, the space-time equivalence inherent in the ergodicity could be interpreted by saying that a randomly selected vertex on a lattice has a probability proportional to n of belonging to a cluster of size n. The familiar asymptotic relation

$$p\langle b\rangle \sim (1-p)\langle n\rangle \quad \text{as} \quad \langle n\rangle \to \infty \tag{6.6}$$

for uncorrelated percolation on a lattice refers to the ergodic distribution conditioned on finite cluster size. However, because of the distinction between the ergodic and empiric distributions of (n, b)-pairs, we must expect that the coefficients c_1 in Table 10 when $\beta = 0$ should be different from $(1-p)/p$ with p given by (6.1). Indeed Tables 7 to 10 serve the quite different purpose of predicting the expected boundary size b of a cluster of given size n on a finite torus of order 100. Moreover, in such a prediction the order of the torus is important because, when n is large, a large cluster tends to wrap itself around the torus: thus a large cluster effectively has a boundary that is composed of the boundaries of the "internal holes" in the cluster (and so a large cluster in the percolative region can have quite a small boundary), whereas a small cluster will have a boundary composed principally of its external boundary. Of course, the two terms "external" and "internal" must here be interpreted loosely and intuitively, since (strictly speaking) there is no vertex at infinity on the torus to afford any rigorous definitional distinction between them. Despite these qualifications, the linear dependence (6.5) of b upon n on the finite torus is very marked: Table 11 quotes $100(1-r^2)$, where r is the correlation coefficient between n and b in the empiric distribution, and so gives the residual variance of b about the regression line expressed as a percentage of the total intra-interrupt variance of b; and it is notable how much of the variation in b is attributable to the concomitant n.

Figure 1 shows the contours for prescribed percentages of black vertices in the left-hand half of the (α, β)-plane. This Figure also shows the approximate thresholds of the region of black percolation and the region of grey percolation. A set of 10 runs at each of the critical points $(\pm 0.36, 0)$ and $(0, +0.88)$ yielded the following estimates for the size of the largest cluster

$$\begin{aligned} &\alpha = +0.36,\ \beta = 0; \quad \hat{n}_W = 3625 \pm 106 \\ &\alpha = -0.36,\ \beta = 0; \quad \hat{n}_B = 3713 \pm 117 \\ &\alpha = 0,\ \beta = +0.88; \quad \hat{n} = 5294 \pm 186. \end{aligned} \tag{6.7}$$

These figures can be used as indicators for the size of the largest black cluster on the black percolation threshold, it being natural to suppose that $\hat{n}_B$ will increase with increasing β; and the behaviour of the threshold can then be roughly sketched in terms of the data in Table 2. The position of this threshold can also be sketched using data from Table 6, since the variance of cluster size has a singularity on the black threshold. We stress, however, that the

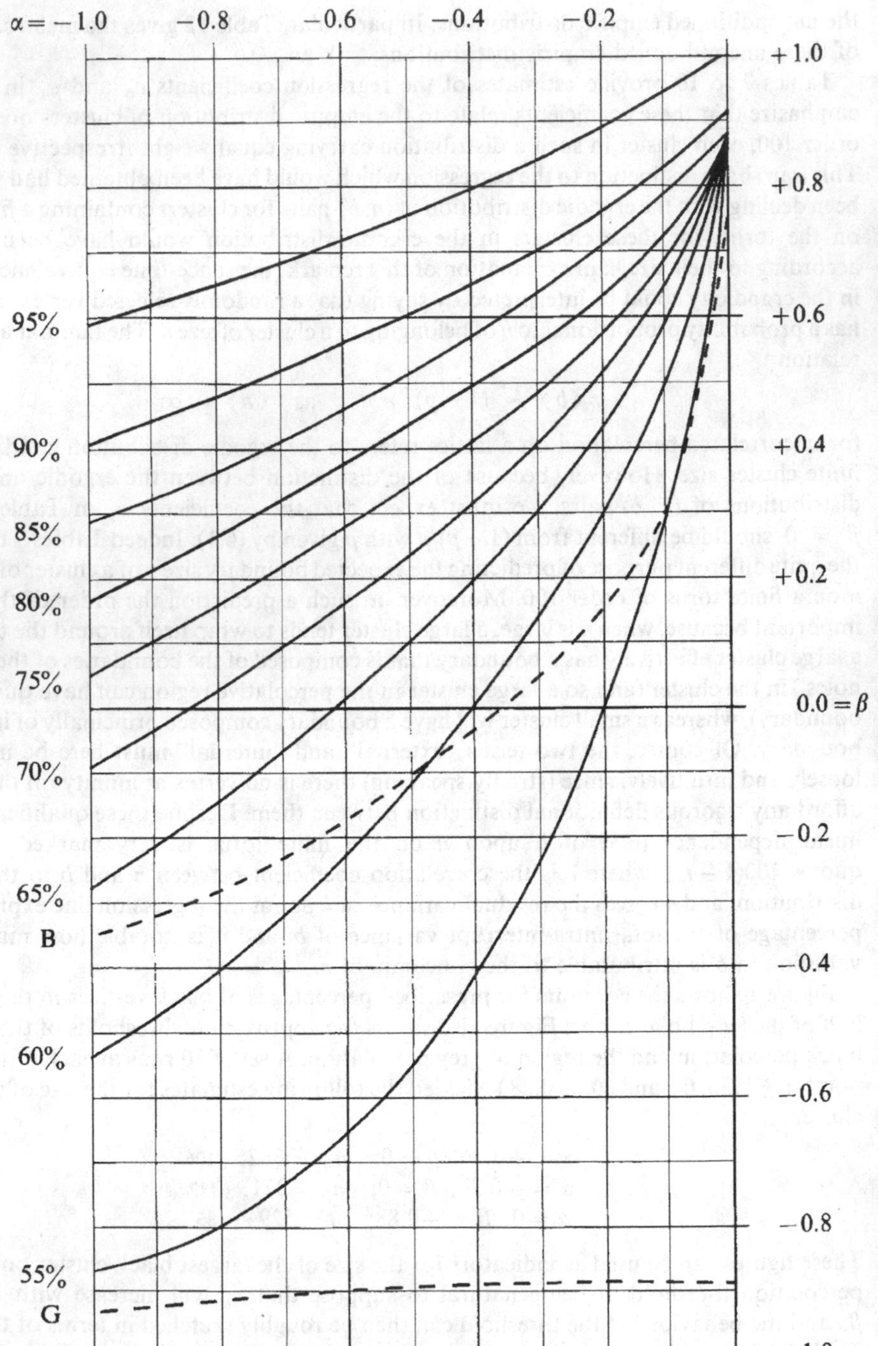

Figure 1. Contours of percentages of black vertices in the region $-1 \leqslant \alpha \leqslant 0$, $-1 \leqslant \beta \leqslant +1$. The curve B is the approximate threshold of black percolation, and the curve G is the threshold of grey percolation given by (6.8).

black threshold in Figure 1 is very approximate. According to a makeshift but plausible argument due to Müller-Hartmann and Zittartz (1977) and Sneddon (1979), the threshold for grey percolation lies along the curve

$$\cosh \tfrac{1}{2}\alpha \;=\; \sinh^2 \beta, \tag{6.8}$$

which is also drawn in Figure 1. Between the two regions of black and grey percolation in Figure 1, there is a band in which no percolation occurs.

On the critical half-line $\alpha = 0$, $\beta \geqslant \beta_0$ above the Onsager point, the Markov field is unstable and the Monte Carlo simulations take much longer to settle down. This can be seen from the diagrams that show time-lapse dynamics for interrupts at $t = 1, 2, 4, 9, 16, 20, 30, 40, ..., 190, 200$. We have ourselves found it very helpful to study these and the other diagrams that show the structure of typical clusters for various values of (α, β). Indeed, one can get quite a good grasp of how the Ising model behaves from prolonged visual inspection, which is more synoptic than Tables 1 to 12. For example, an interesting difference in the structure of the two LR diagrams shows how large clusters become lacier as β decreases along the percolation threshold. (It should be remembered that the graph is a torus; so a single cluster can extend over the top and into the bottom of a diagram, or from one side to the other). Other interesting features can be discerned by careful scrutiny of other diagrams.

The Ising model has always had greater theoretical than practical interest as a model for ferromagnetism. On the other hand, it is a very good practical model for a binary alloy. In this respect the region of grey percolation is especially interesting, in the way it envinces the two phases of binary crystallization. These two phases are separated by boundary lines, on which there is a surface tension that can easily be recognized by looking along the 45° diagonals of the appropriate diagrams with the eye held close to the paper. For further information on polychromatic percolation, consult Halley (1983).

There is, however, a caveat for the model of a binary alloy, namely we have throughout assumed that the structure of the graph $\mathfrak{G}$ remains fixed for all α, β. In practice, this may not be a valid assumption. We began this article with a mention of tin-copper alloy; and in that case the lattice structure changes as the percentage of tin changes:

% tin by weight	*Phase*	*Structure*
0 to 1	α	face-centred cubic
1 to 38	$\alpha + \varepsilon$	
38 to 39	ε	ordered hexagonal
39 to 61	$\varepsilon + \eta'$	
61 to 62	η'	B8 pseudo-hexagonal
62 to 100	$\eta' + \mathrm{Sn}$	(Sn is tetragonal I).

In the mixed phases (e.g. $\alpha + \varepsilon$) there are lumps of α and lumps of ε on a macroscopic scale, these lumps being typically 1 micron to 1 millimetre in size. A single phase lump may be subdivided into grains. So we have three levels of disorder, one on an atomic scale and the other two on a macroscopic scale.

Thus there remains plenty of room for further extensions of percolation theory to include the hitherto unexplored question of critical changes in the structure of the underlying graph.

7. Typographical acknowledgement: OUCS Lasercomp.

This article has been typeset on a Lasercomp at Oxford University Computing Service. We are indebted to Gillian Coates for help with this typesetting.

8. References.

J.E.Besag (1974) "Spatial interaction and the statistical analysis of lattice systems." *J.Roy.Stat.Soc.* **B 36**, 192-236.

K.Binder (1976) "Monte Carlo investigations of phase transitions and critical phenomena." *Phase transitions and critical phenomena* **5 A**, 1-105 (ed. C. Domb and M.S. Green) Academic Press.

A.Coniglio and L.Russo (1979) "Cluster size and shape in random and correlated percolation." *J.Phys.* **A 12**, 545-550.

F.Delyon (1980) "Taille, forme et nombre des amas dans les problèmes de percolation." *Thèse, Université Pierre et Marie Curie, Paris.*

A.G.Dunn, J.W.Essam and J.M.Lovelock (1975) "Scaling theory for the pair-connectedness in percolation models."*J.Phys.***C 8**, 743-750.

A.G.Dunn, J.W.Essam and D.S.Ritchie (1975) "Series expansion study of the pair connectedness in bond percolation models." *J.Phys.* **C 8** 4219-4235.

A.G.Dunn, J.W.Essam and D.S.Ritchie (1975) "Series expansion study of the pair connectedness in bond percolation models." *J.Phys.* **C 8**, 4219-4235.

R. Durrett (1981) "An introduction to infinite particle systems." *Stochastic Processes and Applic.***11**, 109-150.

J.R. Ehrman, L.D. Fosdick and D.C. Handscomb (1960) "Computation of order parameters in an Ising lattice by the Monte Carlo Method." *J.Math.Phys.* **1** 547-558.

J.W. Essam (1972) "Percolation and cluster size." *Phase transitions and critical phenomena* **2**, 197-270 (ed. C.Domb and M.S. Green) Academic Press.

M.E.Fisher (1961) "Critical probabilities for cluster size and percolation problems." *J.Math.Phys.* **2**, 620-627.

A. Flammang (1977) "Percolation cluster sizes and perimeters in three dimensions." *Z.Phys.* **B 28**, 47-50.

C.M. Fortuin and P.W. Kasteleyn (1972) "On the random cluster model. I. Introduction and relation to other models." *Physica* **57**, 536-564.

C.M. Fortuin (1972) "On the random cluster model. II. The percolation model." *Physica* **58** 393-418.

C.M. Fortuin (1972) "On the random cluster model. III. The simple random cluster model." *Physica* **59**, 545-570.

L.D. Fosdick (1959) "Calculation of order parameters in a binary alloy by the Monte Carlo method." *Phys.Rev.* **116**, 565-573.

J.W.Halley (1983) "Polychromatic percolation." *Percolation structure and processes* (ed. G. Deutscher, R. Zallen and J. Adler) *Ann.Israel Phys.Soc.* **5**, 323-352. Adam Hilger, Bristol.

J.M.Hammersley (1957) "Percolation processes. II. The connective constant." *Proc.Camb.Phil.Soc.* **53**, 642-645.

J.M.Hammersley and D.C. Handscomb (1964) *Monte Carlo methods.* Methuen.

J.M.Hammersley (1975) "Rumination on infinite Markov systems." *Perspectives in probability and statistics*, 195-200.

H.Kesten (1982) *Percolation theory for mathematicians.* Birkhäuser, Boston.

R.Kindermann and J.L.Snell (1982) *Markov random fields and their applications. Contemporary Mathematics*, **1**. (Amer.Math.Soc.).

H. Kunz and B. Souillard (1978) "Essential singularity and asymptotic behaviour of cluster size distribution." *J.Stat.Phys.* **19** , 77-106.

D.P. Landau (1977) "Critical behaviour of a bcc Ising antiferromagnet in a magnetic field". *Phys.Rev.* **B 16**, 4164-4170.

P.L.Leath (1976) "Cluster shape and critical exponents near percolation threshold." *Phys.Rev.Lett.* **36**, 921-924.

P.L.Leath (1976) "Cluster size and boundary distribution near percolation threshold."*Phys.Rev.* **B 14**, 5046-5055.

P.L.Leath and G.R.Reich (1978) "Ramification of large clusters near the percolation threshold." *J.Phys.* **C 11**, 1155-1168.

N.Metropolis, A.W.Rosenbluth, N.M.Rosenbluth, A.H.Teller and T.E.Teller (1953) "Equations of state calculations by fast computing machine." *J.Chem.Phys.* **21**, 1087-1092.

J. Moussouris (1974) "Gibbs and Markov random systems with constraints." *J.Stat.Phys.* **10**, 11-33.

E. Müller-Hartmann and J. Zittartz (1977) "Interface free energy and transition temperature of the square lattice Ising antiferromagnet at finite magnetic field." *Z.Phys.* **B 27**, 261-266.

G.F.Newell and E.W.Montroll (1953) "Theory of the Ising model of ferromagnetism." *Rev.Mod.Phys.* **25**, 353-389.

C.M.Newman and L.S.Schulman (1981) "Infinite clusters in percolation models." *J.Stat.Phys.* **26**, 613-628.

L.Onsager (1944) "Crystal statistics I. A two-dimensional model with an order-disorder transition."*Phys.Rev.* **65**, 117-149.

L.Onsager (1949) "The spontaneous magnetization of the Ising model." *Suppl.Nuovo Cimento* **6**, 241-243.

D.C.Rapaport (1978) "Monte Carlo study of the phase boundary of the Ising antiferromagnet." *Phys.Lett.* **65 A**, 147-148.

V.K.S. Shante and S. Kirkpatrick (1971) "An introduction to percolation theory." *Adv.Phys.* **20**, 325-357.

I. Sneddon (1979) "Ising antiferromagnets in a magnetic field." *J.Phys.* **C 12**, 3051-3057.

D.Stauffer (1977) "Exact distribution of cluster size." *Z.Phys.* **B 25**, 391-399.

D.Stauffer (1978) "Scaling assumption for lattice animals in percolation theory." *J.Stat.Phys.* **18**, 125-136.

E. Stoll and C. Domb (1979) "Shape and size of two-dimensional percolation clusters with and without correlations." *J.Phys.* **A 12** 1843-1855.

M.F.Sykes and J.W.Essam (1964) "Critical percolation probabilities by series methods." *Phys.Rev.* **A 133**, 310-315.

E.T.Whittaker and G.N.Watson (1940) *A course of modern analysis*. (4th ed.), Cambridge University Press.

A.P. Young and R.B. Stinchcombe (1975) "A renormalization group theory for percolation problems." *J.Phys.* **C 8**, L535-L540

DIRECTIONALITY EFFECTS IN PERCOLATION

Sidney Redner

Center for Polymer Studies and Department of Physics
Boston University
Boston, Massachusetts 02215

ABSTRACT

The percolation properties of random networks containing resistors (two-way streets) and/or diodes (one-way streets) are considered. The directionality constraints of the diodes are found to lead to novel geometrical behavior. As a simple example, various random cluster models with a preferred direction, such as directed random walks or directed lattice animals, are shown to be anisotropic in character. The critical behavior of directed percolation is then treated and its connection with branching Markov processes is explained. A closely related "reverse" percolation problem, a transition from one-way percolation to isotropic percolation, is introduced. Finally, the geometrical properties of a network containing arbitrarily oriented diodes are treated. Symmetry and duality arguments are applied to yield exact results for certain aspects of its critical behavior.

1. Introduction

Although the classical percolation problem has been extensively studied over the past decade (see, e.g., Stauffer 1979, Essam 1980, Domb 1983, Hammersley 1983 for reviews and comprehensive references), there are still many features of the model which have yet to be addressed. In particular, several interesting and relatively unexplored variations of percolation have been proposed recently, both from the fundamental and applicatory points of view. In this paper, we discuss a particular generalization in which diodes, which connect or permit transport in only one direction along their length, are incorporated into the random network (Fig. 1). In contrast, the conventional bonds of isotropic percolation permit transport in either direction, thus acting as resistors. The directionality constraint gives rise to a broader range of critical phenomena than in classical

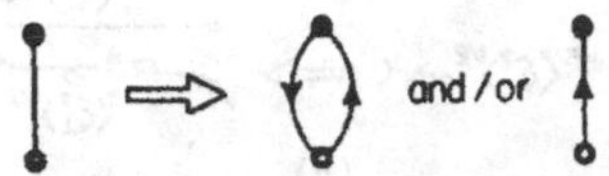

Figure 1

percolation (see, e.g., Kinzel 1983, Redner 1983, and references therein). In particular, if the diodes have a preferred orientation, a random network exhibits anisotropic critical behavior. Furthermore, when the orientation of the diodes is varied, a novel orientational percolation transition occurs.

This paper describes recent work aimed at understanding these new phenomena. First, some simple results for directed walks and lattice animals are outlined to illustrate the effect of a global bias on geometrical models. We then introduce several resistor-diode percolation models of increasing degrees of generality and complexity. The nature of the directed percolation transition and a "reverse" transition -- from one-way to two-way percolation -- are then elucidated. The position-space renormalization group is used to map out the phase diagram of these and more general resistor-diode percolation models and to calculate critical properties. Finally, symmetry and duality are exploited to derive exact results for several critical properties of resistor-diode networks.

2. Simple Phenomenology of Biased Systems

As an example of the effects of a preferred orientation on a geometrical model, consider a discrete random walker who is biased to move preferentially along one direction. For any bias, no matter how small, the average shape of the random walk trajectory is fundamentally changed from the isotropic case (Fig. 2a). With no bias, the root-mean-square displacement scales as N^{ν}, where N is the number of steps in the walk and the correlation length exponent ν has the value 1/2 for any spatial dimension d. On the other hand, the biased walk has an average longitudinal displacement, $\langle\xi_{\parallel}\rangle$, varying as $N^{\nu_{\parallel}}$ with $\nu_{\parallel} = 1$, and an average transverse width $\langle\xi_{\perp}\rangle$ which varies as $N^{\nu_{\perp}}$ with $\nu_{\perp} = 1/2$, again valid for all d.

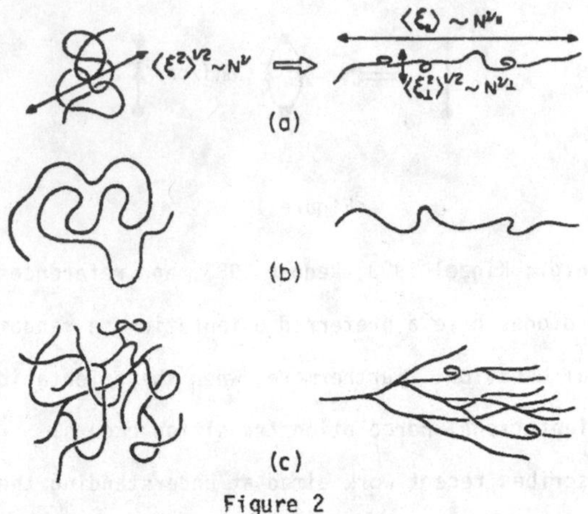

Figure 2

For a self-avoiding walk (Fig. 2b), defined as a random walk in which each site may not be visited more than once, a similar effect occurs. When no bias is present, an excellent approximation for ν is the Flory formula, $\nu = 3/(d + 2)$, valid for $1 \leq d \leq 4$ (see e.g., de Gennes 1979 and references therein). An infinite bias may be represented by a directed self-avoiding walk model. In two dimensions, this problem can be solved by a transfer-matrix method (Redner and Majid 1983). For any dimension, however, one may derive $\nu_\parallel = 1$ and $\nu_\perp = 1/2$. Scaling arguments indicate that these exponent values will also hold for any finite, but non-zero value of the bias. The coincidence of these exponents with those of ordinary directed random walks stems from the fact that for directed self-avoiding walks the excluded-volume effect is no longer fully operative.

More interesting results are obtained for randomly branched structures or lattice animals (Fig. 2c). Early work showed that for an isotropic model with no excluded volume interaction, $\nu = 1/4$ for all d (Zimm and Stockmayer 1949). Once a bias is introduced, heuristic arguments suggest $\nu_\parallel = 1/2$ and $\nu_\perp = 1/4$, again valid for all d (Redner and Yang 1982, Day and Lubensky 1982). With excluded-volume interactions, a Flory-type theory for an isotropic model gives $\nu = 5/\{2(d + 2)\}$, valid for $d \leq 8$ (Issacson and Lubensky 1981). A similar approach for directed animals yields $\nu_\parallel = (d + 11)/\{4(d + 2)\}$ and $\nu_\perp = 9/\{4(d + 2)\}$ for $d \leq 7$ (Redner and Coniglio 1982, Lubensky and Vannimenus

1982). Notice the resemblance between the directed animal in Fig. 2(c) and the shapes of trees in a forest, or the topography of river networks; these two connections have been noted in the literature (Day and Lubensky 1982, Green and Moore 1982). Finally, the Flory method is easily extended to directed percolation, yielding $\nu_\parallel = (d + 9)/\{4(d + 2)\}$ and $\nu_\perp = 7/\{4(d + 2)\}$, valid for $d < 5$.

Despite the extreme simplicity of the Flory approach, these results for $\nu_\parallel$ and $\nu_\perp$ are in excellent agreement with all numerical work. The largest discrepancy is in two dimensions, where the generating function for directed animals can be calculated exactly (Dhar et al. 1982, Nadal et al. 1982, Dhar 1982a, Hakim and Nadal 1983). From these analyses, it is possible to derive $\nu_\perp = 1/2$, but nothing exact for $\nu_\parallel$.

In summary, a preferred orientation strongly affects the large-scale behavior of a wide variety of geometrical models. There is preferential growth along the special axis, and a corresponding transverse shrinking of the structure. This anisotropic behavior can be realized, for example, by diffusion with biased hopping rates (see e.g., Stephen 1981, Derrida and Pomeau 1982, Barma and Dhar 1983, Derrida 1983) or by a purely geometrical problem such as directed animals.

3. Resistor-Diode Percolation Models

The bias discussed in the previous section may also be realized by directed percolation. On the square lattice, for example, each lattice edge is randomly occupied by a bond constrained to "point" upward (if it is vertical) or to the right (if it is horizontal). Such an orientational constraint leads to an anisotropic percolation transition. This behavior and the critical properties of more general networks with arbitrarily oriented diodes are the focus of the remainder of this article.

We therefore introduce a general resistor-diode percolation model which contains directed percolation as a special case. For concreteness, consider again the square lattice whose edges may be occupied by the following bond elements: positive diodes which point either upward or to the right, negative diodes which point in the opposite directions, or resistors (Fig. 3). The occupation probabi-

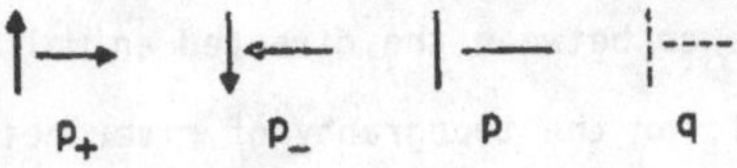

p_+ p_- p q

Figure 3

lities for each of these elements are defined to be p_+, p_- and p respectively, while each bond may be vacant with probability $q \equiv 1 - p - p_+ - p_-$. We shall discuss the features of particular cases of the general model, in increasing order of generality and complexity. These are:

(i) $p, q \neq 0$ and $p_\pm = 0$: classical isotropic percolation.

(ii) $p_+, q \neq 0$ and $p_- = p = 0$: directed percolation

(iii) $p_+, q \neq 0$ and $p_- = q = 0$: "reverse" percolation

(iv) $p_+, p, q \neq 0$ and $p_- = 0$: an "oriented" resistor diode network containing resistors and diodes of one orientation.

(v) $p_\pm, p, q \neq 0$: a "random" resistor-diode network containing resistors and diodes of arbitrary orientation.

(i) Isotropic Percolation: As the concentration of resistors is increased to a critical value, the network undergoes a transition from a non-percolating to an isotropically percolating phase. If a point fluid source is introduced into a percolating network, the fluid will spread isotropically and "wet" a finite fraction of the lattice.

(ii) Directed Percolation: This model exhibits a transition from one-way percolation to no percolation as the concentration of positive diodes is decreased below a critical threshold value (Fig. 4). For $p_+ = 1$, a fluid source at the origin wets the entire first quadrant of the lattice. As p_+ is decreased and q is correspondingly increased, the opening angle ϕ of this wetted region begins to shrink. Near the transition ϕ vanishes as $\xi_\perp/\xi_\parallel$, and fluid flow propagates predominantly along the diagonal. Finally, below the transition, only finite-sized clusters occur. They become increasingly anisotropic in shape as the threshold is approached from below, as discussed in section 2.

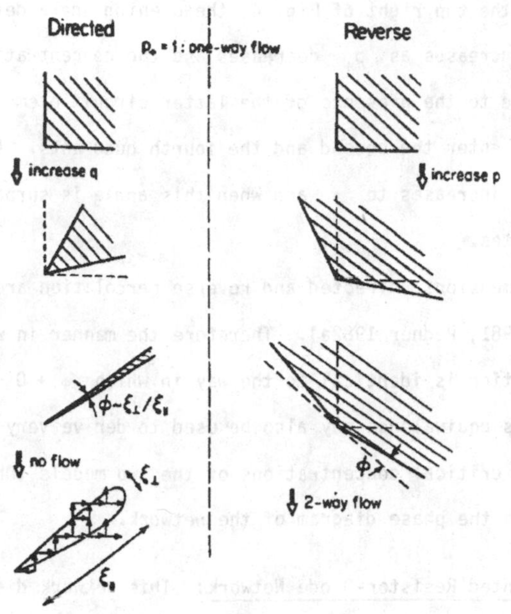

Figure 4

There is a fundamental connection, between directed percolation and branching Markov processes, which is useful in describing diffusion-reaction processes (see e.g., Schlogl 1972, Nicolis and Prigogine 1977, Grassberger and de le Torre 1979, Griffeath 1979). For the anisotropically-shaped cluster at the bottom of Fig. 4, imagine that the diagonal is a "time" axis while one spatial dimension is defined by an axis transverse to the time. Then the equilibrium cluster configuration is equivalent to the trajectory of a diffusing-reacting particle system in one space and one time dimension, where particles may diffuse, split ($\uparrow_{\rightarrow}$) , recombine ($^{\rightarrow}\uparrow$) , or "die". The reaction rates for each of these processes may be arbitrary, and directed percolation is obtained when the rates are chosen to correspond to random bond occupation probabilities (Kinzel 1983). The lifetime of the diffusing particles is equivalent to $\xi_\parallel$ in directed percolation, while the spatial extent of the population is equivalent to $\xi_\perp$.

(iii) <u>Reverse Percolation</u>: There is a transition from one-way to two-way percolation as the concentration of resistors increases to a critical value.

Starting from the top right of Fig. 4, the opening angle defining the region of wetted sites increases as p_+ decreases and the concentration of resistors increases. Due to the presence of the latter circuit element, fluid introduced at the origin may enter the second and the fourth quadrants. As the transition is approached, ϕ increases to π and when this angle is surpassed, the fluid wets all lattice sites.

In two dimensions, directed and reverse percolation are related by duality (Dhar et al. 1981, Redner 1982a). Therefore the manner in which $\phi \to \pi$ in the reverse transition is identical to the way in which $\phi \to 0$ for directed percolation. This equivalence may also be used to derive very accurate rigorous bounds for the critical concentrations of the two models (Dhar 1982b) and qualitatively describe the phase diagram of the network.

(iv) Oriented Resistor-Diode Network: This network displays features from both the directed and reverse transitions, as well as new behavior where the two transitions coalesce. The phase diagram of the network may be described conveniently by a triangle in the composition space spanned by p, q, and p^+ (Fig. 5). This triangle represents the intersection of the half-spaces p, q, $p_+ < 1$, with the plane $p + q + p_+ = 1$. Each corner of the triangle corresponds to a lattice completely filled with one type of bond element, while an interior point corresponds to a lattice with all three elements present. For such a point,

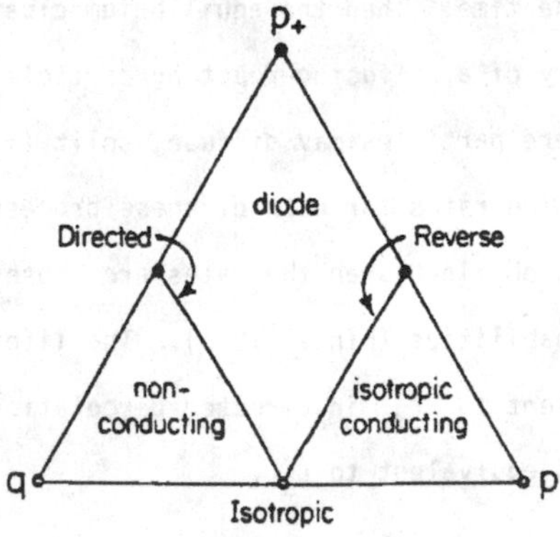

Figure 5

the relative concentration of the ith element is given by the perpendicular distance from the point to the edge opposite the ith corner of the triangle. Directed percolation corresponds to the left edge of the triangle where p_+ and q are non-zero but p is zero, while the right edge of the triangle describes the reverse problem where p_+ and p are non-zero, while q is zero.

A particularly simple, but powerful method to study this model is the position-space renormalization group (see e.g., Reynolds et al 1977, 1980, and Stanley et al. 1982 for a review). This technique was described for isotropic percolation in this author's earlier article of the present volume. For models with directed bonds, the renormalization group can be readily generalized by calculating the probability of traversing a finite cell from bottom to top <u>and</u> vice-versa, and the probability of traversing the cell in one sense only. These probabilities define the recursion relations for p' and $p'_\pm$ respectively, the renormalized probabilities that a cell maps to a resistor in the former case, or to a diode in the latter case.

From these recursion relations, the phase diagram of Fig. 5 is obtained. Two <u>second-order phase transition lines</u> divide the area into three phases characterized by isotropic percolation, one-way percolation, or no percolation. It is useful to think of these situations in terms of forward or reverse conductances, G_+ or G_- respectively, being zero or non-zero. The second-order line on the left is the <u>directed transition</u> where G_+ vanishes with a power-law singularity, while G_- remains zero. The other line is the <u>reverse transition</u> where G_+ is finite and varies smoothly as the line is crossed, while G_- vanishes with a power-law singularity. These two lines meet at the <u>isotropic percolation threshold</u> where G_+ and G_- both vanish. In the theory of critical phenomena, this is a tricritical point as three phases in the system are becoming simultaneously critical (see e.g., Pfeuty and Toulouse 1977). The values of the critical exponents at this point are distinct from the exponents measured anywhere along the two transition lines.

The preceding discussion has centered on geometrical properties of directed percolation problems. Additionally, a network conductivity problem in which the

I-V response of each bond is a general asymmetric function of V may also be treated (Redner 1982b). Two natural choices for this response are $G_+ \neq 0$ and $G_- = 0$, or $G_+ = \infty$ and $G_- \neq 0$, but finite. A wide variety of interesting conductivity properties may be derived.

While the renormalization group method has proved to be of great utility for obtaining global information, the method suffers from the basic flaw that the rescaling procedure is intrinsically isotropic, while the network is geometrically anisotropic along the second-order lines. The construction of a rescaling that correctly treats the anisotropy and gives accurate estimates for both critical exponents, $\nu_\parallel$ and $\nu_\perp$, is still an open problem.

(v) Random Resistor-Diode Network: When negative diodes are included, the three-component triangle discussed above generalizes to a four-component tetrahedron. The renormalization group predicts a symmetric phase diagram (Fig. 6), with a broad range of geometrical transition phenomena. To gain an intuition for the phase diagram, it is helpful to focus on the multicritical line that joins the point marked I to the point marked RM . The former is the isotropic percolation threshold where $p = q = 1/2$ and $p_\pm = 0$, and the latter is "random Manhattan", where $p_\pm = 1/2$ and $p = q = 0$. (This would be the state of midtown Manhattan if one-way street signs were assigned random directions to every block.) This line represents the intersection of the reflection symmetry plane with a dual symmetry plane. The first plane is defined by equal concentrations of positive and negative diodes $(p_+ = p_-)$, while the second is defined by equal concentrations of resistors and vacancies $(p = q)$.

There are two second-order surfaces, emanating upward from the multicritical line, which enclose a wedge-shaped volume. These two surfaces are the continuations of the second-order lines of Fig. 5 into the interior of the tetrahedron. The enclosed volume defines the positive diode phase region of the diagram. An identical structure below the reflection symmetry plane encloses a second volume which defines the negative diode region. The four second-order surfaces meet along the multicritical line, and divide the tetrahedron symmetrically

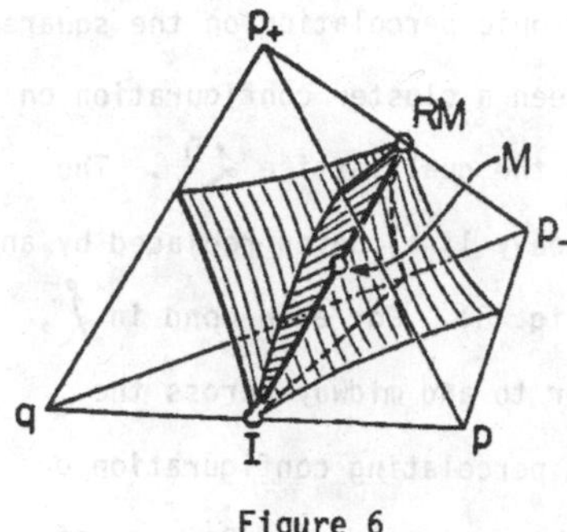

Figure 6

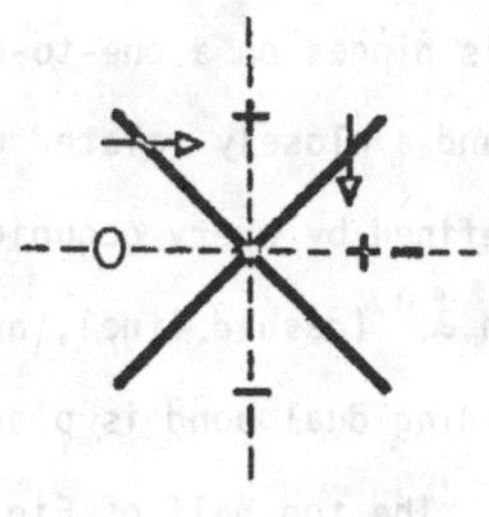

Figure 7

into the positive diode, negative diode, resistor and non-percolating phases.

To help visualize this, consider a two-dimensional projection of the tetrahedron onto a plane perpendicular to the multicritical line (Fig. 7). The four second-order surfaces collapse onto an X-shaped structure that separates the regions of no percolation (0), two-way percolation (+ -), or one-way percolation (either + or -). Moving along a horizontal path in this figure corresponds to exchanging resistor with vacancies, while moving along a vertical path corresponds to exchanging positive and negative diodes. Crossing two of the lines yields a directed transition (0 to +, or 0 to -), while crossing the other two causes a reverse transition (+ to + - or - to + -). The point at the center is a multicritical, fourth-order transition. When this point is approached, there is new critical behavior which is distinct from that of the directed and reverse transitions. The renormalization group predicts that the exponents at this point, and hence along the entire multi-critical line, are just those of isotropic percolation. Thus the critical behavior of random Manhattan is identical to that of isotropic percolation. This striking result is a partial consequence of the duality arguments that follow.

4. Duality for Resistor-Diode Percolation

The dual transformation is an extremely simple but useful tool for deducing exact critical probabilities of a variety of two-dimensional percolation models (see e.g., Essam 1972 and references therein). In this section, we indicate how these arguments can be generalized to random resistor-diode networks.

As a preliminary, we outline duality for isotropic percolation on the square lattice. This hinges on a one-to-one mapping between a cluster configuration on a lattice $\mathcal{L}$ and a closely related configuration on the dual lattice $\mathcal{L}^D$. The mapping is defined by every occupied bond in $\mathcal{L}$ (heavy line) being replaced by an empty bond in $\mathcal{L}^D$ (dashed line), and vice-versa (Fig. 8). For each bond in $\mathcal{L}$, the corresponding dual bond is placed perpendicular to and midway across the original one. The top half of Fig. 8 shows that a percolating configuration on maps into a non-percolating configuration on $\mathcal{L}^D$, and vice-versa. Because of this fact, the critical concentration on the original lattice, p_c, must equal $1 - p_c^D$. Since the square lattice is self-dual by construction, it immediately follows that $p_c = 1 - p_c$, or $p_c = 1/2$.

This mapping can be extended to include networks with diodes (Dhar et al. 1981, Redner 1982a). The rule for transforming resistors and vacancies is kept unchanged, but a diode on $\mathcal{L}$ is replaced by dual diode on $\mathcal{L}^D$ which is rotated by $\pi/2$ clockwise with respect to the original. If the "time" axis on $\mathcal{L}$ is defined to be the upper-right diagonal, then the time axis on $\mathcal{L}^D$ points to the lower right. The dual mapping is therefore defined by $p \leftrightarrow q$ and $p_\pm \leftrightarrow p_\pm$.

Under the extended duality, a non-percolating configuration maps to a two-way percolating configuration just as in pure percolation. However, consider the

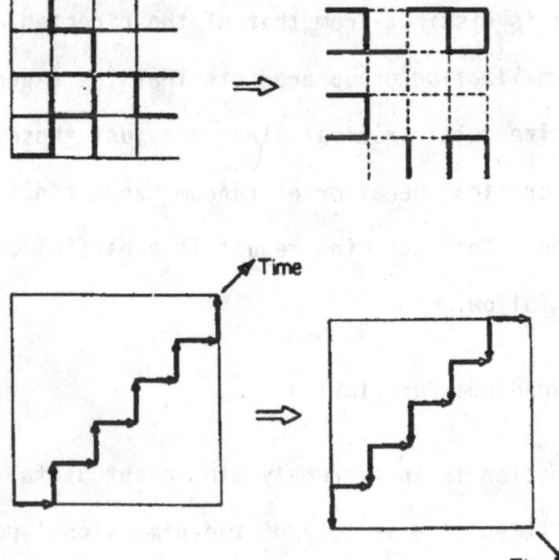

Figure 8

effect of this extended dual transformation on a one-way percolating configuration (bottom row of Fig. 8). On the left, a simple one-way configuration is shown, and its dual counterpart is shown on the right. Since vacancies map to resistors, the empty space on the right-hand side should be completely occupied by resistors. They have not been drawn to illustrate that the one-way path on $\mathcal{L}$ has mapped to a one-way "barrier" on $\mathcal{L}^D$. There may be transport across the barrier to the lower right, but not to the upper left. Thus a one-way percolating configuration maps to another one-way configuration of the same "temporal" sense.

This result implies that the phase diagram must be symmetric across the self-dual plane $p = q$ (the vertical dashed line in Fig. 7). Therefore the intersection of the self-dual plane with the plane of reflection symmetry (the horizontal dashed line in Fig. 7), must be a locus of percolation transitions. Along this line the critical concentration of resistors is given by $p_c = 1/2 - p_+$. This generalizes the result $p_c = 1/2$ for isotropic percolation, to the random resistor-diode network.

Furthermore, the use of duality together with exact results for the functional dependence of the pair-connectedness function shows that the critical exponents anywhere along this line are just those of isotropic percolation (Redner 1982c). This establishes the equivalence between the critical properties of isotropic percolation and random Manhattan.

5. Concluding Remarks

The study of random media has been generally confined to situations where symmetry with respect to reversal of direction holds. Such a system may be described by a random network of resistors. However, when directionality constraints, such as diodes, are introduced, novel geometrical and transport properties result. In the special case where the diodes have an overall orientation, there may be a transition from no percolation to one-way percolation, and also a transition from one-way to two-way percolation. The former is anisotropic in character, and it may be accurately described by a simple Flory-type theory. There is considerably less quantitative information concerning the properties of the latter transition.

A random network containing resistors and arbitrarily oriented diodes displays a wealth of geometrical transition phenomena which may be treated qualitatively by a renormalization group approach. In addition, duality arguments can be formulated to derive exact results for the critical concentration of the network. While some global information has been obtained, there is much less quantitative information available. This may be a promising area for future investigations by rigorous mathematical methods, and by the numerical tools of statistical mechanics such as Monte Carlo and series expansions.

Acknowledgements

I am very grateful to the Institute for Mathematics and its Applications for inviting me to speak at this Conference, and to the members of the Mathematics Institute and the Chemical Engineering Department for their very kind hospitality during the week. This research was supported in part by grants from NSF, ONR and ARO to the Center for Polymer Studies, Boston University.

References

1. Barma, M. and Dhar, D., 1983, J. Phys. C 16, L1451.
2. Day, A.R. and Lubensky, T.C., 1982, J. Phys. A 15, L285.
3. Derrida, B., 1983, preprint.
4. Derrida, B. and Pomeau, Y., 1982, Phys. Rev. Lett. 48, 627.
5. Dhar, D., 1982a, Phys. Rev. Lett. 49, 959.
6. Dhar, D., 1982b, J. Phys. A 15, 1849.
7. Dhar, D., Barma, M., and Phani, M.K., 1981, Phys. Rev. Lett. 47, 1238.
8. Dhar, D., Phani, M.K., and Barma, M., 1982, J. Phys. A 15, L279.
9. Domb, C., 1983, in Percolation Structure and Processes, ed. G. Deutscher, R. Zallen, and J. Adler, Annals of the Israel Physical Society, Vol. 5 (Hilger: Bristol).
10. Essam, J.W., 1972, in Phase Transitions and Critical Phenomena, Vol. 2 (Academic: New York).
11. Essam, J.W., 1980, Rep. Prog. Phys. 43, 833.

12. de Gennes, P.G., 1979, Scaling Concepts in Polymer Physics (Cornell University Press: Ithaca, NY).

13. Grassberger, P., and de la Torre, A., 1979, Ann. Phys. 122, 373.

14. Green, J.E., and Moore, M.A., 1982, J. Phys. A 15, L597.

15. Griffeath, D., 1979, Additive and Cancellative Interacting Particle Systems, Lecture Notes in Mathematics 724 (Springer: Berlin).

16. Hakim, V., and Nadal, J.P., 1983, preprint.

17. Hammersley, J., 1983, in Percolation Structure and Processes, ed. G. Deutscher, R. Zallen, and J. Adler, Annals of the Israel Physical Society, Vol. 5 (Hilger: Bristol).

18. Issacson, J., and Lubensky, T.C. 1980, J. de Physique 41, L469.

19. Kinzel, W., 1983, in Percolation Structure and Processes, ed. G. Deutscher, R. Zallen, and J. Adler, Annals of the Israel Physical Society, Vol. 5 (Hilger: Bristol).

20. Lubensky, T.C. and Vannimenus, J., 1982, J. de Physique 43, L377.

21. Nadal, J.P., Derrida, B., and Vannimenus, J., 1982, J. de Physique 43, 1561.

22. Nicolis, G., and Prigogine, I., 1977, Self Organization in Non-Equilbrium Systems (Wiley: New York).

23. Pfeuty, P., and Toulouse, G., 1977, Introduction to the Renormaliation Group and to Phenomena (Wiley: New York).

24. Redner, S., 1982a, Phys. Rev. B 24, 3242.

25. Redner, S., 1982b, Phys. Rev. B 24, 5646.

26. Redner, S., 1982c, J. Phys. A 15, L685.

27. Redner, S., 1983, in Percolation Structure and Processes, ed. G. Deutscher, R. Zallen, J. Adler, Annals of the Israel Physical Society, Vol. 5, (Hilger: Bristol).

28. Redner, S., and Coniglio, A., 1982, J. Phys. A 15, L273.

29. Redner, S., and Majid, I. 1983, J. Phys. A. 16, in press.

30. Redner, S., and Yang, Z.-R., 1982, J. Phys. A 15, L177.

31. Reynolds, P.J., Klein, W., and Stanley, H.E., 1977, J. Phys. C 10, L167.

32. Reynolds, P.J., Stanley, H.E., and Klein, W., 1980, Phys. Rev. B., 21, 1223.

33. Schlögl, F., 1972, Z. Phys., 253, 14.

34. Stanley, H.E., Reynolds, P.J., Redner, S., and Family, F., 1982 in Real-Space Renormalization, Topics in Current Physics, 30, chap. 7 (Springer: Berlin).

35. Stauffer, D., 1979, Phys. Rep. 54, 1.

36. Stephen, M.J., 1981, J. Phys. C 14, L1077.

37. Zimm, B., and Stockmayer, W., 1949, J. Chem. Phys. 17, 1301.

PERCOLATION AND RELATED SYSTEMS IN EQUILIBRIUM STATISTICAL MECHANICS

J.W. Halley and C. Dasgupta

School of Physics and Astronomy
University of Minnesota
Minneapolis, Minnesota 55455

ABSTRACT

We review the relations of percolation and related geometrical models to equilibrium statistical mechanical models. These relations are of two sorts:
1) Percolation and related geometrical models have been shown to be equivalent to limits of models in equilibrium statistical mechanics in many cases.
2) Many models exist in which a field subject to thermal disorder is added to the geometrically disordered model so that the two kinds of disorder compete.
Throughout this review we attempt to point out places where mathematically exact results might be useful.

1. Introduction

The objective of the workshop which gave rise to this volume was to acquaint the applied mathematics community with problems in disordered media which are of particular interest to technology and physics. Here we review the relationship between percolation and related models of disorder and models of equilibrium statistical mechanics. The disorder models which we will consider are primarily ordinary percolation and the self avoiding walk. In section 2 we define these models and some variations on them, with a review of some of their properties including the empirically observed 'dimensional invariance' of the percolation concentration p_c. The relations of these models to equilibrium statistical mechanics are of two sorts:

1) 'Equivalence relations' establishing models in equilibrium statistical mechanics from which all the desired information about a given geometrical disorder problem can be obtained. Two equivalence relations of this sort are described in

section 3: The bond percolation problem is equivalent in this sense to the statistical mechanical Potts model (as the number of states $s \to 1$.) The self-avoiding walk problem is equivalent to the statistical mechanical n-vector model (as $n \to 0$). We give standard (non-rigorous) arguments for each of these equivalences and cite a large number of similar results for other disorder problems.
2) 'Decorated models' in which one associates a field subject to thermal fluctuations with each site of a disorder model. Such models are discussed in section 4. There we describe notions of 'quenched' and 'annealed' averages, the Harris criterion (establishing when disorder changes critical behavior), scaling in the temperature-concentration plane, the replica trick and localization. We attempt throughout this review to point out places where mathematically exact results might be useful.

2. Percolation and Self-Avoiding Random Walk Models

The standard versions of these models are discussed repeatedly in this volume. To establish notation, consider a lattice (in any integral dimension d). We will generally suppose that the lattice has N sites, periodic boundary conditions and is invariant under discrete lattice translations. These are the standard assumptions concerning lattices studied in solid state physics and crystallography. Such lattices can be characterised by space groups but this will be of no interest here. Little is known about percolation on lattices which do not obey these constraints. We suppose that each site on the lattice is labelled A with probability p and B with probability $1 - p$. One A site is said to be connected to another A site if there is a path in the lattice, connecting the two sites and which contains only A sites. The collection of all the A sites which are connected to any given A site is called a cluster. Suppose that the total number of sites in the lattice is N . Let $P_N(p)$ be the probability that any given site is in the largest cluster. The quantity $\lim_{N\to\infty} P_N(p) = P(p)$ is called the percolation probability. (Here, the limit $N \to \infty$ is taken with the lattice structure unchanged.) Intuitively, $P(p)$ is the probability that a path from one side of the lattice to the other exists. The critical concentration p_c

is the smallest value of p such that $P(p) \neq 0$ for $p > p_c$. One fascination of the percolation problem is that $P(p)$ appears on the basis of numerical evidence to have nonanalytic dependence on p at a nonzero p_c. On the basis of the striking analogy with similar phenomena in phase transitions physicists have defined and estimated critical exponents such as

$$\beta = \lim_{p \to p_c^+} (p - p_c) \frac{d \ln P}{dp} ,$$

so that near and above p_c, $P(p) \propto |p - p_c|^\beta$. To define other critical exponents, let $n_N(s,p)$ be the number of clusters of size s on a lattice of size N. Let $S(p)$ be the mean cluster size in the sense that $S(p) = \lim_{N\to\infty} S_N(p)$ and

$$S_N(p) = \begin{cases} \sum_s' s^2 n_N(s,p) / \sum_s' s\, n_N(s,p) , & p > p_c , \\ \sum_s s^2 n_N(s,p) / \sum_s s\, n_N(s,p) , & p < p_c . \end{cases}$$

Here the prime means that the largest cluster is to be omitted for $p > p_c$. Critical exponents for $S(p)$ are defined by the equations

$$\gamma = - \lim_{p \to p_c^+} (p - p_c) \frac{d}{dp} \ln S(p) ,$$

$$\gamma' = - \lim_{p \to p_c^-} (p - p_c) \frac{d}{dp} \ln S(p) ,$$

so that $S(p) \propto |p - p_c|^{-\gamma}$, $p > p_c$ and $S(p) \propto |p - p_c|^{-\gamma'}$, $p < p_c$. In the analogy with critical phenomena in phase transitions, $P(p)$ behaves like the magnetization, p like the temperature and $S(p)$ as the susceptibility. A connectivity correlation function $g(r_{ij})$ can be defined as the probability that any two sites separated by a distance r_{ij} on the lattice are connected. There is evidence that for a lattice of dimension d this function has the form

$$g(r_{ij}) = K e^{-r_{ij}/\xi} / r_{ij}^{d-2+\eta} \tag{2.1}$$

for r_{ij} much larger than the lattice spacing. K is independent of p and ξ, the coherence length, diverges at $p = p_c$ so that critical exponents

$$\nu = - \lim_{p \to p_c^+} (p - p_c) \frac{d \ln \xi}{dp}$$

$$\nu' = - \lim_{p \to p_c^+} (p - p_c) \frac{d \ln \xi}{dp}$$

can be defined. Estimates of these critical exponents obey scaling relations entirely like those found in phase transitions [1]. The estimated values of the exponents depend only on lattice dimension, suggesting the same type of 'universality' which is found in phase transitions. On the other hand, though values of p_c are known exactly in many cases (see the article by Wierman in this volume), there appear to be no rigorous results confirming this intuitive picture of the percolation transition as a geometrical phase transition. In the next section we will review one approach to establishing this picture on a firmer basis.

So far we have described a model known as the site percolation model. In an equally well known variant, the bond model, one begins with a lattice as before and labels the lines ('bonds') connecting nearest neighbor sites as A (with probability p) or B (with probability 1 - p) . Two sites in the lattice are connected if there is a path containing only A (or only B) bonds between them. Clusters, P(p) , p_c , S(p) , $g(r_{ij},p)$ and other functions can be defined as before. The numerical and other estimates of the critical exponents again depend only on lattice dimension and are the same as those found for site percolation. It is not hard to show that bond percolation on a given lattice is equivalent to site percolation on another lattice, called the covering lattice. This is discussed in the article by Wierman; see also [2].

The site and bond percolation models also display another kind of numerical regularity in the value of p_c for low lattice dimensionalities. (This is sometimes called "dimensional invariance" [2].) For the bond model we form the quantity zp_c where z is the number of nearest neighbors of each site (or coordination number) of the lattice. Numerical results are shown in Table 1. One sees that the product zp_c is approximately the same for all two dimensional and

Lattice	z	$p_c(b)$	$zp_c(b)$	f	$p_c(s)$	$fp_c(s)$
Honeycomb	3	0.6527	1.96	0.61	0.700	0.427
Kagomé	4	--	--	0.68	0.653	0.444
Square	4	0.5000	2.00	0.79	0.590	0.466
Triangular	6	0.3473	2.08	0.91	0.500	0.455
Diamond	4	0.388	1.55	0.34	0.425	0.145
s.c.	6	0.247	1.48	0.52	0.307	0.160
b.c.c.	12	0.119	1.43	0.74	0.195	0.144
h.c.p.	12	0.124	1.49	0.74	0.204	0.151

Table 1. Dimensional Invariants [2]: here $p_c(b)$ $(p_c(s))$ is the percolation concentration for the bond (site problem).

for all three dimensional lattices, but that the two dimensional value differs from the three dimensional one. In the site case one considers the "occupied volume fraction" f at p_c. It is defined as follows: Fill the (d dimensional) lattice with d-spheres centered at each site such that the spheres have the largest possible radius without overlapping. Consider the total volume of the spheres associated with A-sites at p_c. The ratio of this volume to the total volume of the lattice (in the limit of infinite volume) is f. f is shown in another column of Table 1. Again one sees that f depends on d and not on lattice type. (For both types of lattices the values of p_c themselves show no similar regularities). From the point of view of technical applications this dimensional invariance is a very useful property because physical systems with percolation-like properties are often not associated with lattices. Consider, for example, a collection of black (A) and white (B) spheres dumped at random, by a process known as random close packing [3], into a container. Defining nearest neighbor spheres as those in actual contact one can define connectivity and percolation. On the basis of dimensional invariance one can conjecture that percolation will occur when the volume fraction of blacks (A's) is at the value found for lattices of the same dimension. Such a conjecture is confirmed empirically [4]. Zallen [5] has discussed the range of lattice dimensions for which this kind of dimensional invariance can be expected and argues that it cannot extend to arbitrarily large d. We know of no rigorous results on this question.

Recently, many variants on the two basic percolation models have been introduced. Among these are directed percolation, discussed by Redner in this volume, and 'gray' or AB percolation, defined by Hammersley and Mazzarino in this volume. Other variants include polychromatic percolation [6], a variety of A^mB^n percolation models [7], correlated [8] and site-bond models [9]. Space does not permit even definitions of this rich proliferation of models. Generally one can distinguish two directions in modifying the two basic percolation models: One can alter the nature or number of the entities which are identified with lattice sites or bonds or one can alter the definition of connectivity. In the first category of modified models we find directed, polychromatic, correlated and site-bond models while AB and A^mB^n models are in the second category. We will suggest the nature of the possibilities by a brief discussion of 'gray' or AB percolation [7,10]. In this model one considers a lattice with sites labelled A (probability p) or B (probability 1-p) as in site percolation but alters the definition of connectivity: Two sites are said to be connected if a path through alternating A-B-A-B... nearest neighbor sites connects them. One expects that P(p) (defined as before) will be largest at $p = 1/2$ and that there will be two symmetrically positioned critical values $p_c^+(p_c^-)$ above (below) which P(p) will be zero. For lattices which can be divided into two identical sub-lattices, a heuristic argument [7,10] strongly suggests that $P_{AB}(p=1/2) = P_A(p=1/2)$ where the subscripts AB and A identify P as obtained from the AB and site percolation definitions of connectivity respectively. Thus one can conclude that no AB percolation occurs on the square lattice. Numerical results [10] on the triangular lattice suggest that AB percolation exists and has critical properties different from site percolation in two dimensions, whereas AB percolation appears numerically to have the same critical properties as site percolation on the simple cubic lattice [7].

Random walk models may be still more briefly described here. In physics one is particularly interested in self-avoiding walks for which it is non-trivial to calculate the exponent ν associated with the mean square end to end distance $R^2 = \text{constant} \times N^{2\nu}$ in a walk of N steps. We will not be concerned here with

Lévy flights or other types of Markovian random walks with variable step length. (See the papers by Shlesinger and Montroll, and by Hughes and Prager in this volume.)

3. Equivalence to Statistical Mechanical Models

Formulations of non-thermodynamic statistical problems in terms of the thermodynamics of model Hamiltonian systems have played an important role in the theory of geometric critical phenomena. The usefulness of mapping a geometric problem to a Hamiltonian system derives mainly from the fact that a Hamiltonian formulation is, in most cases, the only way of obtaining a continuum field-theoretic description of a geometric problem. A field theoretic description is desirable for several reasons. The structure of the free-energy functional appearing in a field-theoretic formulation determines the universality class of the system and defines the upper critical dimension [11] d_c (which is the dimension above which mean field theory is valid). Furthermore, from a renormalization group [12] (RG) analysis of the free-energy functional, one can identify the relevant scaling variables for the problem, obtain scaling descriptions for most quantities of interest, and determine the exponents appearing in the scaling theory as a power series expansion in $\varepsilon = d_c - d$. For these reasons, Hamiltonian formulations have been very useful in the development of scaling descriptions of geometric critical phenomena. In this section, we briefly discuss Hamiltonian formulations for two of the most widely studied geometric problems: bond percolation and the self-avoiding random walk.

a) Bond percolation

It was pointed out by Kasteleyn and Fortuin [13] that the statistics of clusters in the bond percolation problem are described by the thermodynamics of the s-state Potts model in the $s \to 1$ limit. The s-state Potts model is defined by the Hamiltonian

$$\mathcal{H}(s) = -J \sum_{\langle ij \rangle} (s\,\delta_{\sigma_i \sigma_j} - 1) - H \sum_i (s\,\delta_{\sigma_i 1} - 1) \qquad (3.1)$$

where at each lattice site i, the Potts variable σ_i assumes the values $1,2,\ldots,s$, the sum $\langle ij \rangle$ is over nearest-neighbor pairs of lattice sites, and δ_{mn} represents a Kronecker delta. The partition function of this model can be written as

$$Z = \sum_{\{\sigma\}} \exp[-\mathcal{H}^{(s)}/k_B T] \\ = \exp\left[\frac{s-1}{s}(KN_b + hN)\right] Z' \tag{3.2}$$

where: k_B is Boltzmann's constant and T the absolute temperature; $K = sJ/k_B T$; $h = sH/k_B T$; N_b and N are respectively the total number of bonds and sites of the lattice; and

$$Z' = \sum_{\{\sigma\}} \exp[h \sum_i (\delta_{\sigma_i 1} - 1)] \prod_{\langle ij \rangle} [(1 - e^{-K})\delta_{\sigma_i \sigma_j} + e^{-K}] . \tag{3.3}$$

The connection with the bond percolation problem in which bonds are randomly occupied with probability p is obtained by making the identification

$$p = 1 - e^{-K} . \tag{3.4}$$

Equation (3.3) then reduces to

$$Z' = (1 - p)^{N_b} \sum_{\{\sigma\}} \exp [h \sum_i (\delta_{\sigma_i 1} - 1)] \\ \times \prod_{\langle ij \rangle} [1 + \frac{p}{1 - p} \delta_{\sigma_i \sigma_j}] . \tag{3.5}$$

The product on the right-hand side of Eq. (3.5) can be expanded into a sum of terms each of which corresponds to a graph on the lattice. The factors 1 and $\frac{p}{1-p}\delta_{\sigma_i\sigma_j}$ respectively represent the absence or presence of the bond ij in a graph. The product in Eq. (3.5) generates all possible arrangements of occupied bonds on the lattice. The factors $\delta_{\sigma_i\sigma_j}$ put the Potts variables on all the sites in a connected cluster in the same state. Thus, on taking the trace over the σ variables, a cluster of n_s sites and n_b bonds is weighted by a factor $\left(\frac{p}{1-p}\right)^{n_b} [1 + (s-1)e^{-hn_s}]$. Combining the contributions from the clusters in a graph Y one gets

$$Z' = \sum_Y P(Y) \prod_{n_b, n_s} [1 + (s - 1)e^{-hn_s}]^{G(Y;n_b,n_s)} \tag{3.6}$$

where $G(Y;n_b,n_s)$ is the number of clusters with n_b bonds and n_s sites in the graph Y, and $P(Y)$ is the probability of occurrence of Y. [$P(Y) = p^{N_O(Y)}(1-p)^{N_b - N_O(Y)}$ where $N_o(Y) = \sum_{n_b,n_s} n_b\, G(Y;n_b,n_s)$ is the total number of occupied bonds in Y.] Defining a dimensionless free energy per site in the $s \to 1$ limit as

$$f = \lim_{N\to\infty} \lim_{s\to 1} \frac{1}{N(s-1)} \ln Z \tag{3.7}$$

and using Eq. (3.2) and (3.6), it is easy to obtain the result

$$f = \frac{1}{2} \frac{zJ}{k_B T} + h + \sum_{n_s}{}' K(n_s) e^{-hn_s} \tag{3.8}$$

where z is the number of nearest-neighbor bonds per site of the lattice and $K(n_s) = \lim_{N\to\infty} \frac{1}{N} \sum_{Y,n_b} P(Y) G(Y;n_b,n_s)$ is the average number per site of clusters of size n_s. The $\sum'_{n_s}$ represents a sum over finite clusters.

Equation (3.8) represents the fundamental relation between bond percolation and the $s \to 1$ Potts model. Putting $h = 0$ in Eq. (3.8), we find that the free energy minus $Jz/2k_BT$ represents the average number of finite clusters. The probability $P(p)$ that a particular site belongs to the infinite cluster is given by the 'spontaneous magnetization' of the Potts model

$$\left.\frac{\partial f}{\partial h}\right|_{h=0} = 1 - \sum_{n_s}{}' n_s K(n_s) \equiv P(p) \quad . \tag{3.9}$$

Higher moments of the size distribution of finite clusters are similarly related to higher derivatives of the free energy with respect to the "magnetic field" in the $s \to 1$ limit.

This mapping has been used by many authors [14] to obtain field-theoretic descriptions of the percolation problem by taking the $s \to 1$ limit of continuum generalizations of the s-state Potts model. From these studies, it has been established that $d_c = 6$ for the percolation problem. Using renormalization group methods, the scaling exponents have been calculated to order $\varepsilon^2 (\varepsilon = 6 - d)$. The results are in good agreement with series [15] and Monte Carlo calculations [16] near six dimensions.

b) Self-avoiding random walk

Several years ago, de Gennes [17] showed that the statistics of a single self-avoiding random walk can be obtained from the two-point correlation function of an n-component spin model in the limit $n \to 0$. A brief derivation (following Sarma [18]) of this equivalence is given below.

The n-component spin model on a lattice is defined by the Hamiltonian

$$\mathcal{H} = -J \sum_{<ij>} \vec{S}_i \cdot \vec{S}_j \tag{3.10}$$

where at each lattice site i, there is a spin $\vec{S}_i$ with n components S_i^α, $\alpha = 1,2,\ldots,n$. Each spin has a fixed length $\sqrt{n}$ i.e., $\sum_{\alpha=1}^{n} (S_i^\alpha)^2 = n$. The partition function of this model can be written as

$$\begin{aligned} Z &\equiv \langle \exp(-\mathcal{H}/k_BT) \rangle_A \\ &= \langle \prod_{<ij>} \prod_{\alpha} [1 + \frac{J}{k_BT} S_i^\alpha S_j^\alpha + 1/2 (\frac{J}{k_BT})^2 (S_i^\alpha S_j^\alpha)^2 + \ldots] \rangle_A , \end{aligned} \tag{3.11}$$

where $\langle \ldots \rangle_A$ represents an angular average. In the expansion of the product on the right hand side of Eq. (3.11), one encounters terms of the form $\langle \prod_{\alpha=1}^{n} (S_i^\alpha)^{m_\alpha} \rangle_A$ where m_α's are non-negative integers. To define such averages in the limit $n \to 0$, one introduces a generating function

$$f(k_1,\ldots,k_n) \equiv f(\vec{k}) = \langle \exp\{ i \sum_{\alpha=1} k_\alpha S_i^\alpha \} \rangle_A .$$

Then by the normalization of S_i^α,

$$\nabla^2 f \equiv \sum_\alpha \frac{\partial^2}{\partial k_\alpha^2} f = -nf . \tag{3.12}$$

On the other hand, by direct computation

$$\nabla^2 f = \frac{\partial^2 f}{\partial k^2} + \frac{n-1}{k} \frac{\partial f}{\partial k} \tag{3.13}$$

where $k = \sqrt{(\sum_\alpha k_\alpha^2)}$ and one has used the fact that $f(\vec{k})$ depends only on k. From (3.12) and (3.13)

$$\frac{\partial^2 f}{\partial k^2} + \frac{n-1}{k}\frac{\partial f}{\partial k} = -nf$$

from which the $n \to 0$ limit is obtained:

$$\frac{\partial^2 f}{\partial k^2} - \frac{1}{k}\frac{\partial f}{\partial k} = 0 \ .$$

This has the solution (boundary conditions being $f(0) = 1$, $\partial f/\partial k|_{k=0} = 0$) $f(k) = 1 + ak^2$. Finally $-\partial^2 f/\partial k_\alpha^2|_{k=0} = -2a = \langle (S^\alpha)^2 \rangle = 1$ so that $f(k) = 1 - \frac{1}{2}k^2$. It follows that all the averages of form $\langle \prod_\alpha (S_i^\alpha)^{m_\alpha} \rangle_A$ vanish except $\langle (S^\alpha)_i^2 \rangle_A = 1$ in the $n \to 0$ limit. Thus, if we represent each term in the expansion for Z by a graph, only those terms which correspond to one or more non-intersecting loops will survive in the $n \to 0$ limit. However, since each loop can appear with any one of the indices α , the contribution from these graphs also vanish in this limit, i.e., $Z = 1$ for $n \to 0$. Now, let us consider the two-point correlation function

$$g(R_i,R_j) = \langle S_i^1 \ S_j^1 \rangle = Z^{-1} \ \langle S_i^1 S_j^1 \ \exp\left(\frac{J}{k_B T} \sum_{\langle i,j \rangle} \vec{S}_i \cdot \vec{S}_j \right) \rangle_A \ . \quad (3.14)$$

Since $Z = 1$ in the $n \to 0$ limit, only the numerator needs to be evaluated. Expanding out the exponential as before, we can again represent each term in the expansion by a graph. Since all angular averages except $\langle (S_i^\alpha)^2 \rangle_A$ vanish in the $n \to 0$ limit, the only non-vanishing contributions come from graphs which 1) connect sites i and j ; 2) do not pass through a lattice point more than once; and 3) are labelled by the component index 1 . Also, the weight factor associated with such a graph of N_s steps is $(J/k_B T)^{N_s}$. Thus, it is clear that

$$\lim_{n \to 0} g(\vec{R}_i,\vec{R}_j) = \sum_{N_s} \left(\frac{J}{k_B T} \right)^{N_s} G(N_s,\vec{R}_i - \vec{R}_j) \quad (3.15)$$

where $G(N_s,\vec{R}_i - \vec{R}_j)$ is the total number of self-avoiding walks of N_s steps, connecting sites i and j on the lattice.

This correspondence between the statistics of a single self-avoiding walk and the $n \to 0$ spin model has been generalized by des Cloizeaux [19] to a system of self-avoiding walks that are also mutually exclusive. This mapping has been very useful in the development of scaling theories of dilute and semi-dilute polymer solutions [20]. [Self-avoiding walks provide a good description of the excluded volume effects in the configurational statistics of long polymer chains.] The continuum generalization of the n-component spin model (the n-component Landau-Ginzburg-Wilson free-energy functional [21]) has been extensively studied in the renormalization group literature. Due to the de Gennes - des Cloizeaux equivalence, all the results obtained for the n-component spin model apply directly to the corresponding quantities in the polymer problem if one takes the $n \to 0$ limit. Results obtained from calculations using this equivalence are in good agreement with experiments on polymer solutions [18].

In addition to the two examples discussed above, there are many other geometric problems for which Hamiltonian formulations are available. These include site percolation [22], percolation of trees [23] (i.e. clusters without loops), AB percolation [24], random resistor networks [25], branched polymers [26] and lattice animals [27], self-avoiding rings [28] and polychromatic percolation [29].

We note that the correspondence between a geometric problem and its Hamiltonian counterpart usually involves an analytic continuation of an integer to an apparently unphysical value (e.g. the number of states $s \to 1$ for bond percolation, or the number of components $n \to 0$ for the self-avoiding walks). In most cases, it is not clear whether this analytic continuation is a mathematically well-defined procedure. The results obtained from naive extrapolations usually agree with those obtained from other approaches. Nevertheless, a rigorous analysis of the analytical continuations that appear in these mappings would be very interesting and useful.

4. Dynamical Fields on the Percolation Lattice

In applications to physics and technology one is often interested in more than the geometrical properties of the percolation lattice. There is, in addition, usually a physical variable associated with each site or bond. One such problem is the problem of diffusion of a single particle on the large cluster as discussed by Redner in this volume. Here we will describe another category of problems in which one studies an interacting vector field ϕ^{α} defined at each lattice site (or bond) i if that site is an A-site. In the simplest cases, no field is associated with B-sites. The index α refers to the component of the field, which can be real or complex. A Hamiltonian function $\mathcal{H}(\{\phi\})$ is defined and the equilibrium statistical mechanical problem is to calculate the free energy, schematically written [with $\beta = (k_B T)^{-1}$] as

$$F = -k_B T \,\ell n\, [\operatorname*{Tr}_{\{\phi\}} e^{-\beta\mathcal{H}(\{\phi\})}] .$$

Somewhat more precisely, if the number of sites in the lattice is N, one wishes to calculate the free energy f per site in the limit $N \to \infty$:

$$f = \lim_{N\to\infty} \frac{1}{N} [-k_B T \,\ell n \operatorname*{Tr}_{\{\phi\}} e^{-\beta\mathcal{H}(\{\phi\})}]$$

where $\operatorname*{Tr}_{\{\phi\}}$ is a sum or integral over all allowed values of the field components ϕ_i^{α}. It is the usual practice to assume that f is equal to the value obtained by averaging

$$\frac{1}{N} [-k_B T \,\ell n \operatorname*{Tr}_{\{\phi\}} e^{-\beta\mathcal{H}(\{\phi\})}]$$

over realizations of the percolation model with the limit $N \to \infty$ taken at whatever stage is convenient. Denoting the average over realizations by $\langle \ldots \rangle_r$ one assumes that

$$f = \langle f \rangle_r = \lim_{N\to\infty} \frac{1}{N} \langle -k_B T \,\ell n \operatorname*{Tr}_{\{\phi\}} e^{-\beta\mathcal{H}(\{\phi\})} \rangle_r .$$

We are not aware that these assumptions have been rigorously established for any non-trivial models $\mathcal{H}(\{\phi\})$. Averages of the form $-k_B T \langle \ell n \operatorname{Tr} e^{-\beta\mathcal{H}(\{\phi\})} \rangle_r$ are called 'quenched' averages, in contrast with 'annealed' averages of the form $-k_BT \ln < \mathrm{Tr}\, e^{-\beta \mathcal{H}(\{\phi\})}>_r$. 'Annealed' averages are in general easier to calculate than 'quenched' averages and many of the computational difficulties in the statistical mechanics of these models arise from this fact.

Two particularly straight-forward examples of models of this type are dilute magnets and dilute superconductors. In the magnetic case, the field ϕ_i^α is a spin field whose components at site i are denoted S_i^α . $S_i^1,\ldots,S_i^n$ is a real vector of dimension n and $\mathcal{H}$ is often the n vector model: $\mathcal{H} = -J \sum \vec{S}_i \cdot \vec{S}_j$ where the sum is over nearest neighbor A sites and $J > 0$ is real. (More subtle problems result if couplings extend to large distances or are themselves random.) In the superconducting case ϕ_i^α is the superconducting order parameter (complex) and $\mathcal{H}$ is the Landau-Ginsburg free energy for a superconductor [30].

In these models, one is interested in the interplay between thermally induced phase transitions and the geometrically induced nonanalyticity which occurs in various quantities in the percolation model at $p = p_c$. (It is, of course, important to distinguish between this problem and the discussion of section 3a of this paper, which describes the latter discontinuities in terms of magnetic models.) To illustrate, consider the case in which $\mathcal{H}$ describes the magnetic n-vector model on the A-sites. If the lattice is of high enough dimension ($d > 2$ for Ising models, $d > 2$ for Heisenberg and higher spin dimension models), then one knows that for $p = 1$ a temperature $T = T_c(1)$ exists at which f displays nonanalyticity associated with a magnetic phase transition. For $p < 1$, it seems intuitively possible that this phase transition might still occur at a finite temperature $T_c(p) < T_c(1)$ as long as $p > p_c$ but that no phase transition would occur at any temperature for $p < p_c$. It is, however, also intuitively plausible that the introduction of disorder when $p < 1$ might remove the singularity altogether, rather than merely shift the temperature at which it occurs. Harris [31] presented a physical argument suggesting the conditions under which these two alternatives would occur: He supposed that a local transition temperature $T_c(r)$ could be defined (r labels a site in the lattice) and estimated that the root mean square fluctuation ΔT_c in T_c within a magnetic correlation volume ξ^d

would vary with ξ as $\xi^{-d/2}$. (d is the lattice dimension.) If this is so, then the condition for a well defined coherence length is that $\xi <$ constant $\times (\Delta T_c)^{-\nu} =$ constant $\xi^{+d\nu/2}$ as the temperature T approaches the (spatial average) T_c. (Here ν is the coherence length exponent for the homogeneous magnetic system.) The criterion for a well defined magnetic phase transition in the presence of disorder is thus that $d\nu/2 - 1 > 0$ by Harris' argument. A scaling relation [32], $\alpha = 2 - d\nu$, associates the magnetic specific heat exponent α to this combination of quantities. Thus, Harris' criterion states that a transition continues to exist in the presence of disorder for $\alpha < 0$ and not necessarily for $\alpha > 0$. Turning to common magnetic models, α is believed to be negative for Heisenberg models [33] in 3 dimensions, while for the Ising model, $\alpha = 0$ in a two dimensional lattice [34] and $\alpha > 0$ for three dimensions [35]. Several approximate renormalization group calculations [36-38] also suggest the stronger notion that, when $\alpha < 0$, the critical behavior at the transition is characterized by the same critical exponents which characterize it when $p = 1$ (i.e. in the 'pure' or 'ordered' case). This result occurs because one finds that the variable describing the disorder is 'irrelevant' in the renormalization group sense [39] when α for the pure system is < 0. Systems with α for the pure system greater than zero sometimes appear to exhibit sharp phase transitions in the presence of disorder ($p < 1$) but with exponents different from those of "pure" case and with α for the disordered system less than zero.

When there is a transition, one would like to know the p dependence of $T_c(p)$. Some results for $T_c(p)$ are shown in Figure 1 for the Ising and Heisenberg models. Rigorous bounds on $T_c(p)$ for the diluted Ising model on a two dimensional square lattice have been reported by Bergstresser [40] and by Fisher [41]. A heuristic argument due to Lubensky [42] appears to give the correct dependence of $T_c(p)$ on $p - p_c$ near $p = p_c$. It is based on the de Gennes [43], Skal and Shklovskii [44] picture of the large cluster near p_c as a network of multiply connected 'nodes' tied together by one dimensional 'links'. (See the first paper by Redner in this volume.) In this picture, the 'links' contain an average of $\ell(p)$

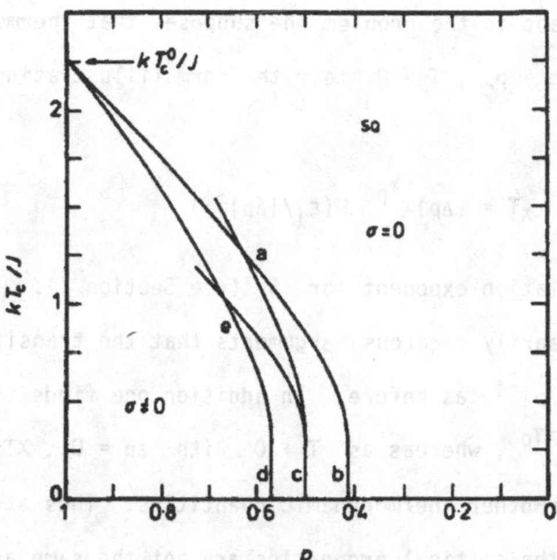

Figure 1: Rigorous Bounds on $T_c(p)$ for the two dimensional dilute ferromagnetic Ising model on a square lattice as found by T. Bergstresser. $T_c(p)$ lies between the lines $k_BT^\circ_C/J$-a-c and $k_BT^\circ_C/J$-e-c. From reference 40.

sites and $\ell(p)$ diverges as $p \to p_c^+$ as $|p - p_c|^{-\phi}$. The Lubensky argument suggests that a magnetic phase transition can occurs only when the distance $\xi_1(T)$ over which spins along the link remain correlated becomes equal to $\ell(p)$. Thus, up to numerical factors, the argument suggests $\xi_1(T_c(p)) = \ell(p)$. Calculation of $\xi_1(T)$ is a problem in the statistical mechanics of one dimensional spin chains and in many cases the answers are well known. At low temperatures, for the Ising model $\xi_1 = \xi_0 e^{J/k_BT}$ and for continuous spin models [45] $\xi_1 = \xi_0' J/k_BT$. Thus the Lubensky argument gives $T_c(p) \propto -1/\ell n|p - p_c|$ for the Ising model and $T_c(p) \propto |p - p_c|^\phi$ for vector spin models. These results are consistent with more detailed estimates [40,41] (if one takes $\phi = 1$ for the Heisenberg case: but see below).

All these results remain, however, non-rigorous and the argument based on the "node" - "link" model may indeed be inadequate, as discussed by Redner in this volume. One may phrase this same argument as a scaling hypothesis for the thermodynamic quantities [46]: Assuming that only the variables $\xi_1(T)$ and

$\Delta p = p - p_c$ are important in the problem one supposes that thermodynamic quantities near the point $p = p_c$, $T = 0$ take the form (illustrating for the susceptibility χ)

$$\chi T = (\Delta p)^{-\gamma_p} F(\xi_1/|\Delta p|^{-\phi})$$

where γ_p is the percolation exponent for S (see Section 1). It follows by standard (but not necessarily rigorous) arguments that the transition occurs when ξ_1 = constant $\times |\Delta p|^{-\phi}$ as before. In addition one finds that, as $\Delta p \to 0$ with $T = 0$, $\chi T \propto |\Delta p|^{-\gamma_p}$, whereas as $T \to 0$ with $\Delta p = 0$, $\chi T \propto (\xi_1)^{\gamma_p/\phi}$, with similar results for other thermodynamic quantities. Thus at $p = p_c$, this argument suggests that the critical properties are not the same as they are in the pure case. This unique behavior at the point $p = p_c$, $T = 0$ in the $T - p$ plane leads to its designation as a multicritical point [46].

This introduction of ϕ through a scaling hypothesis avoids the identification of $|\Delta p|^{-\phi}$ with a geometrical length associated with the Skal-Shklovskii-de Gennes picture of the large cluster. In this context ϕ is called a "crossover exponent". For s-state Potts models, Wallace and Young [47] have shown that $\phi = 1$ for all lattice dimensions and to all orders in perturbation theory. In this formulation, however, ϕ is not necessarily the same for all Hamiltonians $\mathcal{H}(\{\phi\})$ (as suggested by experiments on magnetic systems [48].)

Coniglio [49] has described a renormalization calculation of ϕ in which, for p just below p_c , the bonds of the largest cluster are replaced by a single bond of size $\xi_p \propto |p - p_c|^{-\nu_p}$. At low temperatures, this leads to a different geometrical identification of the length ℓ than that suggested by the Lubensky argument: Qualitatively, in Coniglio's theory, ℓ describes the effective one dimensional length over which thermal correlations need to spread in order to create long range spin correlations across the cluster. In this context, it is plausible that the geometrical identity of ℓ should vary depending on the nature of the spins in $\mathcal{H}$. In particular for Potts spins (including the Ising model as a special case), Coniglio suggests that ℓ is properly identified as the number of bonds which are singly connected at each end in a minimal path across the cluster. For n-vector models (with $n > 2$) he identifies ℓ as the electrical resistance R of the cluster divided by the resistance r per connected bond. Coniglio's calculated values for ϕ are quite close to experimentally estimated ones and his theory suggests resolutions to several previously apparently paradoxical features in this problem.

Still another approach to the problem of the diluted magnet introduces a useful technique, known as the "replica trick" which has been widely used in the statistical mechanics of disordered systems. Basically the idea is to use the identity

$$< \lim_{n \to 0} \frac{Z^n - 1}{n} >_r = < \ell n\, Z>_r$$

to alleviate the problem associated with the calculation of "quenched" averages as discussed earlier. One evaluates $<Z^n>_r$ for positive integer n and then continues the resulting expressions smoothly to $n \to 0$. This is called the "replica trick" because, if n is a positive integer $(\beta \equiv 1/k_B T)$

$$Z^n = \underset{\{\phi^1\}}{\mathrm{Tr}}\ \underset{\{\phi^2\}}{\mathrm{Tr}} \ldots \underset{\{\phi^n\}}{\mathrm{Tr}}\ \exp\{-\beta \sum_{\alpha=1}^{n} \mathcal{H}(\{\phi^\alpha\})\} .$$

Thus Z^n can be regarded as the partition function of the system "replicated" n times. Although it is used widely, the validity of the "replica trick" procedure is still open to mathematical question [50]. Stephen and Grest [51] have used this procedure in the dilute magnet problem we have been discussing. These authors consider the Ising model in a field H on a bond diluted lattice. If the Ising spins are σ_i^α, where i specifies the site and α the "replica", then Stephen and Grest show that with N_B the total number of bonds on the lattice)

$$<Z^n>_r = \mathrm{Tr}\ \exp\{-\beta\mathcal{H}_{eff}\}\ (1 - p)^{N_B} ,$$

where

$$\beta\mathcal{H}_{eff} = -\sum_{<ij>,p} K_p\, \mu^{(p)}(i,j) - H \sum_{i,\alpha} \sigma_i^\alpha$$

and, at low temperatures, the $n \to 0$ limit gives

$$K_p \to \ln\left(\frac{1}{1-p}\right) - 2p \exp(-2J\beta) .$$

Here

$$\mu^{(1)}(i,j) = \sum_{\alpha=1}^{n} \sigma_i^\alpha \sigma_j^\alpha \quad , \quad \mu^{(2)}(i,j) = \sum_{\alpha<\beta} \sigma_i^\alpha \sigma_i^\beta \sigma_j^\alpha \sigma_j^\beta \; ,\ldots$$

Thus, as one approaches $n = 0$ along the line $T = 0$, the Hamiltonian approaches a 2^n ($= 1$ as $n \to 0$) state isotropic Potts model. When $T \neq 0$, the $n \to 0$ limit leads to an anisotropic 2^n-state Potts model. This formulation was used by Wallace and Young [47] to obtain the result $\phi = 1$ cited earlier.

In a similar framework, one can consider some versions of the localization problem, which are model descriptions of electrons on a disordered lattice. One such model ("quantum percolation" [52]) can be described by fermion creation and annihilation operators [53] $c_i^\dagger, c_i$ at each site with

$$\mathcal{H} = \sum_i E_i \, c_i^\dagger \, c_i + \sum_{\langle ij\rangle} t_{ij}[c_i^\dagger c_j + c_j^\dagger c_i] .$$

Here $E_i = 0$ if i is a B site and $E_i = E_0$ if i is an A site; t_{ij} takes the value $t_{AA} = t_{BB}$ if i and j are both A sites or both B sites and another value, t_{AB}, if one of the sites i and j is an A site and the other a B site. The problem in these localization models is to calculate the conductivity $\sigma(\omega)$ as a function of frequency ω. Writing $\mathcal{H} = \sum_{\langle ij\rangle} c_i H_{ij} c_j$, one can write the conductivity in terms of a spectral function $G_{ij}(z)$ given for z outside the spectrum of H_{ij} by the equations

$$z\, G_{ij}(z) - \sum_\ell H_{i\ell}\, G_{\ell j}(z) = \delta_{ij} .$$

If m is the electron mass and $\hbar$ is Planck's constant $\times\, 1/2\pi$, defining

$$p_{ij}^\mu = m/\sqrt{-1}\, \hbar\, H_{ij}(x_i^\mu - x_j^\mu)$$

gives

$$G_{ij}(\omega) = \lim_{\varepsilon\to 0^+} \frac{[G_{ij}(\omega + \varepsilon\sqrt{-1}) - G_{ij}(\omega - \varepsilon\sqrt{-1})]}{2\sqrt{-1}} .$$

With e the electron charge, the conductivity tensor is

$$\sigma^{\mu\nu}(\omega) = \frac{e^2\hbar}{\pi m^2} \sum_{ijk\ell} p^{\mu}_{\ell i} G_{ij}(\omega) P^{\nu}_{jk} G_{k\ell}(\omega) \quad .$$

Here $x^{\mu}_{i} = \langle i|x^{\mu}|i\rangle$ where x^{μ} is the position operator of the electron and $|i\rangle = c_i^{\dagger} |0\rangle$ where $|0\rangle$ is the vacuum state. In this problem, one is interested in the behavior of $\sigma(\omega,p)$ as a function of ω and of the concentration p of A-sites. $\sigma(\omega,p)$ is believed to exhibit a phase transition-like nonanalyticity as a function of ω at one or more frequencies ω_c (the "mobility edges"). For fixed ω, $\sigma(\omega,p)$ will also exhibit a phase transition-like onset at certain values of p. In the case that $t_{AB} = 0$, $\sigma(\omega,p)$ at fixed ω exhibits a transition with the percolation conductivity exponent t (see articles by Redner, this volume). But for $t_{AB} \neq 0$ the onset is characterized by a different exponent, characteristic of the quantum mechanical localization problem. A vast literature on the localization problem exists [54]. Most of it is concerned with models in which the parameters E_i and t_{ij} in $\mathcal{H}$ take a continuum of values. It is believed that the critical properties of the model are insensitive to whether the parameters take a continuum of values or a discrete number of values. We are aware of very few rigorous results on the localization problem [55].

References

1. S.K. Ma, Modern Theory of Critical Phenomena, Benjamin, Reading, Mass. (1976).

2. V.K.S. Shante and S. Kirkpatrick, Adv. Physics 20, 325 (1971).

3. R. Zallen in Fluctuation Phenomena, edited by E.W. Montroll and J.L. Lebowitz, North-Holland, Amsterdam (1979), p. 177.

4. J.P. Fitzpatrick, R.B. Malt and F. Spaepen, Phys. Lett. A 47, 207 (1974); H. Ottavi, J. Clerc, G. Giraud, J. Rousseng, E. Guyon and C.D. Mitescu, J. Phys. C 11, 1311 (1978). See also reference 3.

5. R. Zallen, Phys. Rev. B 16, 1426 (1977).

6. Reference 5. Also, J.W. Halley and W.K. Holcomb, Phys. Rev. Lett. 40, 1670 (1978).

7. J.W. Halley in Percolation Structures and Processes, ed. by R. Zallen, J. Adler and G. Deutscher, Annals of Israel Phys. Soc. (in press); F. Scholl and K. Binder, Z. Physik B 39, 239 (1980).

8. H. Muller-Krumbhaar, Phys. Lett. 48A, 459 (1974); A. Coniglio, J. Phys. A 8, 1773 (1975).

9. H.L. Frisch and J.M. Hammersley, J. Soc. Indust. Appl. Math. 11, 894 (1963); P. Agrawal, S. Redner, P.J. Reynolds and H.E. Stanley, J. Phys. A 12, 2073 (1979).

10. M. Barma and J.W. Halley, Proc. of the Nucl. Phys. and Sol. St. Symposium, Madras (1979); T. Mai and J.W. Halley in Ordering in Two Dimensions, S. Sinha, ed., Elsevier North Holland (1980), p. 369.

11. G. Toulouse and P. Pfeuty, Introduction to the Renormalization Group and to Critical Phenomena, Wiley, N.Y. (1977).

12. Ref. 1: see also Real Space Renormalization, ed. by T.W. Burkhardt and J.M.J van Leeuwen, vol. 30 of Topics in Current Physics, Springer-Verlag, Berlin (1982).

13. P.W. Kasteleyn and C.M. Fortuin, J. Phys. Soc. Japan Suppl. 16, 11 (1969); T. Lubensky in La Matière Mal Condensée, ed. by R. Balian, R. Maynard and G. Toulouse, North Holland, Amsterdam (1979), p. 404.

14. A.B. Harris, T.C. Lubensky, W.K. Holcomb and C. Dasgupta, Phys. Rev. Lett. 35, 327 (1975), M.J. Stephen, Phys. Rev. B 15, 5674 (1977).

15. D.S. Gaunt, M.F. Sykes and H. Ruskin, J. Phys. A 9, 1899 (1976); R. Fisch and A.B. Harris, Phys. Rev. B 18, 416 (1978).

16. S. Kirkpatrick, Phys. Rev. Lett. 36, 69 (1976).

17. P.G. de Gennes, Phys. Lett. A 38, 339 (1972).

18. See M. Daoud et al., Macromolecules, 8, 804 (1975).

19. J. des Cloizeaux, J. Phys. (Paris) 36, 281 (1975). For more refined versions of this correspondence, see P.D. Gujrati, Phys. Rev. A 24, 2096 (1981) and J. Phys. A 14 L345 (1981); J.C. Wheeler and P. Pfeuty, Phys. Rev. A 24, 1050 (1981).

20. P. de Gennes, Scaling Concepts in Polymer Physics, Cornell Univ. Press, Ithaca, N.Y. (1979), Chapter 10.

21. M.E. Fisher, Rev. Mod. Phys. 46, 597 (1974) and reference 1.

22. M.R. Giri, M.J. Stephen, G.S. Grest, Phys. Rev. B 16, 4971 (1977).

23. M.J. Stephen, Phys. Lett. A 56, 149 (1976).

24. Reference 7 and J.W. Halley and M.J. Stephen (unpublished).

25. M.J. Stephen, Phys. Rev. B 17, 4444 (1978); C. Dasgupta, A.B.Harris and T.C. Lubensky, Phys. Rev. B 17, 1375 (1978).

26. T.C. Lubensky and J. Isaacson, Phys. Rev. Lett. 41, 829 (1978), 42, 410(E) (1979).

27. T.C. Lubensky and J. Isaacson, Phys. Rev. A 20, 2130 (1979).

28. P. Pfeuty and J.C. Wheeler, Phys. Lett. A 84, 493 (1981); F. Rys and W. Helfrich, J. Phys. A 15, 599 (1982); P.D. Gujrati, Phys. Rev. B 27, 4507 (1983).

29. A. Coniglio and F. Peruggi, J. Phys. A 15, 1873 (1982).

30. P.J. Scalapino, M. Sears, R.A. Ferrell, Phys. Rev. B 6, 3409 (1972).

31. A.B. Harris, J. Phys. C 7, 1671 (1974).

32. See, e.g., Ref. 21.

33. This is also discussed in Reference 21. Note however, that a sign is wrong in Equation (7.3):

$$a = \frac{4 - n}{2(n + 8)} \varepsilon - \frac{(n + 2)^2 (n + 28)}{4(n + 8)^3} \varepsilon^2$$

34. L. Onsager, Phys. Rev. 64, 117 (1944).

35. Reference 21 and note 33 above.

36. A.B. Harris and T.C. Lubensky, Phys. Rev. Lett. 33, 1540 (1974).

37. A. Aharony, Phys. Rev. B 12, 1038 (1974).

38. G. Grinstein and A. Luther, Phys. Rev. B 13, 1329 (1976).

39. C. Thompson, Contemporary Physics 19, 203 (1978). Also References 1, 11, 21.

40. T. Bergstresser, J. Phys. C10, 3831 (1977).

41. M. Fisher, Phys. Rev. 162, 480 (1967); see also A.B. Harris, J. Phys. C 7, 3082 (1974).

42. T.C. Lubensky, Phys. Rev. B 15, 311 (1972).

43. P.G. de Gennes, J. Phys. Lett. 38, 567 (1977).

44. A.S. Skal and B.I. Shklovskii, Fiz. Tekh. Poluproudn. 8, 1582 (1974) (Sov. Phys. Semicond. 8, 1029 (1975)).

45. H.E. Stanley, Phys. Rev. 179, 570 (1969).

46. H.E. Stanley, R.J. Birgeneau, P.J. Reynolds and J. Nicoll, J. Phys. C 9, L553 (1976).

47. D.J. Wallace and A.P. Young, Phys. Rev. B 17, 2384 (1978).

48. R.J. Birgeneau, R.A. Cowley, G. Shirane and H.T. Guggenheim, Phys. Rev. Lett. 37, 940 (1976); R.J. Birgeneau, R.A. Cowley, G. Shirane, J.A. Tarvin and H.J. Guggenheim, Phys. Rev. B 21, 317 (1980); R.A. Cowley, R.J. Birgeneau, G. Shirane, H.J. Guggenheim and H. Ikeda, Phys. Rev. B 21, 4038 (1980).

49. A. Coniglio, Phys. Rev. Lett. 46, 250 (1981); A. Coniglio in Disordered Systems and Localization, ed. by C. Castellani, C. Di Castro and L. Peliti, vol. 149 of Lecture Notes in Physics, Springer-Verlag, Berlin (1981) pp. 51-55.

50. J.L. van Hemmen and R.G. Palmer, J. Phys. A 15, 3881 (1982); J. Phys. A 12, 563 (1979).

51. M.J. Stephen and G.S. Grest, Phys. Rev. Lett. 38, 567 (1977).

52. A bond version of this model is discussed by A.B. Harris, Phys. Rev. Lett. 49, 296 (1982); see also R. Raghavan and D.C. Mattis, Phys. Rev. B 23, 4791 (1981); Y. Shapir, A. Aharony and A.B. Harris, Phys. Rev. Lett. 49, 486 (1982).

53. A.A. Abrikasov, L.P. Gorkov, I.E. Dzyaloshinski, Methods of Quantum Field Theory in Statistical Physics, Prentice-Hall, Englewood Cliffs, N.J. (1963) Chapter I.

54. See for example, the papers on this subject in Disordered Systems and Localization, ed. by C. Castellani, C. di Castro and L. Peliti, vol. 149 of Lecture Notes in Physics, Springer-Verlag, Berlin (1981).

55. H. Kunz and B. Souillard, pp. 213-218, Ref. 54; H. Kunz and B. Souillard, Comm. Math. Phys. 78 201 (1980).

H. Kunz and B. Souillard, J. Physique Lett. 43, L39, 1982.

F. Bentosela, R. Carmona, P. Duclos, B. Simon, B. Souillard, R. Weder, Comm. Math. Phys. 88, 387, 1983.

H.Kunz and B. Souillard, J. Physique Lett. 44, L411, 1983.

BRANCHED POLYMERS, POLYMER GELS AND PERCOLATION THEORY

S.G. Whittington

Department of Chemistry
University of Toronto
Toronto, CANADA

ABSTRACT

This paper is concerned with the application of percolation theory, and related ideas such as the statistics of lattice animals, to model the behavior of branched polymer molecules and the sol-gel phase transition in polymer systems.

1. Introduction

For polymers in solution, if the interactions of a flexible, linear polymer molecule with itself, with other polymer molecules and with the solvent are neglected, the conformations of the molecule can be described statistically in terms of a random walk model [1]. Deviations from "ideal" random walk behavior caused by the polymer molecule's interactions with itself constitute the excluded volume problem. In a good solvent, the forces of molecular attraction between different segments of the molecule are negligible, and the excluded volume problem is a purely geometrical one: two different segments of the chain molecule cannot occupy the same region of space simultaneously. A self-avoiding walk [1,2] has become the standard model for linear polymer chains in good solvents.

In the case of branched polymers it is only recently that attempts have been made to incorporate excluded volume effects. The classical approach to the statistics of branched polymers has centered on models which investigate the statistics of animals (defined below) on a Bethe lattice, a network lacking closed loops. An interesting set of models which has incorporated both excluded volume effects and the effect of cycles is lattice animals, weakly embeddable in a regular lattice [3].

In section 2 we shall consider a variety of results on the effect of restrictions on the number of cycles and the number of branch points and, in particular, how these restrictions affect the limiting entropy per monomer of a sufficiently large branched polymer molecule. We shall also consider the effect of these restrictions on the values of various critical exponents.

If we consider a polymerization reaction with monomers having a functionality greater than two then the products of the reaction will be a set of branched polymer molecules with a distribution of molecular weights. In the early stages of the reaction, no very large molecules are formed, and the reaction mixture is a sol, or suspension. However, at a certain extent of reaction, the weight-average molecular weight of the reaction mixture will diverge, corresponding to the formation of a very large (mathematically infinite) crosslinked molecule, i.e. a gel. Beyond this point, the reaction mixture will consist of sol and gel components. This gelation "transition" has been modelled both by percolation on a regular lattice and by a more classical model which can be viewed either as percolation on a Bethe lattice or as a branching process. In section 3 we shall review some recent work on the use of lattice decorations to extend the model of random percolation on a lattice to include the effect of bifunctional monomers in the reaction mixture. We also consider how such lattice decorations can be used to relate lattice and continuum percolation models.

2. Lattice Animals and Branched Polymer Molecules

A bond animal is a connected graph, with n vertices, weakly embeddable (see Fig. 1) in a lattice (L) and the number of bond animals is the number of such embeddings per site of the lattice. If we consider the square lattice, the numbers (a_n) of bond animals are $a_1 = 1$, $a_2 = 2$, $a_3 = 6$, $a_4 = 23, \ldots$ With four vertices, eighteen of the animals are simple chains (connected graphs with two vertices of unit degree, all other vertices being of degree two), four are branched structures with one vertex of degree three, and the final one is a square, see Fig. 1. We note that bond animals are sub-graphs of the lattice - two adjacent vertices of the lattice can be members of the vertex set of the animal without the corresponding edge being a member of the edge set. One can also

```
0 —— 0 —— 0 —— 0      (2)          0 —— 0 —— 0      (8)
                                             |
                                             0

0
|
0 —— 0                (4)          0 —— 0 —— 0      (4)
     |                                  |
     0                                  0

0 —— 0                             0    0
|    |                             |    |
|    |                (1)          |    |           (4)
0 —— 0                             0 —— 0
```

Figure 1. Animals with four vertices, embeddable in the square lattice. The six possible "space types" are shown and the figures in brackets indicate the numbers of embeddings obtainable by rotation and reflection. All of these graphs are weakly embeddable (i.e. they are subgraphs of the lattice). The space type shaped like a U is weakly embeddable but not strongly embeddable since it contains a pair of vertices which are not connected by an edge but which are neighbors on the lattice. The other five space types are strongly embeddable.

define animals which are section graphs of the lattice; these are known as site animals. For the square lattice the numbers (A_n) of site animals with n sites are $A_1 = 1$, $A_2 = 2$, $A_3 = 6$, $A_4 = 19,\ldots$ Although bond animals are the more natural model of branched polymers there are good reasons to believe that critical exponents (but not growth constants, defined below) will be the same for site and bond animals.

Not much is known rigorously about the large n behavior of a_n and A_n . The most important result, due to Klarner [4], is the existence of the growth constant, λ . He showed that

$$0 < \lim_{n \to \infty} n^{-1} \log a_n = \sup_{n > 0} n^{-1} \log a_n \equiv \log \lambda < \infty \qquad (2.1)$$

with a similar result for A_n . Like many such results on "thermodynamic limits", the proof rests on showing that a_n is a super-multiplicative function i.e.

$$a_n a_m \leq a_{n+m} . \qquad (2.2)$$

To see the idea, we again consider the square lattice and, for each animal, we define a <u>top vertex</u> as the top-most vertex in the right-most set of vertices, and the <u>bottom vertex</u> as the bottom-most vertex in the left-most set of vertices. For any pair of animals, one having m vertices and the other having n vertices, we translate the animals so that the top vertex of one animal is one lattice spacing to the left of the bottom vertex of the other animal. If we now add an edge to join this pair of vertices, the resulting graph is an animal with $(n + m)$ vertices and each n,m pair gives a unique animal by this construction. Not all animals with $(n + m)$ vertices can be constructed in this way, so that Eq. (2.2) hold and a_n is a super-multiplicative function. One can show that $n^{-1} \log a_n$ is bounded above and this, together with (2.2), implies (2.1). Since (2.1) implies that the limit is approached from below,

$$\lambda \geq a_n^{(1/n)} \tag{2.3}$$

for any n, so that enumerations of animals immediately lead to lower bounds on λ [5].

The sub-dominant asymptotic behavior is not known, but there is good numerical evidence that

$$a_n \sim n^{-\theta} \lambda^n \tag{2.4}$$

and, if this form is accepted, (2.1) implies that $\theta \geq 0$. Using series analysis techniques, Gaunt [6] has estimated the value of θ as a function of dimension (d) and finds that it reaches the mean field value $(\theta = 5/2)$ at $d = 8$, in agreement with predictions of field theory [3]. Parisi and Sourlas [7] have discovered an interesting connection between lattice animals in d dimensions and the Yang-Lee edge singularity in $(d - 2)$ dimensions, which implies that $\theta = 1$ in two dimensions and $\theta = 1.5$ in three dimensions.

Another interesting quantity from the point of view of the polymer problem is the mean-square radius of gyration of the animal, $\langle s_n^2 \rangle$, which is expected to behave as

$$\langle s_n^2 \rangle \sim n^{2\nu} . \tag{2.5}$$

The value of ν has been estimated by scaling arguments [8] and by Monte Carlo methods [9,10].

Lubensky and Isaacson [3] presented a field-theoretic treatment of the animal problem which has had a tremendous influence on the research in this field. One of their important conclusions was that cycles were unimportant in the sense that critical exponents would be independent of cycle fugacity. This led to a number of treatments of lattice trees (animals with no cycles) by a variety of theoretical approaches [11-17]. Klein [12] showed, using an approach similar to that used by Klarner for animals, that

$$0 < \lim_{n \to \infty} n^{-1} \log a_{no} = \sup_{n > 0} n^{-1} \log a_{no} \equiv \log \lambda_o < \infty \qquad (2.6)$$

where a_{no} is the number of trees with n vertices. (The subscript "zero" labels the graphs as having no cycles.) It is probably true [16] that λ_o is strictly less than λ, though this does not appear to have been shown rigorously. The series analysis work [15-17] strongly suggests that the exponent θ_o, defined by

$$a_{no} \sim n^{-\theta_o} \lambda_o^n \qquad (2.7)$$

is equal to θ, the corresponding exponent for animals in agreement with the result of Lubensky and Isaacson. The evidence that ν is the same for trees and animals is less convincing [11,13], and additional series analysis or Monte Carlo work on this question would be useful.

This relationship between trees and animals raises the interesting question of how increasing the number of allowed cycles affects the growth constant and the exponent. If a_{nc} is the number of animals with n vertices and cyclomatic index (a measure of the number of cycles) equal to c, it has been shown [18] that

$$\lim_{n \to \infty} n^{-1} \log a_{nc} = \log \lambda_o , \qquad (2.8)$$

so that the growth constant is <u>independent of c</u>. Assuming that

$$a_{nc} \sim B_c n^{-\theta_c} \lambda_o^n , \qquad (2.9)$$

then

$$\theta_c \geq \theta_{c+1} \geq \theta_c - 1 , \tag{2.10}$$

and series results are consistent with the conjecture that

$$\theta_{c+1} = \theta_c - 1 . \tag{2.11}$$

The interesting feature of this result is that it appears to be possible to change the value of the critical exponent with only an apparently minor change in the imposed constraints. The requirement that the animal contains not more than c cycles leads to the same result, of course.

At first sight this appears to be in disagreement with the argument of Lubensky and Isaacson [3] that the critical exponent is independent of cycle fugacity (z). However, by considering the generating function

$$G(n,z) = \sum_c a_{nc} z^c \sim n^{-\theta(z)} \lambda(z)^n \tag{2.12}$$

it can be shown [18] that (2.8) and (2.11) are consistent with $\theta(z)$ being independent of z, provided that the amplitude B_c in (2.9) depends on c in an appropriate way.

The question of how the exponent ν depends on c does not appear to have been considered in the literature, though ν is probably independent of cycle fugacity [3,11].

The work discussed above focusses on the importance of cycles and on the relationship between trees and animals. Another type of constraint which has been considered is the number of branch points in the tree. If we consider trees weakly embeddable in the d-dimensional hypercubic lattice, the degree of a vertex in the tree can range from unity to $2d$, the lattice coordination number. If we consider trees with n_k vertices of degree k, $k = 1,2,\dots,2d$, then it follows from Euler's theorem that

$$n_1 = 2 + n_3 + 2n_4 + \dots + (2d - 2)n_{2d} \tag{2.13}$$

and the total number of vertices in the tree is simply

$$n = \sum_k n_k \ . \tag{2.14}$$

It is convenient to classify trees by their total number of vertices and the numbers of vertices with degree 3,4,... . Let $t(n,n_3,n_4,\dots,n_{2d})$ be the number, per lattice site, of trees with n vertices, n_3 of degree 3, n_4 of degree 4, etc. We call

$$n^+ = n_3 + n_4 + \dots + n_{2d} \tag{2.15}$$

the number of branch points in the tree and we note that $n^+ = 0$ corresponds to a simple chain. Thus, it follows from a result of Hammersley and Morton [19] that there exists a positive (lattice dependent) constant μ such that

$$\lim_{n\to\infty} n^{-1} \log t(n;0,0,\dots,0) = \inf_{n>0} n^{-1} \log t(n;0,0,\dots,0) = \log \mu \ . \tag{2.16}$$

Suppose that we now allow branch points. The obvious questions which arise are: (i) Does the limit still exist? (ii) If so, is the limit equal to $\log \mu$? Lipson and Whittington [20] have shown that, provided that n_3, n_4, ... , n_{2d} are fixed as n goes to infinity,

$$\lim_{n\to\infty} n^{-1} \log t(n;n_3,n_4,\dots,n_{2d}) = \log \mu \ . \tag{2.17}$$

In the language of the polymer problem, this means that the limiting entropy per monomer is independent of the <u>fixed</u> number of branch points.

The next question is, what happens to the value of the limit when n_3 , n_4, etc. go to infinity as n goes to infinity? One can show that, provided

$$n_k = o(n/\log n) \text{ for all } k > 2 \ , \tag{2.18}$$

the value of the limit is unchanged [20]. If n^+ is allowed to grow linearly with n , that is, if we consider the number $t(n;\alpha)$ of trees subject to the restriction that

$$n^{+} < \alpha n \ , \ \alpha > 0 \tag{2.19}$$

then it can be shown [20] that $\lim_{n\to\infty} n^{-1} \log t(n;\alpha)$ exists and, subject to some formal assumptions, that the value of the limit is strictly greater than $\log \mu$ for all positive α. Roughly speaking, if the number of branch points goes to infinity (as $n \to \infty$) and n^{+}/n can go to a non-zero limit, the limiting entropy per monomer is larger than for a linear polymer. If n^{+} grows less rapidly then the limiting entropy is the same as for a linear polymer.

Gaunt et al. [21,22] have considered a constraint on the type, rather than on the number of branch points. They consider (site) animals with the restriction that no vertex has degree greater than some fixed number v. In this case the growth constant seems to be a strictly monotone function of v but the critical exponent changes from $v = 2$ to $v = 3$ and then remains constant as v is increased further. Stauffer [23] has suggested that this kind of restriction may be a useful model of steric hindrance effects which arise from the presence of a bulky side group attached to each monomer.

3. Percolation Theory and the Sol-Gel Transition

If a vinyl monomer, such as styrene, is polymerized by an addition polymerization reaction, the product is a linear polymer. If a reactant with more than one vinyl group (e.g. divinyl benzene) is copolymerized with styrene then the product will be a branched polymer. In this kind of reaction the product molecules will have a distribution of molecular weights (or equivalently, of degrees of polymerization), and this distribution depends on the extent of reaction $|p|$. The (weight-average) molecular weight of the branched polymer will increase as the extent of reaction increases and will become infinite at a certain stage, well before the reaction has gone to completion. This corresponds to the formation of a gel. The reaction mixture will contain only molecules of finite molecular weight (i.e. it will be a sol) at extents of reaction below this "gel point", and will contain both sol and gel components above the gel point. The probability that a randomly chosen monomer is incorporated into a polymer of infinite molecular weight is the gel fraction, $G(p)$. $G(p)$ will be zero for p less than some critical value p_c , corresponding to the gel point, and will increase rapidly for p just above p_c . As $p \to 1$, i.e. as the reaction goes to completion, $G(p) \to 1$ and all the material will be a gel. The weight-average molecular weight, $M(p)$, of material in the sol will decrease above p_c as molecules are progressively incorporated into the gel.

The connection with percolation theory is the formation of an infinite network of bonds (in the infinite molecular weight molecule) at the gel point. The gel fraction will correspond, in the percolation model, to the percolation probability, $P(p)$, and the weight average molecular weight will correspond to the mean size of finite clusters, $S(p)$. It is clear that the general behavior of $P(p)$ and $S(p)$ will be similar to the behavior of $G(p)$ and $M(p)$; this will be true both for bond percolation on a regular lattice, embeddable in a three dimensional Euclidean space, and for percolation on a Bethe lattice. The classical models of gelation in polymerization reactions can be phrased as percolation on a Bethe lattice [23]. Such models have the disadvantage of ignoring cycles which can form during the polymerization process [24]. (In fact, it is possible to incorporate the effect of small cycles by various approximations [25].)

Although the two models have similar qualitative behavior, the behavior close to the gel point is quantitatively quite different. In each case

$$P(p) \sim (p - p_c)^{\beta} \quad , \quad p \to p_c^+ \tag{3.1}$$

and

$$S(p) \sim (p - p_c)^{-\gamma} \ , \ p \to p_c^+ \ . \tag{3.2}$$

The percolation model gives $\beta \simeq 0.45$ and $\gamma \simeq 1.7$, while the classical model gives $\beta = \gamma = 1$. In spite of the difference in the predicted values of these (and other) exponents it has not yet been possible to distinguish experimentally between the two models [23].

In this paper we shall not pursue the question of whether the percolation model or the classical model should give the better description of gelation. Instead we shall look at several extensions of the percolation model which are attempts to bridge the gap between random bond percolation on a regular lattice and the realities of the chemical situation. In particular, we shall be interested in whether these extensions change the values of critical exponents such as β and γ.

We shall make use of lattice decorations and we first need to describe some results on the invariance of exponents under certain types of decoration [26]. Consider bond percolation on a lattice L. As the bond density p approaches the critical density p_c, we expect that the percolation probability will behave as (3.1). We now consider a finite two-rooted graph G and we replace every bond in L by the graph G. We call the resulting graph the G-decoration of L, L^G. If the bond density on L^G is f then the probability that the pair of roots in G are connected is the transmission function of G, $t(f)$. We also define the association function of G, as the probability that a randomly chosen vertex in G, which is not a root, is connected to at least one root of G. The percolation probability on L^G will be

$$F(f) \sim (f - f_c)^{\beta(G)} , \quad f \to f_c^+ , \tag{3.3}$$

where $\beta(G)$ is the appropriate exponent for the G-decoration of L. If we consider only root percolation, i.e. the probability $F_R(f)$ that a randomly chosen root in L^G is connected to an infinite number of other roots then, since the association function is a finite polynomial,

$$F_R(f) \sim F(f) .$$

We map a configuration (of occupied and unoccupied bonds) on L^G onto a configuration on L by occupying a bond on L if and only if the roots of the corresponding graph G, in L^G, are connected. The bond density on L is related to the bond density on L^G by the transmission function

$$p = t(f) \tag{3.4}$$

so that

$$P(p) \sim (p - p_c)^{\beta}$$
$$= [t(f) - t(f_c)]^{\beta} . \tag{3.5}$$

Since $t(f)$ is a finite polynomial, we can expand $t(f)$ about f_c giving

$$P(p) \sim [t'(f_c)(f - f_c)]^{\beta} \tag{3.6}$$

provided that $t'(f_c) \neq 0$. This condition can be shown [26] to hold for any finite G provided that $p_c \in (0,1)$.

The mapping of configurations from L^G onto L preserves percolation so that

$$P(p) \equiv F_R(f) \tag{3.7}$$

and from (3.6) and (3.7) we see that

$$F_R(f) \sim (f - f_c)^{\beta} . \tag{3.8}$$

Hence, from (3.3), (3.4) and (3.8)

$$\beta(G) = \beta , \tag{3.9}$$

and the exponent is unchanged by such decorations. We note that the above result can be extended to a stochastic decoration in which each bond in L is replaced by a graph G chosen, randomly and independently, from a set of possible decorating graphs [27].

We now examine the relevance of such decorations to the gelation problem. In the typical polymerization reaction discussed at the beginning of this section, a q-functional monomer (e.g. divinyl benzene) is copolymerized with a 2-functional monomer (e.g. styrene). In many cases the mole percentage of 2-functional monomer in the reaction mixture will be large [28], but the standard bond percolation model only considers the q-functional monomers (as vertices of a lattice with coordinate number q). A possible model which incorporates 2-functional monomers is as follows [29]. For any lattice L we construct a stochastic decoration L^S of L by replacing the jth bond in L by a set of n_j bonds in series where the

n_j are discrete random variables independently distributed with probability distribution $\{g(n), n = 1,2,\ldots\}$. The vertices of L will correspond to the q-functional monomers and the inserted vertices in L^S will correspond to the 2-functional monomers.

Any configuration on L^S can be mapped onto a unique configuration on L by occupying a bond on L if and only if every bond is occupied in the simple chain joining the corresponding pair of q-coordinated vertices on L^S. Then a configuration on L^S percolates if and only if its image on L percolates. If the bond densities on L and L^S are p and f respectively then these are related by a generalized transmission function

$$p = \sum g(n) f^n , \qquad (3.10)$$

and the critical densities are related by

$$f_c \approx 1 - (1 - p_c)/\,\overline{n} , \qquad (3.11)$$

where $\overline{n}$ is the mean number of bonds in a simple chain in L^S. As $\overline{n} \to \infty$, $f_c \to 1$. These results will apply for all L, including regular and Bethe lattices.

Using the result described above on the invariance of the exponent β under certain kinds of decoration, it is easy to show [29] that β is unchanged by this stochastic decoration, so that the presence of 2-functional monomers will shift the gel point to higher values of the extent of reaction, but will not change the behavior sufficiently close to the gel point.

One can also show [29], using similar arguments, that the width of the critical region shrinks as the fraction of 2-functional monomers increases. Since measurements of critical exponents involve experiments <u>inside</u> the critical region, this suggests that experimental work should be carried out on systems with a <u>low</u> mole percentage of bifunctional monomers.

This stochastic decoration is related to "chain percolation" which has been used as a model of vulcanization [30-32]. It may also be relevant to gelation occuring in some biopolymeric systems [33].

One of the objections which has been raised to the use of percolation as a model of gelation is that the vast majority of results in percolation theory are for the case of percolation on a lattice. One can argue that gelation occurs in the continuum and that continuum percolation is therefore a more useful model. The question then arises as to whether such quantities as critical exponents are unchanged when one passes from lattice to continuum percolation. Gawlinski and Stanley [34] have presented numerical evidence that lattice and continuum percolation have the same critical exponents. We can also ask whether it is possible to construct sequences of lattices which, when some appropriate limit is taken, become sufficiently good approximations to the continuum.

The first approach of this type is due to Domb [35] who considered percolation extending to first, second, third,... etc. neighbors. In this way the coordination number of the lattice can be increased indefinitely. As this occurs, the critical (site) density (p_c) will decrease to zero but the product of the critical density and the coordination number (q) will go to a finite limit. By using the available numerical data on the critical densities for various lattices with a range of coordination numbers, Domb estimated that the critical parameter [36] for continuum percolation was

$$E_c = \lim_{q\to\infty} p_c q \approx 4.0 \text{ to } 4.5 \tag{3.12}$$

In the same spirit, it is possible to use lattice decorations to form a sequence of lattices for which the limit in (3.12) can be calculated exactly [26]. The idea is as follows. For a lattice L we construct an expanded lattice L^n in which each bond in L is replaced by a set of n bonds in series. We then form the covering lattice of L^n (L^{nc}, say) and the matching lattice of L^{nc} (L^{nc^*}, say). If p_c is the critical density for bond percolation on L, the critical density for bond percolation on L^n is $p_c^{1/n}$. Then the critical density for site percolation on L^{nc} will also be $p_c^{1/n}$ and the critical density for site percolation on L^{nc^*} will be $1 - p_c^{1/n}$. If L is the hexagonal lattice, L^{nc^*} will consist of triangles surrounded by complete graphs on $6n$ vertices, and the coordination number will be $11n - 1$. Using the value given by Sykes and Essam [37] for the critical bond density on the hexagonal lattice gives

$E_c = 4.692...$, in good agreement with the estimate of Domb [35]. The value of E_c depends on the parent lattice L but it may be that $E_c(L)$ is an upper bound on E_c for the continuum, for every L.

The lattices in these sequences have a variety of disadvantages as models of continuum percolation (e.g. the lattice points are countable and are not dense in the plane), but more importantly, there are difficulties in showing that the exponent is invariant when the limit is taken. These problems can be avoided, to some extent, by considering a different type of lattice decoration proposed by Ord [27]. He considers a lattice L and replaces every vertex in L by a complete graph on n vertices, which he calls an "n-pole". Bonds on L are replaced by "transmission lines" such that for every pair of vertices in L incident on a common bond, every pair of vertices in the corresponding pair of n poles is joined by a bond in the transmission line. Alternatively, one can think of every bond, and the pair of vertices incident on this bond, in L being replaced by a complete graph on 2n vertices. The resulting graph L^n is the "n-pole decoration of L". Ord shows that percolation exponents are invariant under this transformation.

To make contact with continuum percolation he considers decorations L^n in which n is a square so that the vertices of the decorated lattice can be arranged in a square array. By choosing L to be the square lattice this yields a sequence of "square lattices" with successively higher connectivity. As $n \to \infty$ the n-pole decorations approach the continuum in that the coordination number becomes infinite <u>and</u> the vertices become dense in the plane. Moreover, Ord shows that the exponents are invariant even in the limit $n \to \infty$.

These decorations form an interesting connection between lattice and continuum percolation and suggest that the critical exponents are the same for the two percolation problems.

To close this section we mention several other extensions of random lattice percolation which make the model more applicable to the gelation problem. Solvent effects have been included by considering site-bond percolation [38,39], correlation effects in the formation of clusters have been considered [38], and dynamical effects have been investigated in the low [40] and high mobility [41] limits.

Whether or not percolation theory is directly relevant to the gelation transition is a question which can only be answered by a comparison between the predictions of the theory and the results of experiments carried out within the critical region. However, theoretical work on gelation has produced interesting results on a variety of extensions of the random percolation model.

4. Discussion

We have reviewed a number of recent results on lattice animal models of branched polymers and on the percolation model of polymer gelation. In particular, we have discussed how various constraints (on the number of cycles and on the number of branch points) affect the growth constant and critical exponent of the animal problem. We have also shown that lattice decorations are useful in extending the percolation problem to include a distribution of vertices of degree 2 (or bifunctional monomers, in the context of gelation) and in making contact between lattice and continuum percolation.

Acknowledgements

The author would like to thank David Gaunt, Jane Lipson, Garnet Ord and Glenn Torrie for numerous helpful conversations on these topics. Much of the research described here has been financially supported by NSERC of Canada.

References

1. M.N. Barber and B.W. Ninham, Random and Restricted Walks, Gordon and Breach, New York, 1970; see also the article by Hughes and Prager in this volume.
2. S.G. Whittington, Adv. Chem. Physics 51, 1 (1982).
3. T.C. Lubensky and J. Isaacson, Phys. Rev. A. 20, 2130 (1979).
4. D.A. Klarner, Can. J. Math., 19, 851 (1967).
5. S.G. Whittington and D.S. Gaunt, J. Phys. A: Math. Gen. 11, 1449 (1978).
6. D.S. Gaunt, J. Phys. A: Math. Gen. 13, L97 (1980).

7. G. Parisi and N. Sourlas, Phys. Rev. Lett. 46, 871 (1981).

8. M. Daoud and J.F. Joanny, J. Physique 42, 1359 (1981).

9. D. Stauffer, Phys. Rev. Lett. 41, 1333 (1978).

10. H.J. Herrmann, Z. Phys. B. 32, 335 (1979).

11. F. Family, J. Phys. A: Math. Gen. 13, L325 (1980).

12. D.J. Klein, J. Chem. Phys. 75, 5186 (1981).

13. W.A. Seitz and D.J. Klein, J. Chem. Phys. 75, 5190 (1981).

14. S. Redner, J. Phys. A: Math. Gen. 12, L239 (1979).

15. J.A.M.S. Duarte and H.J. Ruskin, J. Physique 42, 1585 (1981).

16. D.S. Gaunt, M.F. Sykes, G.M. Torrie and S.G. Whittington, J. Phys. A: Math. Gen. 15, 3209 (1982).

17. H.J. Ruskin and J.A.M.S. Duarte, Phys. Rev. A. 26, 1791 (1982)]

18. S.G. Whittington, G.M. Torrie and D.S. Gaunt, J. Phys. A: Math. Gen. 16, in press (1983).

19. J.M. Hammersley and K.W. Morton, J.R. Stat. Soc. B 16, 23 (1954).

20. J.E.G. Lipson and S.G. Whittington, J. Phys. A: Math. Gen. submitted.

21. D.S. Gaunt, A.J. Guttmann and S.G. Whittington, J. Phys. A: Math. Gen. 12, 75, (1979).

22. D.S. Gaunt, J.L. Martin, G. Ord, G.M. Torrie and S.G. Whittington, J. Phys. A.: Math. Gen. 13, 1791 (1980).

23. D. Stauffer, A. Coniglio and M. Adam, Adv. Polymer Sci. 44, 103 (1982).

24. D. Stauffer, Physica 106A, 177 (1981).

25. Z. Ahmad and R.F.T. Steptoe, Colloid and Polymer Sci. 258, 663 (1980).

26. G. Ord and S.G. Whittington, J. Phys. A: Math. Gen. 13, L307 (1980).

27. G. Ord, Ph.D. Thesis, University of Toronto (1983).

28. R.S. Whitney and W. Burchard, Makromol. Chem. 181, 869 (1980).

29. G. Ord and S.G. Whittington, J. Phys. A: Math. Gen. 15, L29 (1982).

30. P.G. de Gennes, J. de Physique Lettres 40, L355 (1979).

31. M. Daoud, J. de Physique Lettres 40, L201 (1979).

32. A. Coniglio and M. Daoud, J. Phys. A: Math. Gen 12, L259 (1979).

33. D.A. Rees, Chemistry and Industry 630 (1972).

34. E.T. Gawlinski and H.E. Stanley, J. Phys. A 14, L291 (1981).

35. C. Domb, Biometrika 59, 209 (1972).

36. E.N. Gilbert, J. Soc. Ind. Appl. Math. 9, 533 (1961).

37. M.F. Sykes and J.W. Essam, J. Math. Phys. 5, 117 (1964).

38. A. Coniglio, H.E. Stanley and W. Klein, Phys. Rev. Lett. 42, 518 (1979).

39. A.J. Guttmann and S.G. Whittington, J. Phys. A: Math. Gen. 15, 2267 (1982).

40. H.J. Herrmann, D.P. Landau and D. Stauffer, Phys. Rev. Lett. 49, 412 (1982).

41. R.M. Ziff, J. Stat. Phys. 23, 241 (1980).

CRITICAL PROBABILITIES IN PERCOLATION MODELS

John C. Wierman

Department of Mathematical Sciences
The Johns Hopkins University
Baltimore, Maryland 21218

ABSTRACT

A survey of mathematical percolation theory is presented, concentrating on the concept of the critical probability. Various interpretations of critical probability are considered, and the method of rigorous determination of exact critical probability values is outlined.

1. Introduction

Percolation theory provides a mathematical model for the spread of fluid through a medium, where both "fluid" and "medium" may have various interpretations depending on the context. The medium is represented by a graph (or network), from which vertices (or bonds) and edges (or sites) are deleted at random, with fluid flowing in the network of vertices and edges of the resulting graph.

The percolation model may be viewed as an alternative to the diffusion model. In the percolation model the random mechanism is ascribed to the medium, while in a diffusion model the randomness is ascribed to the fluid and the medium is deterministic. In the percolation model the random structure is frozen into the medium for all time. This memory of the medium is responsible for phenomena which do not occur in diffusion models. Fluid may be trapped in a finite region of the medium for all time in a percolation model, while usually in a diffusion model it will eventually reach arbitrarily far, given sufficient time. Trapping of fluid in a percolation model is certain for a range of parameter values, but breaks down when the parameter crosses a threshold, above which infinite flow is possible. Critical phenomena of this sort are typical of percolation processes, but are uncommon in diffusion models. Mathematical percolation theory is in the early stages of development, while an extensive theory of diffusion is currently available. In percolation theory, there are formidable mathematical difficulties to obtain even basic intuitive results which are supported by extensive numerical evidence. This paper will present a survey of mathematically rigorous results in percolation. For additional information the reader is referred to Kesten (1982), Smythe and Wierman (1978), and Wierman (1982).

Percolation models were introduced by Broadbent and Hammersley (1957). They have become increasing popular in recent years, due to the wide variety of applications possible, the ease of visualization, the simplicity of formulation of problems, and the extensive interest in critical phenomena in statistical mechanics research.

2. Preliminaries

There are two standard forms in which randomness may be introduced into the medium. In a <u>bond percolation model</u>, each edge or bond of the graph is open, i.e., allows fluid to flow through in either direction, with probability p, $0 < p < 1$, independently of all other bonds. Every vertex is understood to be open in a bond percolation model, allowing fluid to flow between any pair of bonds of which it is an endpoint. In a <u>site percolation model</u>, each vertex or site is open with probability p, $0 < p < 1$, independently of all other sites, while a bond is open only if both its endpoint vertices are open sites. The connectivity structure of the underlying graph, the random mechanism introduced, and the parameter value p together determine the characteristics of the fluid flow.

Additional variations of the random mechanism have been considered. In <u>mixed percolation models</u>, edges, vertices, and faces are all allowed to be open at random. In multi-<u>parameter models</u>, there exist disjoint subsets of graphical elements, each with a different parameter serving as the probability that each element in the subset is open. An <u>oriented percolation model</u> is one in which some bonds allow fluid to pass through in only one direction if they are open, as discussed in an article by Redner in this volume.

The results presented in this paper will be phrased in terms of site percolation models. This is without loss of generality, since a bond percolation model on a graph G may be converted into an equivalent site percolation model on a different graph, denoted here by G^c, called the covering graph of G, by the bond-to-site transformation introduced by Fisher (1961). The covering graph is constructed by placing a site of G^c at the midpoint of each bond of G, and connecting two sites of G^c by a bond of G^c if the bonds of G corresponding to these sites share a common endpoint in G. Since the bonds of G and the sites of G^c are in one-to-one correspondence, a bond percolation model realization on G is converted to a site percolation model realization on G^c by declaring a site of G^c to be open if and only if the corresponding bond in G is open.

Rigorous results have been obtained for percolation models on certain infinite regular lattice graphs G. The graphs G discussed in the following (unless specifically excepted) are assumed to satisfy:

G is periodic, i.e. G is embedded in R^d for some $d \geq 2$ so that the vertex and edge sets of G are invariant under translations by each of a set of linearly independent vectors $\{v_1,\ldots,v_d\}$ of R^d. (1)

G is connected. (2)

There exists $z < \infty$ such that each vertex is an endpoint of at most z edges. (3)

Each compact set of R^d contains at most finitely many vertices of G. (4)

The square lattice (S), triangular lattice (T), and hexagonal lattice (H) and their covering lattices satisfy these conditions.

For a fixed graph G, let E(G) denote the set of edges of G and V(G) denote the set of vertices of G. A path on G is an alternating sequence of sites $v_i \in V(G)$, $0 \leq i \leq n$, and bonds $e_i \in E(G)$, $1 \leq i \leq n$, of the form $(v_0,e_1,v_1,\ldots,e_n,v_n)$, where e_i has endpoints v_{i-1} and v_i for $1 \leq i \leq n$. A circuit is a path $(v_0,e_1,\ldots,v_n)$ containing n distinct sites with $v_0 = v_n$.

A path or circuit is open if all its bonds and sites are open. The open cluster containing site v, denoted C_v, is the set of all sites which are connected to v by an open path. Let $\#C_v$ denote the number of sites in C_v. Let P_p denote the probability measure corresponding to the parameter value p, and let E_p denote the associated expectation.

3. Critical Probabilities

In a non-oriented bond or site model, consider the spread of fluid from a single source site. If the parameter p is near zero, the fluid spreads only locally, being trapped in a finite region. However, if p is sufficiently large, the fluid may spread throughout the medium. There are several definitions of a critical probability, which separates parameter intervals of local and extensive spread of fluid, corresponding to various interpretations of penetration or extensive spread through the medium.

Define the percolation probability $\theta(p,v) = P_p(\#C_v = \infty)$, the probability that site v is in an infinite open cluster. The cluster size critical probability is defined by

$$p_H = p_H(G) = \inf\{p \in [0,1] : \theta(p,v) > 0\} \tag{5}$$

for a specified $v \in V(G)$. Although $\theta(p,v)$ may depend on the choice of v, p_H is independent of v when G is connected. The mean cluster size critical probability is defined by

$$p_T = p_T(G) = \inf\{p \in [0,1] : E_p(\#C_v) = \infty\}, \tag{6}$$

which again is independent of v for G connected. Since $P_p(\#C_v = \infty) > 0$ implies that $E_p(\#C_v) = \infty$, $p > p_H$ implies that $p \geq p_T$, and thus $p_H \geq p_T$.

In the context of the square lattice bond percolation model, Seymour and Welsh (1978) defined another critical probability, which was modified slightly and extended to periodic graphs by Kesten (1981). For G a periodic graph in R^d, for each n and $i = 1,2,\ldots,d$, let

$$T(n,i) = \{(x_1,\ldots,x_d):\ 0 \le x_i \le n,\ 0 \le x_j \le 3n \ \text{ for } \ j \ne i\}.$$

Define the sponge crossing probability in direction x_i by

$$\tau(n,p;i) = P_p\{\exists \text{ an open path in } T(n,i) \text{ which intersects both } x_i = 0 \text{ and } x_i = n\}. \tag{7}$$

The <u>sponge-crossing critical probability</u> is defined by

$$p_S = p_S(G) = \inf\{p \in [0,1] :\ \limsup_{n\to\infty} \max_{1 \le i \le d} \tau(n,p;i) > 0\}. \tag{8}$$

Although p_S was introduced for technical reasons and did not arise as naturally as the other definitions, it played a crucial role in the rigorous determination of critical probabilities.

For a two-dimensional graph G satisfying (1)-(4), a simple argument shows that $p_T \le p_S$. Considering $T(n,1)$, note that there exist A and B, $0 < A,B < \infty$, such that there are at most An initial sites for a path across the rectangle from left to right, and at least Bn sites in such a path. Therefore, letting I_E denote the indicator random variable for event E, we have

$$\begin{aligned}\tau(n,p;1) &\le An \max_{v\in G} P_p(\#C_v \ge Bn) \\ &\le \frac{An \max_{v\in G} E_p(\#C_v I_{\{\#C_v \ge Bn\}})}{Bn} \\ &= \frac{A}{B} \max_{v\in G} E_p(\#C_v I_{\{\#C_v \ge Bn\}})\end{aligned} \tag{9}$$

where the maximum is actually over the finite set of vertices in a rectangle with side lengths equal to the periods of the graph. If $p < p_T$, the mean cluster sizes are all finite, and thus $\tau(n,p;1) \to 0$ as $n \to \infty$, by applying the dominated convergence theorem in (9). Identical reasoning applies to $\tau(n,p;2)$ so $p < p_T$ implies $p \le p_S$.

Kesten (1981) has proved that in fact $p_T = p_S$ for any periodic graph in R^d satisfying (1)-(4). This is one of the few dimension-free results in mathematical percolation theory.

Sykes and Essam (1964) introduced another concept of critical probability, which is denoted by p_E , in the setting of site percolation models on two-dimensional graphs. For a rectangle R , let $\lambda_R(p)$ denote the mean number of <u>open clusters per site</u> in the region R . As R expands in both directions, $\lambda_R(p)$ converges to a limiting mean open cluster per site function $\lambda(p)$. Sykes and Essam assumed that $\lambda(p)$ has a unique singularity, p_E , which is interpreted as a percolation threshold or critical probability, presumably equal to p_H .

4. Evaluation of Critical Probabilities

Sykes and Essam contributed the important concept of <u>matching pairs of graphs</u>. Let M be a planar graph and $\mathfrak{F}$ be a set of faces of M . A graph G may be constructed by, for each face $F \in \mathfrak{F}$, inserting a bond (diagonal) connecting each pair of non-adjacent vertices of F . The matching graph of G , denoted G^* , is constructed from M by inserting such diagonals in all faces which are not in $\mathfrak{F}$. [Inserting all diagonals in a face F is called "close-packing" the face.] In general, both G and G^* may be nonplanar graphs.

The matching graph of the square lattice S is a square lattice with both diagonals inserted in every square. The triangular lattice T is its own matching graph. Any planar graph in which every face is triangular is self-matching. The covering graph of the square lattice is also self-matching.

The matching property of such pairs of graphs may be interpreted in two ways. For a realization of a percolation model on a rectangular region, either there exists an open path in G across the rectangle from right to left or there exists a closed path in G^* across the rectangle from top to bottom. This provides a relationship between sponge-crossing probabilities on G and G^* , leading to the importance of the p_S version of the critical probability.

Sykes and Essam noted that the random subgraph of open sites and bonds of G contains three types of faces: close-packed faces, empty faces (those which do not contain a closed site), and non-empty faces. Since each non-empty face contains exactly one closed cluster in G^* , a modification of Euler's law relates the number of open clusters in G to the number of closed clusters in G^* . Applying this result to a rectangle R , taking expectation and dividing by the number of sites in R , one derives the relationship that

$$\lambda(p) + \phi(p) = \lambda^*(1 - p) \tag{10}$$

for some polynomial ϕ , where λ^* denotes the mean open clusters per site function on G^* . λ^* is evaluated at 1 - p since closed clusters are counted on G^* . The polynomial ϕ arises as the expectation of the numbers of open sites, bonds, and empty and closed packed faces in G .

The assumption of a unique singularity of the open clusters per site function for each graph implies that

$$p_E(G) + p_E(G^*) = 1 . \tag{11}$$

In particular, Sykes and Essam conclude that for any self-matching graph G , $p_E(G) = 1/2$. This reasoning leads to the conclusions that the critical probabilities for the square lattice bond model, the triangular lattice site model, and the site model on any fully triangulated graph are all equal to 1/2 . By additional arguments, using the star-triangular transformation, Sykes and Essam (1964) obtained critical probabilities for the bond model on the triangular lattice T and the hexagonal lattice H : $p_E(T) = 2 \sin \frac{\pi}{18}$ and $p_E(H) = 1 - 2 \sin \frac{\pi}{18}$.

If G is the square lattice S , its matching graph S^*, or its covering graph S^c , $\lambda(p)$ is analytic on $[0,p_H)$ and $(p_H,1]$ and has two continuous derivatives on [0,1] . It is not known if there is in fact a singularity of $\lambda(p)$.

It is important to note that Van den Berg (1981) has constructed a nonperiodic fully-triangulated graph which has critical probability $p_H = p_T = 1$, rather than 1/2 .

A relationship between critical probabilities similar to that proposed by Sykes and Essam was proved by Seymour and Welsh (1978) for the square lattice bond percolation model. Using the sponge-crossing concept, and the self-matching property of the model, they show that

$$p_T + p_H = 1 , \tag{12}$$

which for generalization to the site percolation matching lattice setting should be interpreted as

$$p_T(G) + p_H(G^*) = 1 . \qquad (13)$$

A relatively simple argument shows that $p_T + p_H \leq 1$. Consider a sequence B_k of rectangles with $B_k = [0,2^k] \times [0,2^{k+1}]$ for k even, and $B_k = [0,2^{k+1}] \times [0,2^k]$ for k odd. As in (9) the probability that an open path crosses the width of B_k may be bounded above in terms of $E_p[\#C I_{\{\#C \geq 2^k\}}]$. If $p < p_T$, the series of bounds converges, so the Borel-Cantelli Lemma applies to show that such crossings occur finitely often with probability one. By the matching property, for all sufficiently large k, B_k is crossed lengthwise by a closed path in G^*. Since the orientations of the B_k alternate, closed paths across B_k and B_{k+1} must intersect, so these closed paths link together to form an infinite closed cluster in G^*. Hence $1 - p \geq p_H(G^*)$, so $1 \leq p + p_H(G^*)$ for all $p < p_T(G)$, which yields the conclusion.

In the proof of the reverse inequality, symmetry plays a crucial role. The main work is to construct an open circuit around the origin when $p > p_S(G)$. Whenever such a circuit exists, the closed cluster in the matching lattice containing the origin is finite. Thus, if such an open circuit exists with probability one, then $1 - p \leq p_H(G^*)$, from which one concludes that $p_S(G) + p_H(G^*) \geq 1$. Recall that $p_T = p_S$ to complete the argument. In the context of the square lattice bond model, Seymour and Welsh proved that if

$$P[\exists \text{ an open path in } [0,n] \times [0,n] \text{ from } x = 0 \text{ to } x = n] > \delta > 0 \qquad (14)$$

then

$$P[\exists \text{ an open path in } [0,3n] \times [0,n] \text{ from } x = 0 \text{ to } x = 3n] > f(\delta) > 0 \quad (15)$$

for a strictly increasing function f, independent of n. Their reasoning used the fact that both coordinate axes are symmetry axes for the square lattice, and relied upon correlation inequalities of Fortuin, Kasteleyn, and Ginibre (1971). Furthermore, open paths crossing such rectangles lengthwise were combined to construct an open circuit in a square annulus, $A_n = [-3n,3n]^2 \setminus [-n,n]^2$, surrounding the center, with probability at least $f^4(\delta)$. If $p > p_S$, there is a sequence $\{n_k; k \geq 1\}$ such that (14) holds and the annuli A_{n_k} are disjoint, and then there exists an open circuit around the origin with probability one.

Kesten (1982, Chapter 6) subsequently showed, in the context of site models on two-dimensional periodic lattices, that one axis of symmetry is sufficient to obtain a lower bound for the probability of existence of an open circuit in an annulus (given a lower bound on the probability of an open crossing of a rectangle, for each direction). From Kesten's results, it is known that $p_T(G) + p_H(G^*) = 1$ for site percolation on a matching pair of periodic graphs with two non-perpendicular symmetry axes. While this is similar to the Sykes and Essam relationship, it involves two concepts of critical probability. Therefore it is not sufficient for determining critical probability values, even for self-matching lattices.

Kesten (1980) supplied a path-cutting methods to accomplish the final step. Suppose one determines a candidate p_0 for the critical probability value, for which the sponge-crossing probabilities of open paths in G and closed paths in G^* are bounded below by $\delta > 0$, for both directions. Then $p_0 \leq p_S(G)$, but a technique of closing sites in independent stages to achieve an arbitrary $p < p_0$ shows that the open sponge-crossing probability converges to zero, so in fact $p_0 = p_S(G)$. Interchanging the roles of G and G^*, we obtain $1 - p_0 = p_S(G^*) = 1 - p_H(G)$, so

$$p_T(G) = p_S(G) = p_H(G) = p_0 . \tag{16}$$

The value $p_0 = 1/2$ for the square lattice bond model and the triangular lattice site model is a consequence of the self-matching property of S^c and T. The determination of p_0 for the triangular and hexagonal lattice bond models employed the star-triangle transformation discussed in Sykes and Essam. In general, determination of the critical probability requires a relationship between sponge crossing probabilities in different directions on the lattice, or sponge crossings in the lattice and its matching lattice. For the most comprehensive results currently known, see Chapter 3 of Kesten (1982).

5. Survey of Results

To summarize the critical probabilities that have been rigorously determined:

In the square lattice bond model: $p_T = p_H = p_S = 1/2$. (17)

In the triangular lattice bond model: $p_T = p_H = p_S = 2 \sin \frac{\pi}{18} \approx .347$. (18)

In the hexagonal lattice bond model: $p_T = p_H = p_S = 1-2 \sin \frac{\pi}{18} \approx .653$. (19)

In the site model on a fully-triangulated periodic planar graph with one coordinate axis as a symmetry axis:

$$p_T = p_H = p_S = 1/2 \ . \quad (20)$$

Russo (1981) shows that $p_T = p_H = p_S$ for the square lattice site percolation model, and Kesten (1982) and Kiguchi (1982) independently improved his techniques to show that $p_H > 1/2$.

In some multiparameter models, the critical surface in the parameter space, which separates the percolative and non-percolative regions, has been determined. For the square lattice bond percolation model where each horizontal bond is open with probability p and each vertical bond is open with probability q , Kesten (1982) showed that the critical surface is $\{(p,q) \in [0,1]^2: p + q = 1\}$. In the triangular lattice, if bonds in each of two directions are open with probability p and each bond in the third direction is open with probability q , the critical surface is $\{(p,q) \in [0,1]^2: p + 2q - pq^2 = 1\}$. The equality of probabilities in two directions is needed to obtain a symmetry axis. The 3-parameter model, with probabilities p, q, and r in the three directions, was conjectured to have critical surface $p + q + r - pqr = 1$ by Sykes and Essam, but this has not been rigorously verified.

Since the rigorous determination of critical probabilities depends crucially on the concept of the matching property, which is not available in higher dimensions, there are no exact values known in dimensions $d \geq 3$.

For the site percolation models on $G = S, S^c, T, T^c$ or H^c , the arguments used to prove $p_H(G) = p_T(G) = p_S(G)$ and $p_T(G) + p_H(G^*) = 1$, which may be rewritten as $p_H(G) + p_H(G^*) = 1$, provide the following description of the cluster sizes on G and G^* (where C_v^* denotes the open cluster in G^* containing v) .

For $p < p_H(G)$,

$$\begin{aligned} &\text{(i)}\quad P_p(\#C_v = \infty) = 0 \\ &\text{(ii)}\quad P_{1-p}(\#C_v^* = \infty) > 0 \\ &\text{(iii)}\quad P_p\ (\exists \text{ a unique infinite closed cluster in } G^*) = 1 . \end{aligned} \tag{21}$$

For $p = p_H(G)$,

$$\begin{aligned} &\text{(i)}\quad P_{p_H}(\#C_v = \infty) = P_{1-p_H}(\#C_v^* = \infty) = 0 \\ &\text{(ii)}\quad E_{p_H}(\#C_v) = E_{1-p_H}(\#C_v^*) = \infty \\ &\text{(iii)}\quad \text{For each } N , \\ &\qquad P_{p_H}\ (\exists \text{ an open circuit in } G \text{ surrounding } [-N,N]^2 \text{ and a} \\ &\qquad \text{closed circuit in } G^* \text{ surrounding } [-N,N]^2) = 1 . \end{aligned} \tag{22}$$

For $p > p_H(G)$

$$\begin{aligned} &\text{(i)}\quad P_p(\#C_v = \infty) > 0 \\ &\text{(ii)}\quad P_{1-p}(\#C_v^* = \infty) = 0 \\ &\text{(iii)}\quad P_p\ (\exists \text{ a unique infinite open cluster in } G) = 1 . \end{aligned} \tag{23}$$

To show the uniqueness of the infinite cluster in (21,iii) or (23,iii), use (22,iii): For (23,iii), suppose the vertices x and y are each in an infinite open cluster. Choose N sufficiently large that $[-N,N]^2$ contains both x and y . Then an open circuit surrounding $[-N,N]^2$ exists in G , and must intersect both C_x and C_y , showing that x and y are in the same open cluster.

Much of the interest of physicists in percolation models is due to the threshold behavior at the critical probability, which is analogous to a phase transition. Usually phase transition is associated with a singularity of some function involved in the physical system's description. Attempts are made to predict the behavior of various quantities in the immediate neighborhood of the threshold, analyzing these quantities as powers of the difference between the parameter value and the threshold. Such powers are called critical exponents. The existence of critical exponents has not been established for any of the periodic graphs discussed in this paper, but some progress has been made in this direction.

For S and S^c, the percolation probability function $\theta(p)$ is a continuous function of $p \in [0,1]$, and is infinitely differentiable at all $p \neq p_H$. It is not known if $\theta(p)$ is analytic for $p > p_H$. Kesten has shown that there exist constants $0 < k_i < \infty$, $0 < \beta_i < \infty$, $1 \leq i \leq 6$, such that

$$k_1(p - p_H)^{\beta_1} \leq \theta(p) \leq k_2(p - p_H)^{\beta_2} \qquad \text{for } p \geq p_H \qquad (24)$$

$$k_3(p_H - p)^{-\beta_3} \leq E_p(\#C_v) \leq k_4(p_H - p)^{-\beta_4} \qquad \text{for } p < p_H \qquad (25)$$

$$k_5(p-p_H)^{-\beta_5} \leq E_p(\#C_v I_{\{\#C_v < \infty\}}) \leq k_6(p-p_H)^{-\beta_6} \qquad \text{for } p > p_H . \qquad (26)$$

For general G, based on numerical evidence, the behavior of θ is expected to be approximately that of $c(p - p_H)^{\beta}$ as $p \to p_H^+$, where β depends only on the dimension of the graph, with similar expectations for the expected cluster size and expected size of finite clusters also.

A crucial step in the proof of these bounds is the derivation of bounds on tail probabilities of the cluster size distribution. These tail probability bounds also illustrate the difference in clustering behaviors between the percolative and non-percolative regions. Kesten's proof that for a periodic graph G, $p_T = p_S$, demonstrates that there is exponential decay of the cluster size distribution in the non-percolative region: For $p < p_T$, there exists a universal constant $k_1 < \infty$ and $k_2(p,G)$ such that

$$P_p(\#C_v \geq n) \leq k_1 \exp\{-k_2 n\} . \qquad (27)$$

At the critical probability p_H, for $G = S$, S^c or S^*, there exist constants

$0 < k_i , \gamma_i < \infty$ such that

$$k_1 n^{\gamma_1 - 1} \le P_{p_H}(\#C_v \ge n) \le k_2 n^{\gamma_2} . \tag{28}$$

[It is relatively simple to obtain an exponent of -1 in the lower bound.] For G periodic in R^d, Aizenman, Delyon and Souillard (1980) showed that for $p > p_H$ there exists $0 < k_1 , k_2 < \infty$ such that

$$P_p(n \le \#C_v < \infty) \ge k_1 \exp\{-k_2 n^{(d-1)/d}\} \qquad (d \ge 2) \tag{29}$$

For $d = 2$ and $G = S$, S^c, or S^* the order of decay is known to be $\exp\{-n^{1/2}\}$ in the percolative region:

$$P_p(n \le \#C_v \le \infty) \le k_3 \exp\{-k_4 n^{1/2}\} \text{ for } p > p_H . \tag{30}$$

In conclusion, current mathematical knowledge of percolation models leaves much to be desired, since many properties which are intuitive and accepted by physical scientists on the basis of numerical evidence, have not been rigorously proved. This paper surveys results at the current frontier of the mathematical research in percolation theory, showing that considerable advances have been made recently. The rapidly growing literature, filled with variations and extensions of the standard model, conjectures, plausibility arguments, and simulation studies, raises many interesting and formidable mathematical questions for future research.

References

1. Aizenman, M., Delyon, G., and Souillard, B., (1980), Lower bounds on the cluster size distribution, J. Stat. Phys. 23, 267-280.

2. Broadbent, S.R., and Hammersley, J.M. (1957) Percolation processes, Proc. Camb. Phil. Soc. 53, 629-641, 642-645.

3. Fisher, M.E. (1961), Critical probabilities for cluster size and percolation problems. J. Math. Phys. 2, 620-627

4. Fortuin, C.M. Kasteleyn, P.W., and Ginibre, J. (1971), Correlation inequalities on some partially ordered sets. Comm. Math. Phys. 22, 89-103.

5. Higuchi, Y. (1982), Coexistence of the infinite (*) clusters; a remark on the square lattice site percolation. Z. Wahrsch. verw. Geb., to appear.

6. Kesten, H. (1980), The critical probability of bond percolation on the square lattice equals ½. Comm. Math. Phys. 74, 41-59.

7. Kesten, H. (1981), Analyticity properties and power law estimates of functions in percolation theory. J. Stat. Phys. 25 717-756.

8. Kesten, H. (1982), Percolation Theory for Mathematicians, Birkhäuser, Boston.

9. Russo, L. (1981), On critical percolation probabilities. Z. Wahrsch. verw. Geb. 56, 229-237.

10. Seymour, P.D. and Welsh, D.J.A. (1978), Percolation probabilities on the square lattice. Ann. Discrete Math. 3, 227-245.

11. Smythe, R.T. and Wierman, J.C. (1978), First-passage Percolation on the Square Lattice. Lecture Notes in Mathematics, vol. 671, Springer-Verlag.

12. Sykes, M.F. and Essam, J.W. (1964), Exact critical percolation probabilities for site and bond problems in two dimensions. J. Math. Phys. 5, 1117-1127.

13. Van den Berg, J. (1981), Percolation theory of pairs of matching lattices. J. Math. Phys. 22, 152-157.

14. Wierman, J.C. (1981), Bond percolation on honeycomb and triangular lattices. Adv. Appl. Prob. 13, 293-313.

15. Wierman, J.C. (1982), Percolation theory. Ann. Prob. 10, 509-524.

CRITICAL EXPONENTS AND THRESHOLDS FOR PERCOLATION AND CONDUCTION

Muhammad Sahimi

Department of Chemical Engineering and Materials Science
University of Minnesota
Minneapolis, MN 55455 U.S.A.

ABSTRACT

We review recent developments in percolation, lattice animals and percolation conductivity theories. We also review recent improvements in numerical methods which are crucial for the study of percolation and related theories.

1. Introduction

Percolation theory continues to receive wide attention. This is no surprise as percolation theory provides a testing ground for theories in phase transitions and critical phenomena, as well as being a good model for a variety of physical phenomena. Recent reviews include those of Stauffer [1], Stanley [2] and Stauffer, Coniglio and Adam [3]. The paper by Hughes and Prager in this volume contains references to some of the earlier reviews. We review in detail only those recent developments in percolation and related theories that have taken place since the publication of these reviews, or were not covered by them; a few earlier results are also mentioned briefly for completeness. In particular, attention is restricted to ordinary (isotropic) percolation, lattice animals and percolation conductivity on one hand and directed percolation and directed lattice animals on the other. These classes of problems belong to different universality classes, and considerable progress has been made in the past two years in understanding the shape and size of directed lattice animals and directed percolation and their relevance to a wide variety of problems. However, we only give the critical exponents of directed percolation and discuss briefly a few rigorous results on the percolation thresholds of directed lattices. The interested reader is referred to the second article by Redner in this volume for a fuller exposition of the directed percolation problem. Many variants of percolation theory have been invented in the past few years to model particular physical phenomena: the interested reader is referred to the excellent article of Stauffer, Coniglio and Adam [3] for a review of the variants of the ordinary percolation problem and their applications.

2. Exact Results for Percolation Thresholds

2.1. Isotropic Percolation

The non-rigorous results of Sykes and Essam [4] for bond percolation thresholds p_c^B of hexagonal ($1-2\sin\pi/18$), square ($1/2$) and triangular ($2\sin\pi/18$) lattices, and site percolation thresholds p_c^S of triangular ($1/2$) and Kagomé ($1-2\sin\pi/18$) lattices have been made rigorous by Kesten [5] and Wierman [6] (See also the paper by Wierman in this volume.) Kondor [7], using a generalized star-triangle transformation and some conjectures of Wu [8] concerning the critical points of 2-dimensional Potts models, had conjectured that $p_c^S = 1/2^{1/2} = 0.707107$ for the hexagonal lattice. This value was very close to the series estimate $p_c^S = 0.698 \pm 0.003$ of Sykes <u>et al</u>. [9], though clearly outside of the error bars. However, Djordjevic <u>et al</u>. [10] invented a new method of estimating the per-threshold (see below) and found that $p_c^S = 0.6962 \pm 0.0006$ which reliably rules out the result of Kondor [7]. In fact Wu's conjectures on the critical points for the Kagome and diced lattices have been shown to be incorrect [11]. Thus the determination of p_c^S for the hexagonal lattice remains an outstanding unsolved problem.

Tsallis [12] proposed a new relationship from which the exact phase diagram of the nearest-neighbor q-state Potts ferromagnet in the fully anisotropic 3-12 lattice can be obtained. His conjecture is based on a star-triangle transformation and recovers all the available exact results concerning particular cases. As a consequence of his conjecture the bond percolation threshold of Kagome lattice is predicted to be $p_c^B = \frac{1}{3}\{1 + 4\sin[\frac{1}{3}\sin^{-1}(1 - \frac{27}{8}\sin\pi/18)]\} = 0.522372$.

For three and higher dimensional Bravais lattices, Sahimi <u>et al</u>. [13] noticed a very curious relationship between bond percolation thresholds and a certain lattice Green function. If by $\{i\}$ we denote the set of nearest neighbors of site i, then a lattice Green function G_i can be defined as the unique solution of the equation [14]

$$zG_i - \sum_{j\epsilon\{i\}} G_j = \delta_{io} \quad (1)$$

which decays to zero as the distance of site i from the origin increases. (Such a decaying solution only exists in three or higher dimensions.) Sahimi et al. [13] observed that

$$p_c^B \cong G_o \quad . \quad (2)$$

They conjectured that this result may in fact be exact for Bravais lattices, based on the excellent agreement that they found between the estimates of p_c^B for three and higher dimensional lattices and the corresponding values of G_o. Since as the dimensionality d of the lattice increases we must have $p_c^B \to 1/(z-1)$, where z is the lattice coordination number, and since $G_o \to 1/(z-1)$ as $d \to \infty$, thus their conjecture has the correct limiting behavior. The lattice Green function G_o figures prominently in the theory of random walks. Hughes et al. [15] also explored the possible relationship between random walk statistics and the percolation thresholds.

We remark that much series, Monte Carlo and renormalization group work remains to be done before the percolation thresholds for three-dimensional lattices are known with the same precision and confidence as they are for the critical points of the three-dimensional Ising model. Since 1961 when Vyssotsky et al. [26] estimated, for the first time, the bond percolation threshold of three-dimensional lattices, various authors have estimates p_c^B for these lattices. All of these estimates supposedly have very tight error bars and yet often these various estimates do not overlap with each other! To give the reader some ideas about how unsettled the situation is, we compile the available estimates of p_c^B for the simple cubic lattice in three dimensions in table 1.

p_c^B	method
0.2465 ± 0.0002 [16]	Series
0.247 ± 0.003 [17,18]	Series
0.2479 ± 0.004 [19]	Series
0.248 ± 0.001 [20]	Monte Carlo
0.2492 ± 0.0002 [21]	Monte Carlo
0.2495 ± 0.0005 [22]	Monte Carlo-Renormalization
0.25 ± ? [23]	Monte Carlo
0.2511 ± 0.0015 [24]	Phenomenological Renormalization of Monte Carlo Data
0.2526 ± 0.00013 [25]	Monte-Carlo Renormalization
0.25273 [13]	G_0 (Eq. 2)
0.254 ± 0.013 [16]	Monte Carlo

Table 1. The available estimates of p_c^B, the bond percolation threshold of the simple cubic lattice in three dimensions. Numbers in brackets denote the reference.

2.2. Directed Percolation

No exact results for the percolation thresholds of directed lattices are known yet, aside from a very special case [27]. However, Dhar, Phani and Barma [28] exploited a relationship between the generating functions for site animals and the percolation probability and obtained the following relationship between the site percolation thresholds of directed hexagonal and square lattices:

$$p_c^h = (p_c^s)^{1/2}. \tag{3}$$

Essam and De'Bell [29] obtained a similar relationship between the site percolation thresholds of directed diamond and simple cubic lattices:

$$p_c^d = (p_c^{sc})^{1/2} \quad . \tag{4}$$

Aside from these two rigorous relationships there are also a few rigorous upper and lower bounds for site and bond percolation thresholds of a fully directed simple cubic lattice in d dimensions. By treating the percolation problem as a Markov process, Mauldon [30] obtained a very strong lower bound for p_c^B of the fully directed square lattice. His results, $p^B > 0.6297$ compares very well with the value $p^B = 0.6447 \pm 0.0001$ that Kinzel and Yeomans [31] obtained by using the transfer-matrix method. This lower bound is identical to that obtained by Dhar [32] very recently and is somewhat better than the lower bound of Gray et al. [33], $p_c^B > 0.6231$. Bishir [34] showed that for the fully directed square lattice $p_c^S > 2/3$ which should be compared with the estimate of Kinzel and Yeomans [31], who obtained $p^S = 0.7058 \pm 0.0001$. Very recently Cox and Durret [35] showed rigorously that $p_c^B(d)$, the bond percolation threshold of a fully directed simple cubic lattice in d dimensions, obeys the inequalities

$$d^{-1} + \frac{1}{2} d^{-3} + O(d^{-3}) < p_c^B(d) < d^{-1} + d^{-3} + O(d^{-4}) \ , \tag{5}$$

while earlier Blease [36] had obtained the (non rigorous) expansion

$$p_c^B(d) = d^{-1} + \frac{1}{2} d^{-3} + d^{-4} + 3d^{-5} + \frac{21}{2} d^{-6} + \frac{479}{12} d^{-7} + O(d^{-8}) \tag{6}$$

which agrees with the result of Cox and Durret. The second paper by Redner in this volume may be consulted for related results about directed percolation.

3. Critical Exponents for Percolation

3.1. Isotropic Percolation

For the definition of critical exponents for percolation see the paper by Hughes and Prager in this volume. Den Nijs [37] proposed a relationship for the thermal exponent y_T (the inverse of the correlation length exponent) between the two-dimensional q-state Potts model and the exactly solved eight-vertex model of Baxter [38], from which one deduces that

$$y_T = \frac{3}{2} \frac{2x-\pi}{x-\pi} \quad , \quad q = 4\cos^2 x \quad , \ 0 < x < \frac{\pi}{2} \quad . \tag{7}$$

Pearson [39] and Nienhuis et al. [40] independently proposed another relationship for the magnetic exponent y_h,

$$y_h = \frac{3}{2} + \frac{3\pi}{8(\pi-x)} - \frac{x}{2\pi} \quad . \tag{8}$$

From these two relationships one can calculate all critical exponents for all $q < 4$ and thus for the percolation limit $q = 1$ through the usual scaling relations

$$\nu = 1/y_T \quad , \quad \beta = (d-y_h)/y_T \quad , \quad 2 - \alpha = \gamma + 2\beta = \beta(1+\delta) = d\nu \quad . \tag{9}$$

Since their discovery three years ago these relationships have been confirmed by a variety of methods, some of which are numerical and others theoretical. We mention briefly these recent developments in support of the conjectures of den Nijs-Pearson-Nienhuis et al. (We note here that the conjecture of Klein et al. [41] that $y_T = \ln 3/2/\ln 3^{1/2} = 0.738$ for percolation at $d = 2$ was reliably ruled out.)

Blöte, Nightingale and Derrida [42] were the first to confirm these conjectures numerically. Using finite-size scaling and transfer matrix methods, they confirmed these conjectures for all continuous values of q in the interval $0 < q < 4$ to within 0.2 percent. The slow convergence of their results around $q \approx 4$ for finite size systems to the infinite system results was attributed to the logarithmic corrections to the critical exponents around $q \approx 4$.

Lobb and Karasek [43] used the large-cell Monte Carlo renormalization method of Reynolds et al. [44] for bond percolation on the square lattice with self-dual cells [45]. Their results strongly supported the conjectures. Eschbach, Stauffer and Herrmann [46] used the same method on triangular site problem with cells up to $b = 10000$, where b is the scale factor of the renormalization, with the stated intention of ruling the den Nijs conjecture out. Their results for y_T confirmed the conjecture! Derrida and De Seze [47], using the transfer matrix method, also confirmed the den Nijs conjecture for the percolation limit. Earlier large-cell Monte Carlo renormalization of Reynolds et al. [44] on the square lattice site problem with cell sizes up to $b = 500$ had lent support to the conjecture of Klein et al. [41].

Duarte and Ruskin [48] reanalyzed the series expansion data of Sykes et al. [9] and Gaunt and Ruskin [17] to obtain the 'susceptibility' exponent $\gamma = 2.41 \pm 0.025$, for percolation, just compatible with the theoretical prediction $\gamma = 43/18 = 2.3888$. Thus they removed a discrepancy between the series estimates [9] $\gamma = 2.43 \pm 0.03$ and [17] $\gamma = 2.425 \pm 0.005$ and the theoretical prediction. Adler and Privman [49] analyzed the three-state Potts model with a series expansion method modified to explicitly account for the effect of confluent corrections to scaling. Their estimate $\beta = 0.1111 \pm 0.0006$ of the magnetization exponent agrees excellently with the conjectured value $\beta = 1/9$. All of the earlier estimates of β had fallen below 0.109, thus casting doubts on the validity of the conjectures. Their estimate $\gamma = 1.449 \pm 0.027$ also agrees well with the theoretical prediction $\gamma = 13/9 = 1.444$.

Black and Emery [50] studied two-dimensional classical models and (1+1)-dimensional quantum systems. They were able to derive Eq. (7) by using a Coulomb-gas representation of the Potts model and showed that it is exact asymptotically. Den Nijs [51] derived the Pearson-Nienhuis et al. conjecture from the spin-spin correlation function in the Coulomb-gas representation. Herrmann [52] took essentially the same approach as that of Black and Emery and derived an exact solution for the one-dimensional quantum q-state Potts model for a finite chain. By combining this solution with finite-size scaling analysis he found $y_T = 1.48 \pm 0.01$ for $q = 4$, in good agreement with the conjectured value $y_T = 3/2$. The largest previously estimated value of y_T had been 1.365 obtained by phenomenological renormalization [42].

Kunz [53] calculated the magnetic exponent y_h for the zero state Potts model. His result $y_h = 2$ agrees wih the conjectured value. It is believed [54] (but not rigorously proven) that the $q = 3$ Potts model is in the same universality class as the hard hexagon lattice gas. Baxter [55] solved the hard hexagon problem exactly and the critical exponents $\alpha = 1/3$ and $\beta = 1/9$ as predicted by his solution agree with the conjectured values. The four-state Potts model is believed [56] to be in the same universality class as the Baxter-Wu [57] model, and the predicted valued [57] of $\alpha = 2/3$ and $\beta = 1/12$ are in agreement with the conjectured values. Nienhuis [58] was also able to give some theoretical (though not rigorous) argument in support of the Pearson-Nienhuis et al. conjecture. Thus there remains little doubt that the den Nijs-Pearson-Neihnuis et al. conjectures are exact for all $0 \leq q \leq 4$ and thus for the percolation limit $q = 1$. We remind the the reader that the Potts model in two dimensions exhibits a discontinuous (first order) phase transition for $q > 4$ as was demonstrated by Baxter and co-workers [59].

Several 'non-critical' exponents can be defined by the following equations,

$$\begin{aligned} n_s(p) &\propto s^{-\theta} \exp(-\text{const. } s^{\zeta}) \\ R_s(p) &\propto s^{\rho} \quad . \end{aligned} \qquad s \to \infty \tag{10}$$

Here $n_s(p)$ is the average number, per lattice site, of clusters of s sites at fraction p of sites and R_s is the average radius of clusters of s sites. Exponents θ, ζ and ρ are called non-critical because they characterize the behavior of $n_s(p)$ and $R_s(p)$ for all values of p. For ordinary percolation one has for ζ, θ and ρ one value for all p such that $0 < p < p_c$, one value for all p such that $p_c < p < 1$ and except for ζ another value for $p = p_c$. For example [1] $\zeta(p < p_c) = 1$, $\zeta(p > p_c) = 1-1/d$ and $\rho(p > p_c) = 1/d$, where d is the dimensionality of the system. There is no scaling relation between these exponents and the critical exponents discussed earlier. We shall discuss the properties of these exponents later in connection with lattice animals.

In contrast to two dimensions there are no (possibly) exact results for the critical exponents or thresholds for three-dimensional systems. The percolation thresholds and the critical exponents can be determined only by numerical means such as series expansion, Monte Carlo, and renormalization group methods. The values obtained by these methods vary greatly. Series expansion methods have not yielded satisfactory results for the critical exponents partly because the method depends heavily on the exact location of the percolation thresholds which are not known yet. In fact very recent work of Gaunt and Sykes [19] seems to contradict the universality principle; e.g. their estimate of β, the percolation probability critical exponent, varies greatly from one lattice to another and it also depends on the percolation problem, i.e. the values are different for site and bond problems. At present there is no general consensus concerning the values of critical exponents obtained from series expansions, Monte Carlo simulations or any other technique. Phenomenological renormalization and transfer matrix methods which proved to be extremely powerful tools of estimating critical properties in two dimensions are not practical at present for three-dimensional systems because of the enormous computer storage and time that are necessary for even relatively small systems. There has been a great deal of work using position-space renormalization group methods (for a review see Stanley et al. [60]) but most of these have been confined to two dimensions. For $d = 3$, this method has been used by several authors but the values vary greatly from one author to another. The most successful attempt was made by Payandeh [61] who calculated all critical exponents in three dimensions. His results are in good agreement with the values listed in Table 2. Table 2 summarizes the accepted values of critical exponents in two and three dimensions. For comparison the mean field values of critical exponents are also included. We mention that site percolation on a continuum (overlapping circles in two dimensions) appears to belong to the same universality class as site percolation on a lattice [62].

Exponent	d = 2	d = 3	Mean Field Value
α	-2/3	-0.64	-1
β	5/36	0.45	1
γ	43/18	1.74	1
δ	91/5	4.87	2
ν	4/3	0.88	1/2
σ	36/91	0.46	1/2
τ	187/91	2.20	5/2

Table 2: The accepted values of the critical exponents in two and three dimensions and the corresponding mean field values. The values for two-dimensional systems are possibly exact. The values for $d = 3$ are from Heermann and Stauffer [20] and Gaunt and Sykes [19].

3.2. Directed Percolation

Series expansion estimates of the critical exponents β and γ for directed percolation in two and three dimensions are given by Blease [63]. The results are $\beta(d = 2) = 0.28 \pm 0.02$, $\gamma(d = 2) = 2.27 \pm 0.04$, $\beta(d = 3) = 0.60 \pm 0.05$, and $\gamma(d = 3) = 1.58 \pm 0.10$. In directed percolation there are two correlation length exponents (see the second paper by Redner in this volume for details). These are $\nu_\parallel$ and $\nu_\perp$ which characterize the divergence of the correlation lengths in longitudinal ("preferred") and transverse directions respectively. Kinzel and Yeomans [31] estimated $\nu_\parallel$ and $\nu_\perp$ in two dimensions by means of the transfer-matrix method. The results are $\nu_\parallel = 1.730 \pm 0.002$ and $\nu_\perp = 1.094 \pm 0.01$, which are in agreement with the estimates of Essam and De'Bell [64] that were obtained by a series expansion method. Redner and Coniglio [65] derived Flory-type approximate formulae for $\nu_\parallel$ and $\nu_\perp$. The results are

$$\nu_\parallel = \Delta(d + 9)/4(d + 2) \tag{11}$$

$$\nu_\perp = 7\Delta/4(d + 2) \tag{12}$$

where $\Delta = \beta + \gamma$ is the "gap" exponent. Obukhov [66] argued that $d = 5$ is the upper critical dimension for directed percolation and derived the critical exponents to leading order in $\varepsilon = 5 - d$. Cardy and Sugar [67] showed that there is an exact mapping between the directed percolation problem and Reggeon field theory which models the creation, propagation and destruction of a cascade of elementary particles.

4. Lattice Animals

Let $A_s(p)$ be the average number (per lattice site) of clusters and a_{sm} be the total number of geometrically different configurations for a cluster with s sites and perimeter m. Thus $A_s(p) = \sum_m a_{sm} p^s (1-p)^m$, and one is not only interested in calculating the total number a_{sm} of geometrically different such clusters, but also its asymptotic behavior for large s, which is usually written as

$$A_s(p) \sim \lambda^s s^{-\theta} \quad , \tag{13}$$

where exponent θ is universal, i.e. it depends only on the dimensionality of the system and not on its details, whereas the growth parameter λ is not universal. It was established by Lubensky and Isaacson [68] that the properties of lattice animals cannot be determined through the percolation problem at the percolation threshold, i.e. animals have their own critical points and critical exponents and there is no simple relationship between their exponents and those of percolation. This was also shown by Family and Coniglio [69] (with a position-space renormalization group approach) and by Harris and Lubensky [70]. These authors showed that the isotropic lattice animals are the q = 1 limit of a special case of the polychromatic Potts model. In fact the non-critical exponents of percolation are the same as the critical exponents of lattice animals. The upper critical dimensionality d_u for the two problems is different: while mean-field exponents become valid at $d_u = 6$ for isotropic percolation [71], for lattice animals the upper critical dimensionality is [68,70] $d_u = 8$. For directed percolation the upper critical dimensionality is [66] $d_u = 5$, whereas for the directed animals, animals that can grow only into a single 'quadrant', it is [72] $d_u = 7$. For isotropic and directed animals the most important recent developments are as follows.

4.1. Isotropic Animals

We define the exponent ν through

$$\xi \sim s^{\nu} \tag{14}$$

where ξ is a cluster radius for animals. Employing a field theoretic model for the polymer problem developed by Lubensky and Isaacson [68], Parisi and Sourlas [73] demonstrated that the universal asymptotic properties of large lattice animals (dilute branched polymers) in d dimensions are the same as the universal critical properties of the Yang-Lee edge singularity of the Ising model in (d-2) dimensions. As an immediate consequence one can conclude that the upper critical dimension of isotropic lattice animals is $d_u = 8$ as mentioned above and that the two animal exponents θ and ν are not independent but uniquely determined by the single exponent σ, which controls the behavior of the magnetization M near the Yang-Lee edge in the presence of an imaginary external field H: $M \sim (H-H_o)^{\sigma}$ (see,

Fisher [74]). Parisi and Sourlas [73] showed that

$$\theta = \nu(d-2) + 1 . \qquad (15)$$

Family [75] succeeded in deriving (15) by using Ginzburg criteria and without reference to Yang-Lee problem. The Parisi-Sourlas-Family relation is valid for $p < p_c$ and in particular it yields the exact values $\theta(d = 2) = 1$, $\theta(d = 3) = 3/2$, and $\nu(d = 3) = 1/2$. In two dimensions the best estimate of ν seems to be $\nu = 0.64$ obtained by the transfer matrix method and by renormalization group method [76]. This value of ν was obtained for site animals. For bond animals the only available estimate in $\nu \approx 0.63$ which was obtained [77] by renormalization group method with small cells. From the universality principle one expects these two exponents (for site and bond animals) to be equal. Lubensky and McKane [78] made a connection between percolation problem and the Ising droplets and obtained (for $p > p_c$)

$$\theta = (1 + 4d - d^2)/2d \qquad (16)$$

This equation is not valid at $d = 3$ and $d = 5$, there $\theta\ (d = 3, p > p_c) = - 1/9$ and $\theta(d = 5, p > p_c) = - 449/450$. Isaacson and Lubensky [79] and Daoud and Joanny [80] have also derived various approximate formulae for animal exponents using Flory-type approximations. These formulae are in excellent agreement with the exact results mentioned above. We mention that the mean field values for θ and ν (i.e. their values at $d_u = 8$) are 5/2 and 1/4 respectively. In addition to the statistics of dilute branched polymers, the configurational statistics of isotropic lattice animals have applications in a variety of diverse problems including the cell growth problem [81], homogeneous turbulent flow [82] percolation [1] and spinodal decomposition [83].

4.2. Directed Animals

For directed animals we have to define two cluster radii, one for the longitudinal direction, and one for transverse direction,

$$\xi_{\parallel} \sim s^{\nu_{\parallel}} \quad , \quad \xi_{\perp} \sim s^{\nu_{\perp}} . \qquad (17)$$

Family [75], using Ginzburg criteria, derived the following relationship which is the analog of the Parisi-Sourlas-Family relation for directed animals, viz.

$$\theta = \nu_{\perp}(d-1) \quad . \tag{18}$$

Breuer and Janssen [84] and Cardy [85], by employing the field theoretic model developed by Day and Lubensky [72], showed that the universal asymptotic properties of large d-dimensional directed animals, too, are equivalent to the universal critical properties of the Yang-Lee singularity in (d-1) dimensions and derived the same relationship as in Eq. (18). This connection had been conjectured earlier by Day and Lubensky [72]. Numerically, Stanley, Redner and Yang [86] studied the lattice animal problem by series expansions method and established Eq. (18). In particular Eq. (18) implies the exact results $\nu_{\perp} = \theta = 1/2$ for $d = 2$ and $\theta = 0$ for $d = 1$. We therefore have an interesting correspondence betweeen the Yang-Lee problem and those of isotropic and directed animals. This is very intriguing because it means that the singularities of an Ising model in an imaginary magnetic field are related to the exponents characterizing a purely geometrical problem. Shapir [87] showed that the Yang-Lee edge singularity belongs to the same universality class as the negative dimer problem! Redner and Coniglio [65] and Lubensky and Vannimenus [88] derived various Flory-type approximate formulae for directed lattice animals (and percolation) exponents with excellent accuracy.

The critical exponent $\nu_{\parallel}$ of directed animals requires knowledge of the dynamics of the Yang-Lee problem. This problem seems to be nontrivial, even in one dimension. Breuer and Janssen [84] showed that

$$\nu_{\parallel}/\nu_{\perp} = z_{YL}(d-1) \quad . \tag{19}$$

The dynamical exponent z_{YL} characterizes critical slowing down at the Yang-Lee edge according to the Glauber [89] - kinetic Ising model. Eqs. (18) and (19) imply the exact results $\theta = 3/2$, $\nu_{\parallel} = 1/2$ and $\nu_{\perp} = 1/4$ at $d = 7$ (the upper critical dimensionality for directed animals) and $\nu_{\parallel} = 1$ at $d = 1$. Series expansions estimates of $\nu_{\parallel}$ and $\nu_{\perp}$ for all $2 < d < 7$ are given by Redner and Yang [90], and Stanley, Redner and Yang [86], in particular $\nu_{\parallel}(d = 2) = 0.8$, $\nu_{\parallel}(d = 3) = 0.7$,

$\nu_\perp(d = 3) = 0.46$ and $\theta(d = 3) = 0.94$. Series expansion methods for directed lattice animals (and percolation) are supposed to be more accurate than the accuracy that can be reached for the isotropic problem. This is because the directional constraint greatly simplifies the problems, at least in numerical studies.

Perhaps the most important development concerning the directed lattice animals has been the exact solution for the directed-site animal problem on the square and triangular lattices derived by Dhar [91]. Dhar showed that this problem is equivalent to Baxter's [92,93] hard-square lattice-gas model with anisotropic next-nearest neighbor interaction at negative fugacity. Thus Dhar was able to derive exact formulae for the animal numbers as function of their size s for both square and triangular lattices, confirming the formulae that had been guessed earlier by Dhar, Phani and Barma [28]:

$$A_s = \frac{1}{2\pi} \int_0^{2\pi} dk(e^{-ik} + 1)(1 + 2\cos k)^{s-1} \qquad \text{square} \qquad (20)$$

$$A_s = \frac{1}{2\pi} \int_0^{2\pi} dk(e^{-ik} + 1)(2 + 2\cos k)^{s-1} \qquad \text{triangle} \qquad (21)$$

The exact solution of Dhar also implies that $\lambda = 3$ exactly for the square lattice and $\lambda = 4$ for the triangular lattice animals. Nadal, Derrida and Vannimenus [94] independently obtained $\lambda = 3$ for square lattice animals. These authors also conjectured that $\nu_\parallel = 9/11$ for two dimensional directed lattice animals. They also obtained very accurate estimates of critical properties of some variations of the two dimensional directed lattice animals. Directed lattice animals may model dilute branched polymers in a suitably flowing solvent, and are also closely related to branching Markov processes with a single source. Green and Moore [95] proposed that the exponents $\nu_\parallel$ and $\nu_\perp$ describe the large-scale properties of river networks!

5. Improvements in Numerical Methods

Djordjevic, Stanley and Margolina [10] invented a new method for calculation of critical percolation thresholds. Their starting point is the scaling assumptions of Stauffer [1]. According to this scaling law $n_s(p)$, the mean number of s-sites (s-bonds) clusters per lattice site, obeys the asymptotic relation

$$n_s(p) \sim s^{-\tau} f(xs^{\sigma}) \qquad x \to 0,\ s \to \infty , \tag{22}$$

where $x = (p_c-p)/p$ and τ and σ are critical exponents related to the usual percolation exponents through $\tau = 2 + 1/\delta$ and $\sigma = 1/\beta\delta$ (see also Table 2). Moreover the function f has been found to be universal (see e.g. Nakanishi and Stanley [96]) in that it depends only on system dimensionality. If we define p_{max} to be the value of p for which $n_s(p)$ achieves its maximum value for fixed s and define $R_s \equiv n_s(p_{max})/n_s(p_c)$, then for the limiting value $R = \lim_{s\to\infty} R_s$ as $s \to \infty$ and $p_{max}(s) \to p_c$ one expects from (22) that

$$R = f(x_{max})/f(0) \tag{23}$$

and that R be universal. Here x_{max} is the value of x for which f achieves its maximum. Thus if R can be estimated to a good degree of accuracy for a lattice whose p_c is known exactly (since R is sensitive to p_c), then p_c for a lattice whose critical point is not known can be estimated by a much better accuracy by varying p_c^{trial} until $R(p_c^{trial}) \approx R$. This method was tried for triangular bond problems: p_c^B was estimated to be $p_c^B = 0.3478 \pm 0.0007$ in excellent agreement with the exact value $p_c^B = 0.3473$. For the hexagonal lattice they found $p_c^B = 0.6522 \pm 0.0007$, in excellent agreement with the exact value $p_c^B = 0.6527$. The site percolation threshold of the square lattice was estimated to be $p_c^S = 0.5923 \pm 0.0007$, in agreement with transfer-matrix method estimate [47] $p_c^S = 0.5927$.

Finite-size scaling combined with the transfer-matrix method has proven to be a powerful method for the investigation of critical behavior of two-dimensional systems. The number of systems to which this method has been applied makes a very long list and the reader is referred to the very recent review article by Nightingale [97] for more details. The large-cell Monte Carlo renormalization method appears to be somewhat less accurate than the transfer matrix method, although the latter has been restricted to two dimensions up to now. The large-cell Monte Carlo renormalization method has the advantage that it works for any problem for which connectivity is important. Barber and Selke [98] developed a phenomenological renormalization that can be used with Monte Carlo data. This is based on calculating any quantity $P_\ell(x)$ of interest, where ℓ is the linear dimension of system, and consequently determining the quantity $\zeta_{\ell,\ell'}(x)$ defined by

$$\zeta_{\ell,\ell'}(x) = \ln[P_\ell(x)/P_{\ell'}(x)] \;/\; \ln(\ell/\ell') \qquad (24)$$

If $P_\ell(x)$ scales with ℓ as

$$P_\ell(x) \sim \ell^{\omega} F(\ell^{1/\nu} y) \quad , \qquad (25)$$

where $y = (x-x_c)/x_c$, $\omega = \rho/\nu$, and ρ is the critical exponent of $P_\ell(x)$ as $\ell \to \infty$, i.e.

$$P_\infty(x) \sim y^{-\rho} \quad , \qquad (26)$$

and ν is the correlation length exponent, then from (24) the intersection of $\zeta_{\ell,\ell'}(x)$ and $\zeta_{\ell',\ell''}(x)$ as function of x is approximately (x_c,ω). Barber and Selke [98] applied this method to the analysis of the structure factor of the two-dimensional axial next-nearest neighbor Ising model with considerable success. It would be of considerable interest to apply this method to percolation problems in order to obtain more accurate results for three-dimensional systems.

Monte Carlo methods are still useful tools in the study of percolation and related topics. The earliest Monte Carlo calculations for percolation were made [26] with only 2000 sites. But the algorithm of Hoshen and Kopelman [99] opened the way to fast simulation of larger systems. They used the fact that the occupation of each site of the lattice is not influenced by other sites. Thus one can use techniques which require storage only of one row of a two-dimensional lattice or one plane of a three-dimensional one. Thus it became possible to simulate a $400 \times 400 \times 400$ simple cubic lattice [100] and a 35000×35000 triangular lattice [101]. Currently a $10^6 \times 10^6$ triangular lattice is under investigation by the Boston University group for correction to scaling laws (H. Nakanishi, Private Communication, April 1983).

6. Percolation Conductivity

Ten years ago Kirkpatrick [102] introduced the concept of percolation conductivity; the conductivity of a network in which a fraction of bonds do not conduct. Since then this subject has observed intense activity. Straley [103] argued that the conductivity g of the network near the percolation threshold must obey a power law similar to percolation quantities

$$g(p) \sim |p-p_c|^u \quad . \tag{27}$$

Then depending on the distribution of the bond conductances one may have two distinct cases. In the first case a fraction p of bonds are conducting and the rest (1-p) not. Then as the percolation threshold is neared the conductivity of the network <u>vanishes</u> as

$$g(p) \sim (p-p_c)^t \qquad p \to p_c^+ \tag{28}$$

where it is believed that $t > 1$ for $d > 2$. On the other hand if a fraction p of the bonds are 'superconductors', i.e. have zero resistance, and the rest (1-p) fraction of bonds are ordinary conductors with finite conductance, then the conductivity of the network <u>diverges</u> as the percolation threshold is approached from below

$$g(p) \sim (p_c-p)^{-s} \qquad p \to p_c^- \tag{29}$$

where it is believed that $s < 1$ for three and higher dimensional systems. In two dimensions a simple duality argument by Straley [104] establishes that $s = t$. The exponent t has received considerable attention and many authors have tried to estimate it. Both the exponents s and t are believed to be universal, although there is no rigorous proof of this in the literature. The exponent s might have relevance to the problem of calculating the viscosity of a polymer (see the review article by Stauffer, Coniglio and Adam [3]), although there are several sets of exper-imental data that are not compatible with this assertion [3].

The exponents s and t (which are believed to be independent of each other

[103]) are not related to the percolation exponents by scaling laws similar to the usual scaling laws of critical phenomena. Although Fortuin and Kasteleyn [105] established that a linear random resistor network is the limit $q \to 0$ of a q-state Potts model, this correspondence has not helped much because this result holds for a fully connected network and not the one in which a fraction of bonds (sites) has been removed at random. Stephen [106] was able to construct a Hamiltonian and to define an order parameter for the percolation conductivity problem and derived $\varepsilon = 6-d$ expansion for t and s, but his results were not quantitatively useful for the realistic dimensions d = 2 and 3. The two percolation conductivity exponents cannot be connected to the percolation exponents unless some assumptions are made about the structure of the backbone near the percolation threshold. Kirkpatrick [107] defined the backbone of a resistor network as the current-carrying part of the network. Thus the backbone is the largest biconnected part of the infinite percolation cluster. The backbone fraction is assumed to obey a power law near the percolation threshold:

$$B(p) \sim (p-p_c)^{\beta'} \quad , \tag{30}$$

where the exponent β' is not related to other percolation exponents in a simple manner, and $\beta < \beta'$ according to Kirkpatrick [107]. The value of β' at d = 2 is about 0.52 [108], obtained by large-cell Monte Carlo renormalization method. No reliable value of the backbone exponent β at d = 3 is available yet. It is thus obvious that insight about the backbone structure is crucial to understanding the percolation conductivity phenomena. Thus various models have been proposed for the structure of the backbone near p_c. The most prominent of these are as follows.

<u>Nodes and Links Model</u>: This model was proposed by Skal and Shklovskii [109] and independently by de Gennes [110]. In their model the backbone of the infinite cluster above the percolation threshold is a super-lattice consisting of nodes separated by a distance of the order of correlation length ξ, connected by macrobonds. Each macrobond is made of L links where $L \sim |p-p_c|^{-\zeta}$. The backbone of the incipient infinite cluster, the very large cluster just below the percolation threshold, is also assumed to have the same structure as the macrobonds. The

advantage of this model is the relative ease with which it can be handled. A simple scaling argument with this model of backbone yields [110]

$$t = (d-2)\nu + \zeta . \quad (31)$$

This yields $t = \zeta$ in two dimensions. De Gennes [110] assumed that $\zeta = 1$ and thus $t = 1$ at $d = 2$ which is not consistent with the most accurate estimates of t [111,112] $t \cong 1.27$. This result for t at $d = 2$ also shows that the conjecture of Levinshtein et al. [113] that $t = (d-1)\nu$ is incorrect. The nodes and links model works well in high dimensions where closed loops do not play a dominating role, but in low dimensions these loops are important and cannot be ignored. At low dimensions this description of the backbone does not agree with Monte Carlo simulations [107] nor with simultaneous resistivity and superconducting critical current measurements [148] or with electron microscope pictures of granular superconductors [150]. Coniglio [114] has discussed several other shortcomings of this model. He also showed rigorously that within the context of this model $\zeta = 1$ independent of dimension, as had been assumed by de Gennes [110].

Sierpinski Gasket Backbone: This is a fractal model and has been advocated by Gefen et al. [115] although it was first studied by Kirkpatrick [107]. It is a model of the backbone of the infinite cluster at the percolation threshold and is made of nodes and only multiply connected bonds lumping together in blobs. Thus it is the opposite extreme of the nodes and links model. It has a self-similar structure and the advantage of the model is that it can be solved exactly. For the exponent t Gefen et al. obtained the following relation

$$t = \nu\{d-2 + \ln[(d+3)/(d+1)]/\ln 2\} \quad , \quad (32)$$

where ν is the correlation length exponent. The predictions of Eq. (32) are rather poor, for example it yields $t = 0.98$ at $d = 2$. This model neglects links which are very important for percolation conductivity as discussed below. Coniglio [114] has discussed other shortcomings of this model.

Nodes, Links, and Blobs Model: This model was suggested by Stanley [116]. He observed that the backbone of the incipient infinite cluster below p_c can be partitioned in two classes: (1) the links (cutting bonds), such that if one is cut, the cluster breaks into two parts, and (2) the remaining multiply connected bonds (blobs). Pike and Stanley [117] studied this model by Monte Carlo methods on the square lattice and Coniglio [118] studied it in connection with dilute Heisenberg and Ising systems. It was pointed out that both links and blobs are important. In fact Monte Carlo simulations of Sahimi et al. [119,120] showed that conduction near p_c is controlled by links whereas the dispersion of a contaminant in flow through a random network near the percolation threshold is controlled by blobs (Sahimi et al. used maze in place of blob). The available experimental data show that there is large dispersion near p_c, but there is negligible dispersion near p_c if blobs are neglected, as was demonstrated by Monte Carlo simulations of Sahimi et al. [119,120]. Coniglio [114] also suggested that the infinite cluster above p_c can be described as nodes connected by links and blobs. Thus from the work of Stanley [116], Coniglio [114], and Sahimi et al. [119,120] it appears that the nodes, links, and blobs model seems to be the most appropriate model of the backbone. But because of the complexity of this model, it has not yet been possible to derive a relationship between t and percolation critical exponents, though Coniglio [114] derived several exact relationships for the topological structure of the backbone within the context of this model.

The principal 'experimental' method of obtaining the conductivity of random conductance networks is Monte Carlo simulation, which was initiated by Kirkpatrick [102]. This approach has been taken by many authors a list of whom is too long to be given here. Then the conductivity of the network is fitted to a power law like Eq. (27) in order to obtain t and s. Another approach has been renormalization-group method which was pioneered by Stinchcombe and Watson [121], and has been used by several authors [122]. This method is successful in two dimensions but it does not yield good results in three dimensions mainly because small renormalization cells do not yield good estimates of p_c and consequently overestimate or underestimate the length of the critical region; therefore the estimates of t and s are not realistic. The large-cell Monte Carlo renormalization method [123] in practice is not appreciably easier to apply than the Monte Carlo method alone.

Fisch and Harris [124] used the relationship between the q-state Potts model and random resistor network [105] and combined it with a scaling relation similar to Eq. (31) in order to estimate t by the series expansion method. Their results for lower dimensions are in good agreement with the best estimates of t and they improve as the dimensionality of the system increases. They also obtained the correct t = 3 for six dimensions which is the mean-field value of t [125] (s = 0 (log) at [125] d = 6). These methods combined have caused a very large spread in the reported values of t in two and three dimensions: at d = 2, t has been reported to be anywhere from [126] t = 1 to [124] t = 1.43, at d = 3 it is from [102] t = 1.5 t [127] t = 2.36. Figure 1 present the distribution of reported values of t at d = 2 and d = 3 obtained by various methods!

Self-consistent methods have been used to estimate the conductivity of a random network over the entire concentration range of the conductors. These include the effective-medium approximation derived by Kirkpatrick [102] its cluster

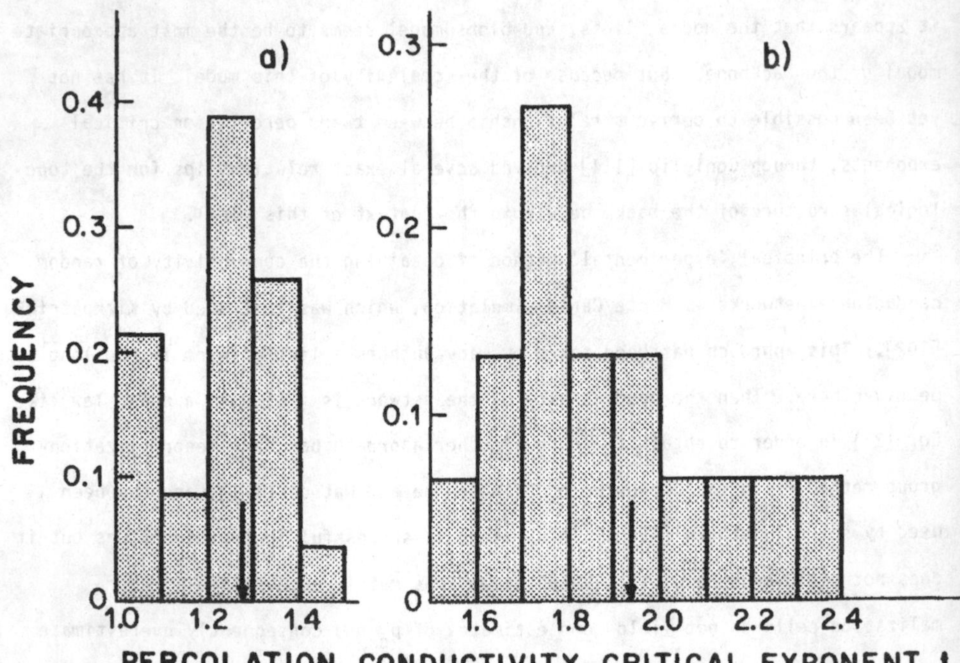

Fig. 1. The distribution of the reported values of percolation conductivity critical exponent t. (a) values for t(d=2), (b) values for t(d=3). The arrows indicate the currently accepted values of t.

extension [128] and the cumulant theory of Hori and Yonezawa [129]. All of them provide excellent prediction of the conductivity outside the critical region, but none of them performs well within the critical region and their predictions for three-dimensional systems are particularly poor; they yield $t = s = 1$ at all dimensions. Recently Sahimi _et al._ [130] combined the effective-medium approximation with the renormalization group approach to develop a new method which yields excellent predictions of conductivity over the _entire_ concentration range of conductors at _all_ dimensions. The method is based on the fact that renormalizing a network creates a new network which is (usually) topologically equivalent to the old network (aside from a scaling factor) but it is farther from criticality because the correlation length in the renormalized lattice is smaller than the one in the old one by a factor of b, where b is the renormalization scale factor. Thus the effective-medium approximation, if applied to the renormalized network should perform better. In particular Sahimi _et al._ produced $p_c^B = 1/2$ (exact) for the square lattice, $p_c^B \approx 0.265$ for the simple cubic lattice in three dimensions which is much closer to the best estimate of p_c^B of the simple cubic lattice (see table 1) than the prediction of the simple EMA, $p_c^B = 1/3$, and $p_c^B = 0.157$ for the four-dimensional simple cubic lattice, in excellent agreement with the series estimate of Gaunt and Ruskin [17] $p_c^B \approx 0.161$ and the conjecture of Sahimi _et al._ [13] $p_c^B \approx 0.156$.

Derrida and Vannimenus [111] used the transfer-matrix method to estimate that $t \approx 1.28 \pm 0.03$ in two dimensions. Sahimi _et al._ [112] and Jerauld _et al._ [131] used the finite-size scaling approach to obtain $t = 1.27 \pm 0.05$ and $t = 1.27 \pm 0.03$ respectively. Thus $t \approx 1.27$ seems to be the most accurate estimate of t at $d = 2$. Sahimi _et al._ [112] also estimated that $t = 1.87 \pm 0.04$ at $d = 3$. This seems to be an accurate estimate of t at $d = 3$ since it is in very good agreement with the series estimate [124] $t = 1.95 \pm 0.04$ and experimental data [132] $t = 1.9$ and [133] $t = 2$. The best estimate of s at $d = 3$ seems to be [104] $s = 0.70$. Very recently Derrida _et al._ [134] estimated t for three-dimensional systems by the transfer-matrix method (the first of such calculation for a three dimensional percolation problem). They obtained $t = 1.94 \pm 0.1$ in agreement with the data mentioned above.

Alexander and Orbach [135] studied some transport problems on self-similar structures such as fractals. Scaling considerations led them to a new dimension which they called the fracton dimension, which may be thought of as a fractal dimension in reciprocal space. From examining the available numerical data, it appears that the fracton dimension of percolating clusters at the percolation threshold is 4/3, independent of the spatial dimension of the system. It is thus a 'superuniversal' value. Alexander and Orbach [135] assumed this to be exact and exploited the connection between diffusion and conductivity (see below) and proposed that

$$t = 1/2[(3d-4)\nu - \beta] \quad . \tag{33}$$

This yields $t = 91/72 \cong 1.264$ and $t = 1.975$ in two and three dimensions respectively, in good agreement with the results mentioned above. Eq. (33) also yields the exact value $t = 3$ at $d = 6$. Very recently Hughes and Sahimi [136] noted that to a high degree of accuracy

$$t = 1 + 2\beta \tag{34}$$

for all dimensions. Eq. (34) yields $t = 92/72 \cong 1.277$ at $d = 2$, $t = 1.9$ at $d = 3$ and $t = 3$ at $d = 6$, in excellent agreement with the best available data. Eqs. (34) was originally derived implicitly by Harris and Fisch [137], but it was not appreciated mainly because accurate estimates of t were not available at the time.

De Gennes [138] posed a variation of the percolation problem in an attempt to probe the structure of percolation clusters. He asked the question: if a random walker (the 'ant') starts at an arbitrary origin of a random network (the 'labyrinth') and executes a random walk, what will be the time dependence of $\langle R^2(t)\rangle$, its mean-squared displacement from the origin of the walk at time t? Of course below the percolation threshold, $\langle R^2(t)\rangle$ must approach a p-dependent constant value R_∞^2 which obeys a certain scaling law

$$\langle R^2(t)\rangle = R_\infty^2\,(1 - ae^{-bt}) \qquad p < p_c \, , \tag{35}$$

where a and b are some constants and

$$R^2 \sim (p_c - p)^{-2\nu + \beta} \, . \tag{36}$$

Eq. (36) was first derived by Straley [139]. Above the percolation threshold $\langle R^2(t)\rangle$ grows linearly in time like a diffusion process with diffusion coefficient D : $\langle R^2(t)\rangle \sim Dt$. However, this diffusive behavior holds only above the percolation threshold in very long times such that the average span $R_s \equiv (\langle R^2(t)\rangle)^{1/2}$ of the walks is much larger than the correlation length ξ of the percolation problem. If $R_s \ll \xi$, then one expects a behavior which is similar to that of the motion of the 'ant' in 'the labyrinth' at $p = p_c$. In such a regime

$$R_s \sim t^{1/d_f} \, , \tag{37}$$

where d_f is the fractal dimensionality of the random walk at $p = p_c$. It is obvious that with Eq. (37) one obtains

$$D \sim t^{(2 - d_f)/d_f} \tag{38}$$

Clerc <u>et al</u>. [140] have very recently reveiwed the progress and results on this problem through 1981. It is now possible to derive a relation between d_f and other percolation exponents. The diffusion constant D and the dc conductivity g are related via the Einstein relation, $g = ne^2D/kT$, where n is the density of electrons. Since on a percolation cluster only those electrons that are in the infinite cluster contribute to g we must have: $n \sim (p - p_c)^\beta$. If we combine this with equation Eq. (28) we obtain $D \sim (p - p_c)^{t-\beta}$. Thus if $R_s \sim \xi$, the percolation correlation length (which scales as $(p - p_c)^{-\nu}$) , then combining Eq. (37) with $D \sim (p - p_c)^{t-\beta}$, $R_s \sim \xi \sim (p - p_c)^{-\nu}$, and $R_s^2 \sim Dt$, we obtain

$$d_f = 2 + (t - \beta)/\nu \, . \tag{39}$$

Thus if one can calculate d_f, then t can be calculated accurately since β and ν are known to a good degree of accuracy. Ben-Avraham and Havlin [142] presented computer simulations results in two and three dimensions for d_f. However, their results do not seem to be very accurate since the exponent t which is calculated based on their results does not agree well with the best available estimates, e.g. they obtained $t \approx 1.05$ in two dimensions. Sahimi and Jerauld [143] developed a two-parameter renormalization group method and estimated d_f to a high degree of accuracy. They then obtained $t \approx 1.27$ and $t \approx 1.8$ in two and three dimensions respectively, in good agreement with the data mentioned above. There is no analog of Eq. (39) for percolation conductivity in directed lattices. For related matters, the reader is referred to the paper by Hughes and Prager and the first article by Redner in this volume.

Acknowledgements

I am very grateful to Professors H. Ted Davis and L.E. Scriven and Dr. Barry D. Hughes for their constant encouragement and stimulating discussions. I am also grateful to Drs. Hughes and Fereydoon Family for a critical reading of the manuscript. I would also like to thank the Institute for Mathematics and its Applications of the University of Minnesota for providing me the opportunity to speak. This work was supported by the U.S. Department of Energy.

Note added in Proof: Since the completion of this review article several new results have been developed which we briefly mention here. Coniglio [144] has presented a Potts model formulation of branched polymers (lattice animals) in a solvent. Besides the three critical behaviors expected in analogy with a linear chain in a poor solvent, a percolation critical behavior is also found in some special cases. This percolation critical point corresponds to a higher-order critical point, where the effect of the solvent on the branched polymer is the same as the screening effect of the other clusters in percolation. Hakim and Nadal [145] proved a conjecture of Nadal et al. [94] giving the exact number of directed lattice animals of s sites with any root (a given set of occupied sites), on a strip of finite width of a square lattice. Carton [146] has formulated the random resistor network problem as a theory of interacting Bose and Fermi fields. He has derived an effective medium approximation which is identical in form to the one derived by Kirkpatrick [102], but it predicts a bond percolation threshold given by: $p_c = 1 - \exp(-2/z)$ (instead of $p_c = 2/z$ as in Kirkpatricks' case). This p_c is identical to the one predicted by the cumulant expansion theory of Hori and Yonezawa [129]. Kapitulnik and Deutscher [147] have demonstrated, by experiment, a percolation crossover from two to three dimensions in their resistivity measurement of co-evaporated Al-Ge random mixture films. The conductivity critical exponent t is found to the 2.1 ± 0.5 for three dimensional systems. The interested reader is also referred to the review article by Deutscher [148] for other experimental relevance of percolation. Sahimi and Hughes [149] have proposed that the percolation conductivity critical exponent s obeys the following relation

$$s = \nu - \frac{1}{2}\beta .$$

This equation satisfies the exact results $s(d = 1) = 1$ and $s(d = 6) = 0$. If we combine this equation with the Alexander-Orbach relation for t we obtain

$$t - s = \frac{3}{2}\nu(d - 2)$$

which satisfies the duality requirement $t = s$ at two dimensions. The predictions of this simple equation for s are in excellent agreement with the available numerical data at all dimensions.

REFERENCES

1. D. Stauffer, in: Disordered Systems and Localization, p. 9, edited by C. Castellani, C. Di Castro and L. Peliti (Springer-Verlag, Berlin 1981).
2. H. E. Stanley, in: Disordered Systems and Localization, p. 59, edited by C. Castellani, C. Di Castro and L. Peliti (Springer-Verlag, Berlin 1981).
3. D. Stauffer, A. Coniglio and M. Adam, Adv. Poly. Sci. 44, 103 (1982).
4. M. F. Sykes and J. W. Essam, J. Math. Phys. 5, 1117 (1964).

5. H. Kesten, Comm. Math. Phys. 74, 41 (1980); and in H. Kesten, Percolation Theory for Mathematicians (Birkhauser, Boston 1982).

6. J. C. Wierman, Adv. Appl. Prob. 13, 298 (1981).

7. I. Kondor, J. Phys. C 13, L531 (1980).

8. F. Y. Wu, J. Phys. C 12, L645 (1979).

9. M. F. Sykes, D. S. Gaunt and M. Glen, J. Phys. A 9, 97 (1976).

10. Z. V. Djordjevic, H. E. Stanley and A. Margolina, J. Phys. A 15, L405 (1982).

11. F. Y. Wu, Rev. Mod. Phys. 54, 235 (1982).

12. C. Tsallis, J. Phys. C. 15, L757 (1982).

13. M. Sahimi, B. D. Hughes, L. E. Scriven and H. T. Davis, J. Phys. A 16, L67 (1983).

14. M. Sahimi, B. D. Hughes, L. E. Scriven and H. T. Davis, J. Chem. Phys. 78, 6849 (1983).

15. B. D. Hughes, M. Sahimi and H. T. Davis, Physica A (Utrecht) (1983) in press.

16. R. Fisch, Ph.D. Thesis, University of Pennsylvania (1977).

17. D.S. Gaunt and H. Ruskin, J. Phys. A11, 1369 (1978).

18. M.F. Sykes, D.S. Gaunt and M. Glenn, J. Phys. A9, 1705 (1976).

19. D.S. Gaunt and M.F. Sykes, J. Phys. A16, 783 (1983).

20. D.W. Heermann and D. Stauffer, Z. Phys. B44, 339 (1981).

21. S. Wilke, Phys. Lett. 96A, 344 (1983).

22. S. Kirkpatrick, in Ill-Condensed Matter, edited by R. Balian, R. Maynard and G. Toulouse, 339 (North-Holland, Amsterdam 1978).

23. H. Nakanishi and H.E. Stanley, Phys. Rev. B22, 2466 (1980).

24. M. Sahimi and B.D. Hughes, preliminary estimate.

25. A.C.N. de Magalhaes, C. Tsallis and G. Schwachheim, J. Phys. C13, 321 (1980).

26. V.A. Vyssotsky, S.B. Gordon, H.L. Frisch and J.M. Hammersley, Phys. Rev. 123, 1566 (1961).

27. E. Domany and W. Kinzel, Phys. Rev. Lett. 47, 5 (1981); see also: F. Y. Wu and H. E. Stanley, Phys. Rev. Lett. 48, 775 (1982).

28. D. Dhar, M. K. Phani and M. Barma, J. Phys. A 15, L279 (1982).

29. J. W. Essam and K. De'Bell, J. Phys. A 15, L601 (1982).

30. J.G. Mauldon, Proc. 4th Berkeley Symp. on Mathematical Statistics and Probability, 2 337 (1961).

31. W. Kinzel and J.M. Yeomans, J. Phys. A14, L163 (1981).

32. D. Dhar, J. Phys. A15, 1849 (1982).

33. L. Gray, J.C. Wierman and R.T. Smythe, J. Appl. Prob. 17, 979 (1980).

34. J. Bishir, Royal Stat. Soc. J. 25B, 401 (1963).

35. J.T. Cox and R. Durret, Math. Proc. Camb. Phil. Soc. 93, 151 (1983).

36. J. Blease, J. Phys. C10, 925 (1977).

37. M. P. M. den Nijs, J. Phys. A 12, 1857 (1979).

38. R. J. Baxter, Phys. Rev. Lett. 26, 832 (1971).

39. R. B. Pearson, Phys. Rev. B 22, 2579 (1980).

40. B. Nienhuis, E. K. Riedel and M. Schick, J. Phys. A 13, L189 (1980).

41. W. Klein, H. E. Stanley, P. J. Reynolds and A. Coniglio, Phys. Rev. Lett. 41, 1145 (1978).

42. H. W. J. Blöte, M. P. Nightingale and B. Derrida, J. Phys. A 14, L45 (1981).

43. C. J. Lobb and K. R. Karasek, Phys. Rev. B 25, 492 (1982).

44. P. J. Reynolds, H. E. Stanley and W. Klein, Phys. Rev. B 21, 1223 (1980).

45. P. J. Reynolds, W. Klein and H. E. Stanley, J. Phys. C 10, L167 (1977); J. Bernasconi, Phys. Rev. B 18, 2185 (1978).

46. P. D. Eschbach, D. Stauffer and H. J. Herrmann, Phys. Rev. B 23, 422 (1981).

47. B. Derrida and L. De Seze, J. Physique (Paris) 43, 475 (1982).

48. J. A. M. S. Duarte and H. J. Ruskin, Z. Phys. B 46, 225 (1982).

49. J. Adler and V. Privman, J. Phys. A 15, L417 (1982).

50. J. L. Black and V. J. Emery, Phys. Rev. B 23, 429 (1981).

51. M. P. M. den Nijs, Phys. Rev. B 27, 1674 (1983).

52. H. J. Herrmann, Z. Phys. B 43, 55 (1981).

53. H. Kunz, unpublished work as referred to in reference 11.

54. S. Alexander, Phys. Lett. A54, 353 (1975).

55. R.J. Baxter, J. Phys. A13, L61 (1980).

56. I.G. Enting, J. Phys. A8, L35, 727, 1690 (1975); E. Domany and E.K. Riedel, J. Appl. Phys. 49, 1315 (1978).

57. R.J. Baxter and F.Y. Wu, Phys. Rev. Lett. 31, 1294 (1973).

58. B. Nienhuis, J. Phys. A 15, 199 (1982).

59. R.J. Baxter, J. Phys. C6, L445 (1973); R.J. Baxter, H.N.V. Temperly and S.E. Ashley, Proc. Roy. Soc. Lond., Ser. A358, 535 (1978).

60. H.E. Stanley, P.J. Reynolds, S. Redner and F. Family, in Real-Space Renormalization, p. 169 edited by T.W. Burkhardt and J.M.J. van Leeuwen, (Springer-Verlag Topics in Current Physics, 1982).

61. B. Payandeh, Riv. Nuovo Cimento 3 (3), 1 (1980).

62. E. T. Gawlinski and H. E. Stanley, J. Phys. A 14, L291 (1981).

63. J. Blease, J. Phys. C10, 917 (1977).

64. J.W. Essam and K. De'Bell, J. Phys. A14, L459 (1981).

65. S. Redner and A. Coniglio, J. Phys. A15, L273 (1982).

66. S.P. Obukhov, Physica (Utrecht) 101A, 145 (1980).

67. J.L. Cardy and R.L. Sugar, J. Phys. A13, L423 (1980).

68. T. C. Lubensky and J. Isaacson, Phys. Rev. Lett. 41, 829 (1978) and (E) 42, 410 (1978), see also: D. S. Gaunt, J. Phys. A 13, L97 (1980).

69. F. Family and A. Coniglio, J. Phys. A 13, L403 (1980).

70. A. B. Harris and T. C. Lubensky, Phys. Rev. B 23, 3591 (1981); A. B. Harris and T. C. Lubensky, Phys. Rev. B 24, 2656 (1981).

71. G. Toulouse, Nuovo Cimento 23, 234 (1974).

72. A. R. Day and T. C. Lubensky, J. Phys. A 15, L285 (1982).

73. G. Parisi and N. Sourlas, Phys. Rev. Lett. 46, 871 (1981).

74. M. E. Fisher, Phys. Rev. Lett. 40, 1610 (1978).

75. F. Family, J. Phys. A 15, L583 (1982).

76. F. Family, J. Phys. A 16, L97 (1983).

77. F. Family, J. Phys. A 13, L325 (1980).

78. T. C. Lubensky and A. J. McKane, J. Phys. A 14, L157 (1981).

79. J. Isaacson and T. C. Lubensky, J. Physique (Paris) 41, L469 (1980).

80. M. Daoud and J. F. Joanny, J. Physique (Paris) 42, 1359 (1981).

81. H. P. Peters, D. Stauffer, H. P. Holters and K. Loewenich, Z. Phys. B 34, 399 (1979).

82. H. G. Hentschel and I. Procaccia, Phys. Rev. Lett. 49, 1158 (1982); Phys. Rev. A27, 1266 (1983).

83. W. Klein, Phys. Rev. Lett. 47, 1569 (1981).

84. N. Breuer and H. K. Janssen, Z. Phys. B 48, 347 (1982).

85. J. L. Cardy, J. Phys. A 15, L593 (1982).

86. H. E. Stanley, S. Redner and Z. R. Yang, J. Phys. A 15, L569 (1982).

87. Y. Shapir, J. Phys. A 15, L433 (1982).

88. T. C. Lubensky and J. Vannimenus, J. Physique-Lettres 43, L377 (1982).

89. R. J. Glauber, J. Math. Phys. 4, 294 (1963).

90. S. Redner and Z. R. Yang, J. Phys. A 15, L177 (1982).

91. D. Dhar, Phys. Rev. Lett. 49, 959 (1982).

92. R. J. Baxter, J. Phys. A 13, L61 (1981); see also A.M. Verhagen, J. Stat. Phys. 15, 219 (1976).

93. R. J. Baxter and P. A. Pearce, J. Phys. A 15, 897 (1982).

94. J. P. Nadal, B. Derrida and J. Vannimenus, J. Physique (Paris) 43, 1561 (1982).

95. J. E. Green and M. A. Moore, J. Phys. A 15, L597 (1982).

96. H. Nakanishi and H. E. Stanley, J. Phys. A 14, 693 (1981).

97. M. P. Nightingale, J. Appl. Phys. 53, 7927 (1982).

98. M. N. Barber and W. Selke, J. Phys. A 15, L617 (1982).

99. J. Hoshen and R. Kopelman, Phys. Rev. B14, 3438 (1976).

100. J. Hoshen, R. Kopelman and E.M. Manberg, J. Statist. Phys. 19, 219 (1979).

101. A. Margolina, H.E. Stanley, D. Stauffer and Z. Djordjevic, paper presented at the Spring meeting, New England section of American Physical Society, May 21, 1982.

102. S. Kirkpatrick, Rev. Mod. Phys. 45, 574 (1973).

103. J.P. Straley, J. Phys. C 9, 783 (1976).

104. J. P. Straley, Phys. Rev. B 15, 5733 (1977).

105. C. M. Fortuin and P. W. Kasteleyn, Physica (Utrecht) 57, 536 (1972), see also: M. J. Stephen, Phys. Lett. 56A, 149 (1976).

106. M. J. Stephen, Phys. Rev. B 17, 4444 (1978).

107. S. Kirkpatrick, in: AIP Conference Proceedings 40, p. 99, edited by J. C. Garland and D. B. Tanner (American Institute of Physics, New York, 1978).

108. G. Shlifer, W. Klein, P. J. Reynolds and H. E. Stanley, J. Phys. A 12, L169 (1979).

109. A. S. Skal and B. I. Shklovskii, Fiz. Tekh. Poluprov. 8, 1586 (1974) (Sov. Phys. Semicond. 8, 1029, 1975).

110. P. G. de Gennes, J. Physique-Lettres 37, L1 (1976).

111. B. Derrida and J. Vannimenus, J. Phys. A 15, L557 (1982).

112. M. Sahimi, B. D. Hughes, L. E. Scriven and H. T. Davis, J. Phys. C 16, L521 (1983).

113. M. E. Levinshtein, M. S. Shur and A. L. Efros, Zh. Eksp. Teor. Fiz. 69, 2203 (1975) (Sov. Phys. JETP 42, 1120, 1976).

114. A. Coniglio, J. Phys. A 15, 3829 (1982).

115. Y. Gefen, A. Aharony, B. B. Mandelbrot and S. Kirkpatrick, Phys. Rev. Lett. 47, 1771 (1981).

116. H. E. Stanley, J. Phys. A 10, L211 (1977).

117. R. Pike and H. E. Stanley, J. Phys. A 14, L169 (1981).

118. A. Coniglio, Phys. Rev. Lett. 46, 250 (1981).

119. M. Sahimi, H. T. Davis and L. E. Scriven, Chem. Eng. Comm. (1983) in press.

120. M. Sahimi, A. A. Heiba, B. D. Hughes, H. T. Davis and L. E. Scriven, submitted to Chem. Eng. Sci.; M. Sahimi, B. D. Hughes, A. A. Heiba, L. E. Scriven and H. T. Davis, submitted to Chem. Eng. Sci.

121. R. B. Stinchcombe and B. P. Watson, J. Phys. C 9, 3221 (1976).

122. R. Rosman and B. Shapiro, Phys. Rev. B 16, 5117 (1977); J. P. Straley, J. Phys. C 10, 1903 (1977); H. Kunz and B. Payandeh, Phys. Rev. B 20, 2185 (1979), see also references 61 and 123.

123. C. J. Lobb and D. J. Frank, J. Phys. C 12, L827 (1979).

124. R. Fisch and A. B. Harris, Phys. Rev. B 20, 1285 (1978).

125. J. P. Straley, J. Phys. C 10, 3009 (1977).

126. C. D. Mitescu, H. Ottavi and J. Roussenq, in: AIP Conference Proceedings 40, p. 377, edited by J. C. Garland and D. B. Tanner (American Institute of Physics, New York, 1978).

127. S. Kirkpatrick, Phys. Rev. B 15, 1533 (1977).

128. J. A. Blackman, J. Phys. C 9, 2049 (1976); G. Ahmed and J. A. Blackman, J. Phys. C 12, 837 (1979).

129. M. Hori and F. Yonezawa, J. Math. Phys. 16, 352 (1975).

130. M. Sahimi, B. D. Hughes, L. E. Scriven and H. T. Davis, Phys. Rev. B 28, 307 (1983); M. Sahimi, B.D. Hughes, L.E. Scriven and H.T. Davis, submitted to J. Phys. C. (July 1983).

131. G. R. Jerauld, J. C. Hatfield, L. E. Scriven and H. T. Davis, J. Phys. C., (1983) in press.

132. B. Abeles, H. L. Pinch and J. I. Gittleman, Phys. Rev. Lett. 35, 247 (1975).

133. D. Adler, L. P. Flora and S. D. Senturia, Solid State Comm. 12, 9 (1973).

134. B. Derrida, D. Stuaffer, H.J. Herrmann and J. Vannimenus, J. de Physique-Lettres (June 1983).

135. S. Alexander and R. Orbach, J. Physique-Lettres 43, L625 (1982).

136. B. D. Hughes and M. Sahimi, unpublished.

137. A.B. Harris and R. Fisch, Phys. Rev. Lett. 38, 796 (1977).

138. P.G. de Gennes, La Recherche 7, 919 (1976).

114. A. Coniglio, J. Phys. A 15, 3829 (1982).

115. Y. Gefen, A. Aharony, B. B. Mandelbrot and S. Kirkpatrick, Phys. Rev. Lett. 47, 1771 (1981).

116. H. E. Stanley, J. Phys. A 10, L211 (1977).

117. R. Pike and H. E. Stanley, J. Phys. A 14, L169 (1981).

118. A. Coniglio, Phys. Rev. Lett. 46, 250 (1981).

119. M. Sahimi, H. T. Davis and L. E. Scriven, Chem. Eng. Comm. (1983) in press.

120. M. Sahimi, A. A. Heiba, B. D. Hughes, H. T. Davis and L. E. Scriven, submitted to Chem. Eng. Sci.; M. Sahimi, B. D. Hughes, A. A. Heiba, L. E. Scriven and H. T. Davis, submitted to Chem. Eng. Sci.

121. R. B. Stinchcombe and B. P. Watson, J. Phys. C 9, 3221 (1976).

122. R. Rosman and B. Shapiro, Phys. Rev. B 16, 5117 (1977); J. P. Straley, J. Phys. C 10, 1903 (1977); H. Kunz and B. Payandeh, Phys. Rev. B 20, 2185 (1979), see also references 61 and 123.

123. C. J. Lobb and D. J. Frank, J. Phys. C 12, L827 (1979).

124. R. Fisch and A. B. Harris, Phys. Rev. B 20, 1285 (1978).

125. J. P. Straley, J. Phys. C 10, 3009 (1977).

126. C. D. Mitescu, H. Ottavi and J. Roussenq, in: AIP Conference Proceedings 40, p. 377, edited by J. C. Garland and D. B. Tanner (American Institute of Physics, New York, 1978).

127. S. Kirkpatrick, Phys. Rev. B 15, 1533 (1977).

128. J. A. Blackman, J. Phys. C 9, 2049 (1976); G. Ahmed and J. A. Blackman, J. Phys. C 12, 837 (1979).

129. M. Hori and F. Yonezawa, J. Math. Phys. 16, 352 (1975).

130. M. Sahimi, B. D. Hughes, L. E. Scriven and H. T. Davis, Phys. Rev. B 28, 307 (1983); M. Sahimi, B.D. Hughes, L.E. Scriven and H.T. Davis, submitted to J. Phys. C. (July 1983).

131. G. R. Jerauld, J. C. Hatfield, L. E. Scriven and H. T. Davis, J. Phys. C., (1983) in press.

132. B. Abeles, H. L. Pinch and J. I. Gittleman, Phys. Rev. Lett. 35, 247 (1975).

133. D. Adler, L. P. Flora and S. D. Senturia, Solid State Comm. 12, 9 (1973).

134. B. Derrida, D. Stauffer, H.J. Herrmann and J. Vannimenus, J. de Physique-Lettres (1983) in press.

135. S. Alexander and R. Orbach, J. Physique-Lettres 43, L625 (1982).

136. B. D. Hughes and M. Sahimi, unpublished.

137. A.B. Harris and R. Fisch, Phys. Rev. Lett. 38, 796 (1977).

138. P.G. de Gennes, La Recherche 7, 919 (1976).

139. J.P. Straley, J. Phys. C13, 2991 (1980).

140. J.P. Clerc, G. Giraud, J. Roussenq, R. Blanc, J.P. Carton, E. Guyon, H. Ottavi and D. Stauffer, Annales de Physique 8, 3 (1983).

141. R. Rammal and G. Toulouse, J. Physique-Lettres 44, L13 (1983).

142. D. Ben-Avraham and S. Havlin, J. Phys. A15, L691 (1982).

143. M. Sahimi and G.R. Jerauld, J. Phys. C. (1983) in press.

144. A. Coniglio, J .Phys. A16, L187 (1983).

145. V. Hakim and J.P. Nadal, J. Phys. A16, L213 (1983).

146. J.P. Carton, J. Phys. A16, L219 (1983).

147. A. Kapitulnik and G. Deutscher, J. Phys. A16, L225 (1983).

148. G. Deutscher, in: Disordered Systems and Localization, p. 26, edited by C. Castellani, C. Di Castro and L. Peliti (Springer-Verlag, Berlin, 1981).

149. M. Sahimi and B.D. Hughes, submitted to J. Physique-Lettres (July 1983).

150. A. Kapitulnik and G. Deutscher, Phys. Rev. Lett. 49, 1444 (1982).

MODELS OF DISORDERED MEDIA: SOME NEW RESULTS, INCLUDING SOME NEW CONNECTIONS BETWEEN COMPOSITE-MEDIA, FLUID-STATE, AND RANDOM-FLIGHT THEORIES

George Stell

Departments of Mechanical Engineering and Chemistry
State University of New York
Stony Brook, New York 11794

ABSTRACT

Some new theoretical results on the microstructure of models of two-phase disordered media are given, as well as the new quantitative bounds on the thermal conductivity that follow for one such model (randomly centered spherical inclusions). A second set of results is then given for random flights, including random flights with hit expectancy prescribed in a unit ball around the flight origin. Finally, some interesting correspondences are demonstrated, via the Ornstein-Zernike equation, between random-flight results, liquid-state results and percolation-theory results.

1. Introduction

In these notes we describe some results of two lines of research that began quite separately but prove to share several fundamental features, some of which we had not anticipated.

The first set of results, summarized in Section 2, comes from a study of the microstructure of certain models of two-phase disordered media and the relation of that microstruture to effective transport properties. The work, begun with Salvatore Torquato in connection with his Ph.D. thesis research, has continued as an ongoing collaboration with Torquato on various extensions of his thesis results. An interesting feature of this work is its very close connection with the results of liquid-state theory in two ways: (i) it involves equations identical in form to the Mayer-Montroll and Kirkwood-Salsburg equations of liquid-state theory; (ii) it reveals the utility of a class of models (randomly centered particles, impenetrable spheres, permeable spheres, adhesive spheres) to which the sophisticated approximation techniques of liquid-state physics are immediately applicable.

Our second set of results, summarized in Section 3, draws on a random-flight study that we have made in collaboration with Peter Cummings. In that work we exploit the fact that equations of Ornstein-Zernike form, much used in liquid-state theory and recently introduced in percolation problems, are fundamental in random-walk theory too. The ubiquitous presence of equations of Ornstein-Zernike form illuminates the well-known key role of the lattice Green's function in both the theory of random walks on a lattice and in the theory of the spherical model of a ferromagnet and its variants. It also leads to novel and somewhat more subtle isomorphisms among results for a certain class of non-lattice random flight models, liquid-state models, and percolation-problem models. Moreover, the percolation-problem models that lend themselves to analysis via this Ornstein-Zernike approach are the same models that are natural to consider using the techniques of Section 2.

2. Some new results for two-phase disordered systems

A. Specific results

The microstructure of a two-phase composite medium consisting of identical particles suspended or embedded in a matrix (which may be fluid, solid, or void) is often characterized in terms of a hierarchy of n-point "matrix functions", $S_n(\vec{r}_1,\dots,\vec{r}_n)$, where S_n is the probability of simultaneously finding n points at $\vec{r}_1$, $\vec{r}_2$,..., $\vec{r}_n$, respectively, all in the matrix phase. The questions that naturally arise concerning these functions are fundamental to the study of disordered media. One wants to know the connection between the S_n and the n-particle probability densities $\rho_n(\vec{r}_1,\dots,\vec{r}_n)$ that characterize the probability of simultaneously finding a particle centered in volume element $d\vec{r}_1$, another particle centered in $d\vec{r}_2$, etc. One further wants to evaluate the S_n for models of interest. Finally one wants to know the extent to which the information contained in the S_n for low n yields sharp information about the transport properties of such models. In collaboration with Sal Torquato, we have investigated these questions and have obtained the following results, almost all of which are to be found in Sal Torquato's Ph.D. thesis [1] and some of which have subsequently appeared in articles by Torquato and Stell [2].

1. For media such that the location of each particle is fully defined by a position vector $\vec{r}$, we find

$$S_n(\vec{r}_1,\vec{r}_2,\ldots,\vec{r}_n) = 1 + \sum_{s=1}^{\infty} \int \cdots \iint \rho_s(\vec{r}_{n+1}, \vec{r}_{n+2},\ldots,\vec{r}_{n+s}) \times \prod_{j=n+1}^{n+s} m^{(n)}(\vec{r}_{j1},\vec{r}_{j2},\ldots,\vec{r}_{jn})d\vec{r}_j \ , \quad (2.1a)$$

where

$$m^{(n)}(\vec{r}_{j1},\vec{r}_{j2},\ldots,\vec{r}_{jn}) = 1 - \prod_{i=1}^{n}[1 - m(\vec{r}_{ij})] \quad (2.1b)$$

and

$$\vec{r}_{ij} = \vec{r}_j - \vec{r}_i \ .$$

Here $m(\vec{r})$ is an indicator function. For spherical particles of unit radius, for example,

$$\begin{aligned} m(\vec{r}) &= 1 \quad \text{for} \quad r < 1 \\ &= 0 \quad \text{for} \quad r > 1 \ , \end{aligned} \quad (2.1c)$$

where $r = |\vec{r}|$. We should stress that the particle distribution described by the ρ_n does not have to be that of thermal equilibrium (although computationally, this is perhaps the easiest non-trivial distribution to work with) nor does it have to be statistically uniform (i.e., homogeneous and isotropic).

2. We have also found a somewhat different representation of the S_n in terms of probability densities $\rho_{i,j}(\vec{r}_1,\ldots,\vec{r}_i;\vec{r}'_1,\ldots,\vec{r}'_j)$ that characterize a mixture of point particles at infinite dilution and the non-point particles (e.g. spheres) that actually constitute the inclusions of the composite medium. Letting $\rho_{i,j}(\vec{r}_1,\ldots,\vec{r}_i;\vec{r}'_1,\ldots,\vec{r}'_j)$ refer to the density associated with i point particles centered at $\vec{r}_1,\ldots,\vec{r}_i$, respectively, and j non-point particles centered at $\vec{r}'_1,\ldots,\vec{r}'_j$, respectively, we have

$$S_n(\vec{r}_1,\ldots,\vec{r}_n) = \prod_{i=1}^{n-1} [S_1(\vec{r}_i)/\rho_1(\vec{r}_i)] \sum_{s=0}^{\infty} \frac{(-1)^s}{s!} \int \cdots \iint \rho_{n-1,s}(\vec{r}_2,\ldots,\vec{r}_n;\vec{r}'_1,\ldots,\vec{r}'_s) \times \prod_{j=1}^{s} m(\vec{r}_j - \vec{r}_i)d\vec{r}'_j \,. \tag{2.2}$$

In a statistically uniform system, the functions in (2.1) and (2.2) simplify somewhat in form; we have (adjusting notation in an obvious way for simplicity)

$$S_1(\vec{r}) = \phi \ , \ \rho_1(\vec{r}) = \rho \ , \ \rho_{1,0}(\vec{r}_1,\vec{r}_2) = \rho \ , \ m(\vec{r}) = m(r) \ ,$$

$$S_2(\vec{r}_1,\vec{r}_2) = S_2(r_{12}) \ , \ \rho_2(\vec{r}_1,\vec{r}_2) = \rho_2(r_{12}) \ , \ \rho_{1,1}(\vec{r}_1,\vec{r}_2) = \rho_{1,1}(r_{12}) \ ,$$

$$S_3(\vec{r}_{12},\vec{r}_{13}) = S_3(r_{12},r_{13},r_{23}) \ , \text{ etc.}$$

where ϕ is the volume fraction of the matrix phase and ρ is the particle number density. If the particles are randomly centered spheres (referred to as "fully penetrable spheres" in some of our discussion below), then the ρ_n and $\rho_{i,j}$ of (2.1) and (2.2), respectively, are just ρ^n and ρ^{i+j}, respectively, and both (2.1) and (2.2) yield

$$S_n(\vec{r}_1,\ldots,\vec{r}_n) = \exp\{-\rho V_n(\vec{r}_1,\ldots,\vec{r}_n)\} \tag{2.3}$$

where V_n is the union volume of n spheres.

3. If the particles are mutually impenetrable spheres, (2.1) and (2.2) each have only a finite number of non-zero terms. In particular, in the uniform case, (2.1) yields

$$S_1 = \phi = 1 - \rho V_1 \tag{2.4}$$

$$S_2(r_{12}) = 1 - \rho V_2(r_{12}) + \iint \rho_2(r_{34})m(r_{13})m(r_{24})d\vec{r}_3 d\vec{r}_4 \tag{2.5}$$

$$S_3(\vec{r}_{12},\vec{r}_{13}) = 1 - \rho V_3(\vec{r}_{12},\vec{r}_{13}) + \frac{1}{2!} \int \rho_2(r_{45})m^{(3)}(r_{14},r_{24},r_{34})m^{(3)}(r_{15},r_{25},r_{35})d\vec{r}_4 d\vec{r}_5 - \frac{1}{3!} \int \rho_3(\vec{r}_{45},\vec{r}_{46}) \prod_{i=4}^{6} m^{(3)}(r_{1i},r_{2i},r_{3i})d\vec{r}_i \tag{2.6}$$

while (2.2) yields the same S_1 and

$$S_2(r_{12}) = \phi - (\phi/\rho) \int \rho_{1,1}(r_{23})m(r_{13})d\vec{r}_3 \tag{2.7}$$

$$S_3(\vec{r}_{12},\vec{r}_{13}) = S_2(r_{23}) - (\phi/\rho)^2 \int \rho_{2,1}(\vec{r}_{23};\vec{r}_{24})m(r_{14})d\vec{r}_4 \,. \tag{2.8}$$

[Eq. (2.5) and (2.6) were known to Weissberg and Prager [3]. Eq. (2.7) and (2.8) appear to be new, as are the more general (2.1) and (2.2).]

For particles with positions given by a uniform equilibrium distribution we have evaluated S_2 exactly through $O(\rho^2)$ and find for spheres of unit radius

$$S_2(r) = 1 - \rho V_2(r) + \rho^2 M_0(r) + O(\rho^3) \tag{2.9a}$$

$$M_0(r) = \begin{cases} [-\frac{16}{9} + \frac{r^3}{3} - \frac{r^4}{10} + \frac{r^6}{1260}]\pi^2 + \frac{16\pi^2}{9} & 0 < r < 2 \\ [\frac{256}{35r} - \frac{128}{9} + \frac{32r}{5} - \frac{5r^3}{9} + \frac{r^4}{10} - \frac{r^6}{1260}]\pi^2 + \frac{16\pi^2}{9}, & 2 < r < 4 \\ \frac{16\pi^2}{9}, & r > 4 \end{cases} \tag{2.9b}$$

In the context of the Percus-Yevick [4] or Mean Spherical Approximations [5] to $\rho_2(r)$ (identical approximations for this distribution of particles) one can use (2.5) to obtain an approximation for $S_2(r)$ that is in closed form through each order in ρ. In (2.5) we have also used the Verlet-Weis [6] fit to $\rho_2(r)$ (which appears to be of high accuracy for all ρ corresponding to the disordered state) to compute S_2 for a variety of such ρ. Figure 1 compares the smooth decay of S_2 for randomly centered spheres with the oscillating behavior of that function for impenetrable spheres at a representative high particle concentration of $1 - \phi = 4\pi\rho R^3/3 = 1/2$. This is slightly above the transition density at which hard-sphere ordering is expected on the basis of computer-simulation studies and hence corresponds to a metastable disordered (i.e., glassy) state for which close packing is generally agreed to occur at about $1 - \phi = 0.62$. Table 1 gives further results. (At $1 - \phi = 0.4$ impenetrable spheres in thermal equilibrium are in a fully stable disordered state.)

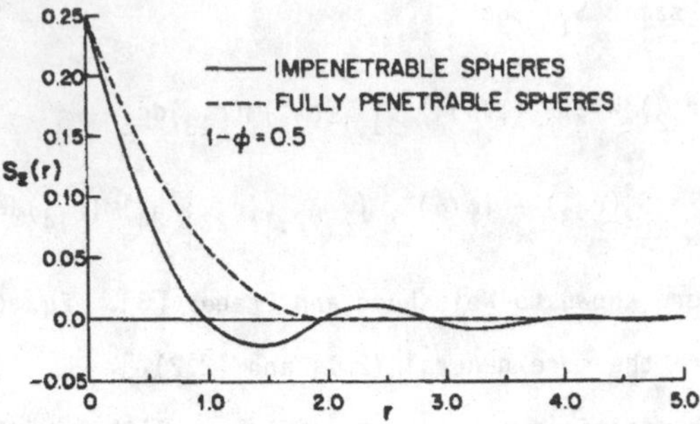

Figure 1. The 2-point void function for impenetrable and randomly-centered ("fully penetrable") spheres at a particle concentration of 0.5.

TABLE 1

S_2 as a Function of r for an Impenetrable Sphere System at 1-ϕ = 0.4, 0.5, 0.62

r	$S_2(r)$		
	1-ϕ = 0.4	0.5	0.62
0.0	0.60000	0.50000	0.38000
0.4	0.48278	0.35455	0.20529
0.8	0.40030	0.27086	0.14157
1.2	0.35444	0.23598	0.13087
1.6	0.34165	0.23468	0.14053
2.0	0.35597	0.25480	0.15872
2.4	0.36931	0.26314	0.15542
2.8	0.36519	0.25183	0.14201
3.2	0.35840	0.24514	0.14171
3.6	0.35671	0.24766	0.14653
4.0	0.35963	0.25270	0.14794
4.4	0.36171	0.25232	0.14297
4.8	0.36081	0.24882	0.14175
5.2	0.35940	0.24823	0.14519
5.6	0.35928	0.25020	0.14635
6.0	0.36005	0.25128	0.14422
6.4	0.36045	0.25024	0.14274
6.8	0.36015	0.24919	0.14442
7.2	0.35980	0.24960	0.14567
7.6	0.35983	0.25043	0.14454
8.0	0.36005	0.25041	0.14343
8.4	0.36013	0.24982	0.14413
8.8	0.36002	0.24965	0.14516
9.2	0.35993	0.25002	0.14470
9.6	0.35996	0.25025	0.14383
10.0	0.36002	0.25006	0.14409

4. We have used Eq. (2.3) to compute variational bounds on the effective thermal conductivity K^* of a system of randomly centered ("fully penetrable") spheres of conductivity K_2 in a matrix of conductivity K_1. The best such bounds that require only knowledge of S_1 (i.e., of ϕ) were given by Hashin and Shtrikman some time ago [7]. Various hierarchies of bounds involving more and more detailed knowledge of the microstructure, as contained in S_2, S_3, etc. were shortly thereafter developed and investigated by Prager [8], Brown [9] Beran [10] and others, and subsequently used and refined by numerous workers. We have evaluted for randomly centered spheres the Beran and Brown bounds that involve S_n through $n = 3$ and have compared the results with the Hashin-Shtrikman bounds as well as somewhat different bounds given by DeVera and Strieder [11] for randomly centered spheres. (Torquato has subsequently extended these latter bounds to a more general class of models.) Some results are shown in Figs. 2 through 4. Over a wide range of K_2/K_1 and ϕ the Beran bounds are sharpest; at low sphere concentration (Fig. 2) the best bounds are very sharp for $K_2/K_1 \approx 10$ and remain useful for $K_2/K_1 \approx 100$. In our figures and tables, K_2/K_1 is denoted as α.

For intermediate sphere concentrations (Fig. 3) the best bounds are still useful at $K_2/K_1 \approx 10$ but are far apart at $K_2/K_1 \approx 100$. At higher sphere concentration (Fig. 4) the best bounds become sharper at $K_2/K_1 \approx 10$ but remain far apart for $K_2/K_1 \approx 100$.

Table 2 and Figs. 5 and 6 suggest that the lower bounds for randomly centered spheres provide a surprisingly useful (albeit rough) guide to the effective conductivities of a variety of experimentally investigated composites in which the spherical inclusions in an insulating matrix are mutually impenetrable, even when they are good conductors (so that K_2/K_1 is very large as in Table 2).

The bounds we have considered behave in the same overall way for values of K_2/K_1 between 1 and 0.01 as they do between 1 and 100 (i.e., they are reasonably sharp at 0.1 but far apart at 0.01). There are certain two-phase systems of physical interest for which the upper bounds appear to be useful approximants in this range (see Table 3). Table 4 gives representative bounds for $K_2/K_1 < 1$.

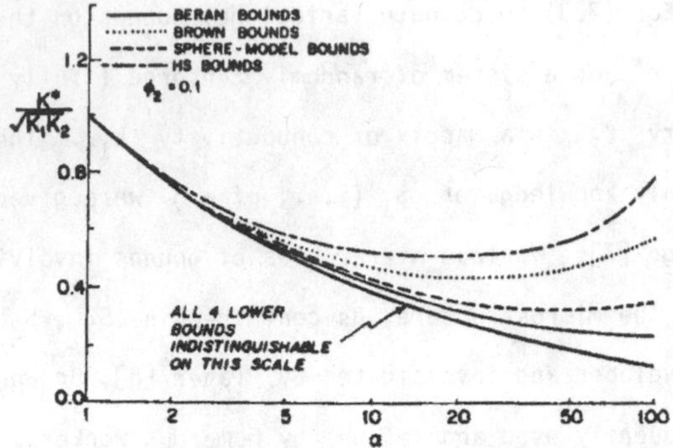

Figure 2. Comparison of the theoretical upper and lower bounds on $K^*/\sqrt{K_1K_2}$ as a function of α for randomly centered spheres at $\phi_2 = 0.1$, where ϕ_2 is the volume fraction of particle phase, $\phi_2 = 1 - \phi$. K_1 = conductivity of matrix, K_2 = conductivity of particles, K^* = effective conductivity, and $\alpha = K_2/K_1$. Sphere-model bounds due to DeVera and Strieder [11]. HS bounds due to Hashin and Shtrikman [7].

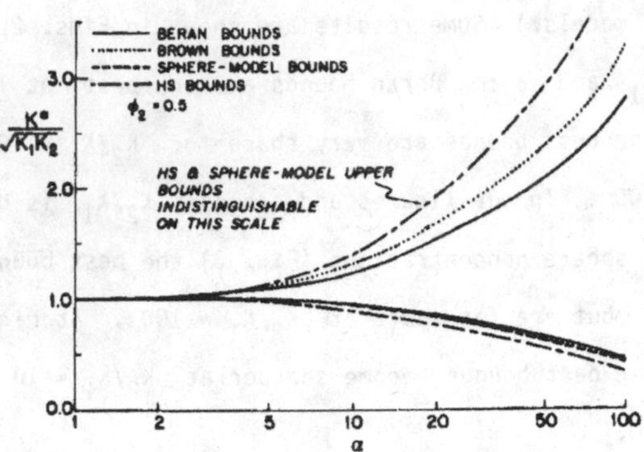

Figure 3. As in Fig 2 for $\phi_2 = 0.5$.

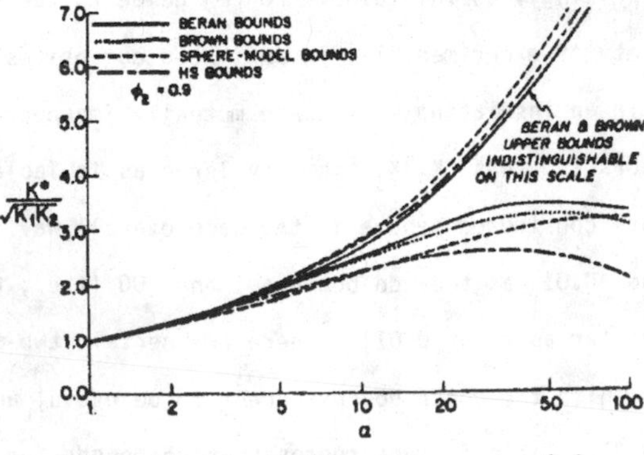

Figure 4. As in Fig. 2 for $\phi_2 = 0.9$.

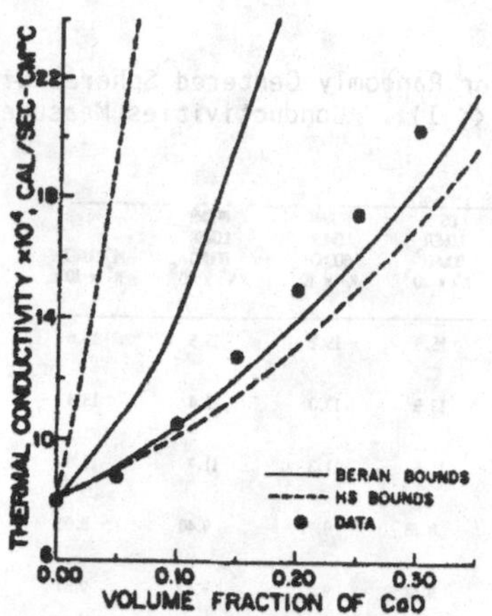

Figure 5. Comparison of the HS and Beran bounds on K^* with thermal conductivity data for CaO particles dispersed in polyethylene. Data from Sundstrom and Lee [12].

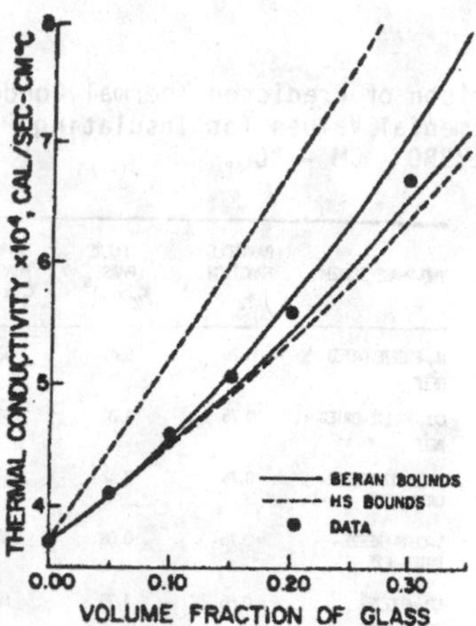

Figure 6. Comparison of the HS and Beran bounds on K^* with thermal conductivity data for glass particles dispersed in polysterene. Data taken from Ref. [12].

Table 2

Comparison of Predicted Thermal Conductivities for Randomly Centered Spheres with Experimental Values for Conducting Spheres ($\alpha >> 1$). Conductivies Measured in CAL/SBC - CM - °C.

TWO-PHASE SYSTEM	PARTICLE FRACTION ϕ_2	SPHERE PHASE $K_2 \times 10^5$	MATRIX PHASE $K_1 \times 10^2$	HS LOWER BOUND[a] $K^* \times 10^2$	DS LOWER BOUND[b] $K^* \times 10^2$	BERAN LOWER BOUND $K^* \times 10^2$	BROWN LOWER BOUND $K^* \times 10^2$	MEASURED $K^* \times 10^2$
COPPER-EPOXY SUSPENSION	0.45	91.0	54	186.0	215.0	218.0	216.0	240[c]
COPPER-EPOXY SUSPENSION	0.60	91.0	54	296.0	380.0	389.0	380.0	390[d]
H_2O - LIQUID LEAD SHOT PACKED BED	0.60	8.43	153	758.0	907.0	949.0	931.0	831[e]
GLYCERIN-LIQUID LEAD SHOT PACKED BED	0.51	8.43	68	270.0	318.0	325.0	321.0	489[e]
H_2 - LEAD SHOT PACKED BED	0.60	8.43	46	245.0	308.0	317.0	310.0	335[e]
AIR-LEAD SHOT PACKED BED	0.69	8.43	6	47.6	66.5	67.5	64.6	162[e]

[a] HS = Hashin-Shtrikman
[b] DS = DeVera-Strieder
[c] From Delmonte, Metal Filled Plastics, Reinhold, NY, 1961
[d] From Colletti and Rebori, Insulation (Libertyville, Ill.), 27 (1965).
[e] From Schumann and Voss, Fuel, 13, 249 (1934).

Table 3

Comparison of Predicted Thermal Conductivities for Randomly Centered Spheres with Experimental Values for Insulating Particles ($\alpha \ll 1$) . Conductivities Measured in CAL/SBC - CM - °C .

TWO-PHASE SYSTEM	PARTICLE FRACTION ϕ_2	SPHERE PHASE $K_2 \times 10^5$	MATRIX PHASE $K_1 \times 10^5$	HS LOWER BOUND[a] $K^* \times 10^5$	DS LOWER BOUND[b] $K^* \times 10^5$	BERAN LOWER BOUND $K^* \times 10^5$	BROWN LOWER BOUND $K^* \times 10^5$	MEASURED[c] $K^* \times 10^5$
N_2-FREEZE-DRIED BEEF	0.76	6.45	62	16.5	15.9	15.2	15.5	15.6
CO_2-FREEZE-DRIED BEEF	0.76	4.20	62	14.5	13.5	13.0	13.4	13.4
FREON 12-FREEZE-DRIED BEEF	0.76	2.42	62	12.9	11.5	11.3	11.7	11.9
VACUUM-FREEZE-DRIED-BEEF	0.76	0.08	62	10.9	8.78	8.91	9.40	8.93
CO_2-FREEZE-DRIED APPLE	0.86	4.35	62	10.2	9.58	9.12	9.41	9.59
FREON 12-FREEZE DRIED APPLE	0.86	2.57	62	8.5	7.55	7.32	7.66	7.90
VACUUM-FREEZE-DRIED APPLY	0.86	0.23	62	6.29	4.68	4.94	5.34	5.37
FREON 12-POLYURETHANE FOAM	0.85	3.76	23	5.0	4.93	4.65	4.74	5.08
VACUUM-POLYURETHANE FOAM	0.85	0.42	23	2.82	2.30	2.29	2.45	2.15

[a] Hs = Hashin-Shtrikman

[b] DS = DeVera-Strieder

[c] As given in ref. [11]. Raw data from Harper and El Sahrigi, Ind. Eng. Chem. Fundam. 3, 318 (1964).

Table 4

Tabulation of the Beran Upper and Lower Bounds on $K^*/\sqrt{(K_1K_2)}$ for Overlapping Spheres. Here $\phi_2 = \rho V_1 = 1 - \phi$.

ϕ_2	$\alpha \to 0$	$\alpha = 0.01$	$\alpha = 0.1$
0.1	$0.855\ \alpha^{-1/2}$	8.57	2.77
	$329.0\ \alpha^{1/2}$	6.87	2.73
0.3	$0.593\ \alpha^{-1/2}$	5.98	2.05
	$30.6\ \alpha^{1/2}$	2.11	1.84
0.5	$0.370\ \alpha^{-1/2}$	3.78	1.42
	$8.90\ \alpha^{1/2}$	0.769	1.16
0.7	$0.188\ \alpha^{-1/2}$	1.97	0.899
	$3.49\ \alpha^{1/2}$	3.28	0.703
0.9	$0.049\ \alpha^{-1/2}$	0.598	0.480
	$1.51\ \alpha^{1/2}$	0.148	0.415

Table 5

Tabulation of the Integral $I(P_2)$ as a Function of the Particle $\phi_2 = 1 - \phi$.

$(1 - \phi)$	$I(P_2)$
0.00	0.0
0.10	0.0189
0.20	0.0315
0.30	0.0387
0.40	0.0411
0.50	0.0395
0.60	0.0346
0.70	0.0273
0.80	0.0184
0.90	0.00883
1.00	0.0

The integral involving S_3 appearing in the Beran and Brown bounds considered above is the double-volume integral

$$I(P_2) = \frac{1}{8\pi^2} \iint S_3(r_{12},r_{13},r_{23}) \frac{P_2(\cos\theta)}{r_{12}^3 r_{13}^3} d\vec{r}_{12}\, d\vec{r}_{13} \qquad (2.10)$$

where θ is the angle opposite r_{23} and

$$S_3(r_{12},r_{13},r_{23}) = S_3(r_{12},r_{13},r_{23}) - S_2(r_{12})S_2(r_{13})\phi^{-1} . \qquad (2.11)$$

For a macroscopic sample, one integrates over all of space in (2.10).

This integral can be expected to appear in a variety of bounds for conductivities that involve S_3 ; hence we give its tabulation in Table 5.

B. General observations

1. Equations (2.1) and (2.2) have the same structure as a certain limiting case (discussed in [1] and [2]) of equations (the Mayer-Montroll and Kirkwood-Salsburg equations, respectively) which were first derived to describe the systems of particles in thermal equilibrium [13]. However, a study of our derivations of (2.1) and (2.2) shows that they have a more general domain of validity; they give the S_n corresponding to a set of ρ_n or $\rho_{i,j}$, respectively, whether they give the latter describe thermal equilibrium or not [14]. At the same time, we have found that the Mayer-Montroll and Kirkwood-Salsburg equations also reveal [15] that the ρ_n of thermal equilibrium characterize a less special microstructure than might first be apparent. For example, we have found in the spatially uniform case that for a system of impenetrable particles the single condition

$$\rho_n(\vec{r}_1,\ldots,\vec{r}_n)/\rho^n \to \prod_{1\le i<j\le n} [1 - m(\vec{r}_{ij})] \quad \text{as} \quad \rho \to 0 \qquad (2.12)$$

yields a set of ρ_n identical to that of thermal equilibrium. Put in words, (2.12) simply means that the ρ_n for a system of n isolated particles is geometrically uniform in three-space for all configurations that do not involve interpenetration, which has zero probability.

2. In the summary of Section A above, we have focused exclusively on two systems of included particles -- randomly centered particles and impenetrable particles. Neither Eqs. (2.1) and (2.2) nor the variational techiques to bound transport coefficients are limited to these two special cases, however; they are applicable for example to the permeable-sphere model of the author [16] and to the adhesive-sphere model of Baxter [17], both of which were originally introduced in the context of liquid-state theory. The connectivity properties of both models (as models of two-phase media) were recently studied by Chiew and Glandt [18] in work upon which we shall again touch at the end of Section 3.

3. Equations of Ornstein-Zernike Form: A Connecting Link Between Liquid-State Theory, Random-Flight Theory, and Percolation Theory

Here we draw heavily upon work we have been doing in collaboration with Peter Cummings, which is discussed in detail in Ref [19]. (Part B of this Section follows closely the Appendix of that reference.)

There is a remarkable similarity between the generating function for an n-step random walk on a periodic lattice (the $E(\vec{r},t)$ we discuss below) and the spin-spin correlation function of the spherical model (SM) of a ferromagnet [20] and its variants -- the mean spherical model (MSM) [21], the Gaussian model (GM) [20], and the mean spherical approximation (MAS) and Gaussian approximation (GA) for the Ising model [22]. Moreover, the SM and its variants all have lattice-gas versions, in which the particle-particle correlation function also looks like $E(\vec{r},t)$ except for trivial factors. This connection between these two-point correlation functions and $E(\vec{r},t)$ has already been noted by a number of workers [23]. Here we shall make clear how the connection follows from the existence of an underlying equation of Ornstein-Zernike type in both the random walk (on a lattice) and lattice-gas problems. (We use lattice-gas language rather than spin-spin language to maximize correspondence between lattice systems and fluid results and we lean heavily on our earlier work [24] in discussion of the lattice gas.) We shall further discuss the connection between the generating function for a random flight not restricted to a lattice (the $e(\vec{r},t)$ of this paper) and the two-particle correlation function for a fluid in the MSA. (The SM, GM and MSM are not well-defined for a fluid.) Here again the connection is via the fact that there is an equation of Ornstein-Zernike (OZ) form underlying both the random-flight and fluid problems. Finally, we extend the connection induced by the Ornstein-Zernike equation to percolation theory and demonstrate an isomorphism between the exact analysis of a particular random flight problem given by Cummings and Stell [19] and a corresponding percolation problem solved in the context of the MSA (equivalently, for this problem, in the PY [4] approximation) [18].

In a fluid of monatomic particles, the Ornstein-Zernike equation that relates the two-particle correlation function $h(\vec{r},\rho)$ and the direct correlation function $c(\vec{r},\rho)$ has the form [25]

$$h(\vec{r},\rho) = c(\vec{r},\rho) + \rho \int d\vec{s}\; c(\vec{s},\rho)h(\vec{r} - \vec{s},\rho) . \tag{3.1}$$

Here ρ is the expected number density of the fluid system, which we assume to be uniform and infinite in volume.

A. **Some random flight results**

We shall now summarize four sets of random-flight results set forth in detail recently by Cummings and Stell [19] (who also obtain a number of results that we shall not touch upon here at all).

1. The relation between the jump probability density $\tau(\vec{r})$ in a random flight and the hit expectance density $e(\vec{r})$ for such a flight is given by an equation of OZ form

$$e(\vec{r}) = \tau(\vec{r}) + \int d\vec{s}\; \tau(\vec{s})e(\vec{r} - \vec{s}) . \tag{3.2}$$

Here

$$\tau(\vec{r})d\vec{r} = \text{probability of the jumper making a jump from the origin to within the volume element } d\vec{r} \text{ centered on the point } \vec{r}, \tag{3.3}$$

while $e(\vec{r})$ is such that

$$1 + E(V) = \int_V [\delta(\vec{r}) + e(\vec{r})]d\vec{r}$$

= expected number of times that the volume V enclosing the origin is visited by the jumper during a random flight of arbitrary length. (3.4)

The delta-function corresponds to the contribution of a jumper who makes no jumps, i.e., who remains at the origin. The quantity $E(V)$ is the "hit expectancy", i.e., the expected number of times the volume is visited once the jumper has left the origin.

2. In random flights a natural generating function to consider is $e(\vec{r},t)$, the generating function for the n-jump probability densities $w_n(\vec{r})$, which is defined by

$$e(\vec{r},t) = \sum_{n=1}^{\infty} t^{n-1} w_n(\vec{r}) \tag{3.5}$$

where t is an arbitrary parameter. The function $e(\vec{r},t)$ has two important properties:

$$w_{n+1}(\vec{r}) = \frac{1}{n!} \frac{\partial^n}{\partial t^n} e(\vec{r},t)\Big|_{t=0} \; , \tag{3.6a}$$

$$e(\vec{r}) = e(r, t = 1) \; ; \tag{3.6b}$$

$e(\vec{r},t)$ and $\tau(\vec{r})$ are also related by an equation of OZ form,

$$e(\vec{r},t) = \tau(\vec{r}) + t \int d\vec{s}\; \tau(\vec{s})e(\vec{r} - \vec{s},t) \tag{3.7}$$

with t formally corresponding to the parameter ρ of (3.1). For $t = 1$, (3.7) reduces to (3.2).

3. Suppose one wishes to consider flights with prescribed hit expectance density $e(\vec{r})$ in a unit sphere about the origin. Perhaps the simplest problem to contemplate is the problem of a jump of zero probablity outside the sphere and a

uniform expectance density within the sphere. (Then $\tau(\vec{r})$ and $e(\vec{r})$ clearly depend on $\vec{r}$ only through $r = |\vec{r}|$.) We have

$$\tau(\vec{r}) = 0 \ , \quad r > 1 \ , \tag{3.8a}$$

$$e(\vec{r}) = \varepsilon \ , \quad r < 1 \ , \tag{3.8b}$$

where ε is a constant. Cummings and Stell solve this problem by Wiener-Hopf techniques [26] and find that in three dimensions

$$\begin{aligned} \tau(\vec{r}) &= r/\pi \ , \quad 0 \leq r \leq 1 \\ \varepsilon &= 3/\pi \ . \end{aligned} \tag{3.9}$$

For $r > 1$ the function $e(\vec{r})$ may be found analytically in a "zonal" representation. For example, for $1 \leq r \leq 2$

$$e(\vec{r}) = \frac{3}{\pi}\,\frac{1}{r} - \frac{4}{\pi}\,\frac{e^{1-r}}{r}\,[\cos(r-1) + \sin(r-1)] \ . \tag{3.10}$$

4. Cummings and Stell go on to solve (3.2) using (3.8) in five dimensions and to solve (3.2) with a generalization of (3.8) in three dimensions that involves nonzero $\tau(\vec{r})$ for $r > 1$. We refer the reader to ref. [19] for explicit results.

B. The OZ connection

We focus now on the relation between (3.1) and (3.7). The similarity is obvious. The differences are that $c(\vec{r},\rho)$ in general depends nontrivially on ρ , while $\tau(\vec{r})$ is independent of t , and that $\int \tau(\vec{r})d\vec{r} = 1$ and $\tau(\vec{r}) \geq 0$ because $\tau(\vec{r})$ is a probability density, whereas $c(\vec{r},\rho)$ has no sign restriction and must satisfy $\int c(\vec{r})d\vec{r} = 1/\rho$ only at singular points associated with the existence of a phase transition. In the lattice-gas analog of (3.1) the volume integral $\int d\vec{s}$ is simply replaced by (or interpreted as) a sum over all lattice sites. Similarly (3.5)-(3.7) have immediate lattice-walk analogs relating a generating function $E(\vec{r},t)$ for n-step lattice-walk probabilities $W_n(\vec{r})$ to a step probability $T(\vec{r})$.

In Fourier space, (3.1) becomes

$$\hat{h}(\vec{k},\rho) = \hat{c}(\vec{k},\rho) + \rho\hat{c}(\vec{k},\rho)\hat{h}(\vec{k},\rho) \tag{3.11}$$

for both the fluid and lattice gas. In the fluid case, the Fourier transform is given by

$$\hat{f}(\vec{k}) = \int d\vec{r}\; f(\vec{r})\,\exp(i\vec{k}\cdot\vec{r}) \tag{3.12a}$$

while in the case of a lattice gas, one sums over all the vectors $\vec{r}$ corresponding to lattice sites instead of integrating, so that

$$\hat{f}(\vec{k}) = \sum_{\vec{r}} f(\vec{r})\exp(i\vec{k}\cdot\vec{r}) \;. \tag{3.12b}$$

In the lattice-gas case, we shall choose a simple hypercubic lattice in d dimensions as a concrete example. Then

$$f(\vec{r}) = (2\pi)^{-d} \int d\vec{k}\; \hat{f}(\vec{k})\exp(-i\vec{k}\cdot\vec{r}) \tag{3.13}$$

where the integration extends from $-\pi$ to π for each of the d components of the vector $\vec{k} = (x_1,\ldots,x_d)$. It immediately follows from (3.1) and (3.13) that

$$\frac{1}{(2\pi)^d} \int d\vec{k}\; \frac{\exp(-i\vec{k}\cdot\vec{r})}{1-\rho\hat{c}(\vec{k},\rho)} = \delta_{\vec{r},\vec{0}} + \rho h(\vec{r},\rho) \tag{3.14}$$

where $\delta_{\vec{r},\vec{0}}$ is a Kronecker delta.

Similarly, the corresponding relation between Fourier transforms of the lattice-walk generating function $E(\vec{r},t)$ and the single step probability $T(\vec{r})$ on the same lattice is given by

$$\frac{1}{(2\pi)^d} \int d\vec{k}\; \frac{\exp(-i\vec{k}\cdot\vec{r})}{1-t\hat{T}(\vec{k})} = \delta_{\vec{r},\vec{0}} + tE(\vec{r},\rho) \tag{3.15a}$$

which follows immediately from the lattice walk analog of (3.7)

$$E(\vec{r},t) = T(\vec{r}) + t \sum_{\vec{s}} T(\vec{s})E(\vec{r} - \vec{s},t) \tag{3.15b}$$

which, like (3.7), is of the Ornstein-Zernike form. The formal similarity between

(3.14) and (3.15a), like that between (3.7), and (3.1), is clear. We shall next show that the differences one finds in functional forms between $c(\vec{r},\rho)$ and $T(\vec{r})$ do not destroy the essential correspondence between $E(\vec{r},t)$ and the lattice-gas $h(\vec{r},\rho)$.

Although $T(\vec{r}) = 0$ for $\vec{r} = \vec{0}$ in the simple random walk (to which we restrict ourselves here), $c(\vec{r},\rho)$ at $\vec{r} = \vec{0}$ is not zero in the SM, MSM, GM or MSA. Moreover, it is only at $\vec{r} = \vec{0}$ that $c(\vec{r},\rho)$ is ρ-dependent in the context of these descriptions of the lattice gas. For both these reasons, it is useful to decompose $c(\vec{r},\rho)$ into two terms,

$$\begin{aligned} c(\vec{r},\rho) &= c_0(\rho)\,\delta_{\vec{r},\vec{0}} + c^1(\vec{r}) , \\ c^1(\vec{r}) &= c(\vec{r},\rho) \qquad \text{for } \vec{r} \neq \vec{0} , \\ c^1(\vec{r}) &= 0 \qquad \text{for } \vec{r} = \vec{0} . \end{aligned} \tag{3.16}$$

Then we can write (3.14) as

$$\frac{1}{(2\pi)^d} \int d\vec{k}\, \frac{\exp(-i\vec{k}\cdot\vec{r})}{1 - zc^1(\vec{k})} = \left(\frac{\rho}{z}\right) [\delta_{\vec{r},\vec{0}} + \rho h(\vec{r},\rho)] \tag{3.17a}$$

where (dropping the argument ρ of c_0 for notational simplicity)

$$z = \rho(1 - \rho c_0)^{-1} . \tag{3.17b}$$

In the SM and its variants, $c^1(\vec{r})$ can be identified with the pair potential $\phi(\vec{r})$ between particles on lattice sites displaced from one another by the vector $\vec{r} \neq 0$, divided by k_BT, k_B = Boltzmann's constant, T = absolute temperature [24]. Thus

$$c^1(\vec{r}) = -\phi(\vec{r})/k_BT , \quad \vec{r} \neq \vec{0} . \tag{3.18}$$

A typical $\phi(\vec{r})$ of interest is a nearest-neighbor interaction, such that $\phi(\vec{r})$ is nonzero only for $|\vec{r}| = 1$, where it is ϕ_1, so that

$$\begin{aligned} c^1(\vec{r}) &= c_1 \qquad \text{for } |\vec{r}| = 1 \\ &= 0 \qquad \text{otherwise} , \\ c_1 = -\phi_1/k_BT &, \quad c^1(\vec{0}) = 2dc_1 . \end{aligned} \tag{3.19}$$

In (3.15)

$$T(\vec{0}) = 1 \tag{3.20}$$

because $T(\vec{r})$ is a probability, so $\sum_{\vec{r}} T(\vec{r}) = 1$. Thus, $t = 1$ is a singular value of the integrand, at which $E(\vec{r},t)$ becomes infinite for $d = 1$ and 2 and long-ranged for $d \geq 3$. (We recall that $E(\vec{r},1)$ is the expected number of visits to the lattice site described by vector $\vec{r}$.) In (3.17), $c(\vec{0})$ is not necessarily unity, but for $c^1(\vec{r}) > 0$, an identical singularity will occur for z such that

$$1 - zc^1(\vec{0}) = 0 \ . \tag{3.21}$$

This is associated with a phase transition of the model and the details vary [24] among the models described by (3.17). In the Gaussian approximation, for example, $c_0 = (\rho - 1)^{-1}$ so $z = \rho(1 - \rho)$ and the singularity occurs when $\rho(1 - \rho)c^1(\vec{0}) = 1$ for all d. In the SM, MSM and MSA, c_0 is instead determined by a "core condition" [24] $h(\vec{0},\rho) = -1$, which from (3.17) requires the satisfaction of

$$\frac{1}{(2\pi)^d} \int d\vec{k}\, [1 - zc^1(\vec{k})] = \rho(1 - \rho)/z \ . \tag{3.22}$$

This relation prevents (3.21) from being realized for nonzero temperature T when $d = 1$ or 2. We refer to ref. [24] for details but note here that despite the non-positive value of $h(\vec{0},\rho)$ the full right-hand side of (3.17a) remains positive for all $\vec{r}$ when $\phi(\vec{r}) > 0$ for the physically realizable range of ρ, $0 < \rho < 1$.

The upshot of the above analysis is that the left-hand sides of (3.15a) and (3.17a) can be identified as the same function if one identifies $T(\vec{r})$ with $c^1(\vec{r})/c^1(\vec{0})$ and t with $zc^1(\vec{0})$. We have thus shown how the Ornstein-Zernike structure induces a correspondence (but not quite an identity) between $E(\vec{r},t)$ and the lattice-gas $h(\vec{r},\rho)$ for the models under consideration despite some differences between $T(\vec{r})$ and $c(\vec{r},\rho)$.

If we were looking at the case in which $c^1(\vec{r}) < 0$, the analysis would be different. For example, using (3.19) with $c_1 < 0$, we would find the integrand of (3.17a) becoming singular at $\vec{k} = (\pm\pi,\pm\pi,\ldots,\pm\pi)$, rather than at $\vec{k} = \vec{0}$, for physically realizable densities ρ . But if we let z be negative, corresponding to unphysical (negative) ρ , we would then again find a simple identity between the integrals of (3.17) and (3.15) for certain pairs of t and z. The inequality $c^1(\vec{r}) \leq 0$ corresponds to a repulsive interaction between particles for $\vec{r} \neq \vec{0}$ in the SM and its variants. (In spin-system language, it corresponds to an antiferromagnetic exchange interaction.)

We return now to the fluid case, in which the mean spherical approximation [5] is typically of useful accuracy for hard-core molecules. Outside the core region, defined by the core diameter σ , one has the fluid analog of (3.18)

$$c(\vec{r},\rho) = -\phi(\vec{r})/k_BT \quad , \; r > \sigma \; , \tag{3.23}$$

while for $r < \sigma$, one has a core condition reflecting the impenetrability of hard-core molecules:

$$h(\vec{r},\rho) = -1 \quad \text{for} \quad r < \sigma \; . \tag{3.24}$$

(We shall take the hard-core diameter to be unity without loss of generality.) In contrast to the lattice-gas case, there is a lack of correspondence between the $h(\vec{r},\rho)$ in the MSA and the random flight $e(\vec{r},t)$ within the physically realizable range of ρ and T for $\phi(\vec{r})$ of interest. This is a result of the negative that both $c(\vec{r},\rho)$ and $h(\vec{r},\rho)$ typically have over the whole range $0 \leq r \leq 1$, as well as the highly non-trivial ρ dependence that $c(\vec{r},\rho)$ typically assumes over this range, in contrast to the independence of $\tau(\vec{r})$ on t . One sees this already for the case of a hard-sphere fluid in which $\phi(\vec{r}) = 0$ for $r > \sigma$, where (3.23) becomes (for hard spheres of unit diameter)

$$c(\vec{r},\rho) = 0 \quad \text{for} \quad r > 1 \; . \tag{3.25}$$

If one solves (3.1) with conditions (3.24) and (3.25) for <u>negative</u> (and hence unphysical) ρ , however, one finds a distinguished ρ value, $\rho = -3/\pi$, at which

$$1 - \rho \int c(\vec{r},\rho)d\vec{r} = 0 \tag{3.26}$$

and for which $\rho h(\vec{r},\rho)$ and $\rho c(\vec{r},\rho)$ are <u>identical</u> to $e(\vec{r},1)$ and $\tau(\vec{r})$ respectively, of Eqs. (3.7) and (3.8). Put another way, the solution of (3.1) subject to conditions (3.24), (3.25) and (3.26) is exactly equivalent to the solution of (3.7) subject to (3.8). In the first case, one finds a solution at $\rho = -3/\pi$. In the second case, equivalently, one finds the solution $\varepsilon = 3/\pi$.

Similarly, one finds that each of the further cases solved by Cummings and Stell in which $e(\vec{r})$ is prescribed inside and $\tau(\vec{r})$ outside a unit ball corresponds exactly to a solution of (3.1) for the distinguished negative value of ρ that satisfies (3.26), subject to (3.24) and (3.25) for a prescribed $\phi(\vec{r})/k_B T$. These cases, which correspond to purely repulsive potentials in the MSA, are fluid generalizations of the lattice case in which $c^1(\vec{r}) < 0$, discussed between (3.22) and (3.23).

We shall end with one more remarkable application of the Ornstein-Zernike equation that leads again to our solution of (3.1) subject to conditions equivalent to (3.24), (3.25) and (3.26), but this time involves a physically realizable number density. It is an application to the theory of percolation and gelation and involves the pair connectedness function $P(\vec{r},\rho)$ that yields the probability of finding two points displaced by $\vec{r}$ that are in the same connected cluster of particles. $P(\vec{r},\rho)$ is related to a "direct" connectivity function $C^{\dagger}(\vec{r},\rho)$, just as $h(\vec{r},\rho)$ is related to $c(\vec{r},\rho)$, by the Ornstein-Zernike equation (3.1). Moreover, the divergence of the mean cluster size is given by the condition

$$1 = \rho \int C^{\dagger}(\vec{r},\rho)d\vec{r} \qquad (3.27)$$

exactly analogous to the infinite compressibility condition (3.26). Condition (3.27) occurs at the percolation threshold (or gelation) density ρ_c . We refer readers to Coniglio et al [27] for details of the general formalism.

Suppose the particles whose clustering properties we are considering are the randomly centered spheres of Section 2. Using the obvious definition of connectedness for such particles (which is <u>not</u>, however, the connectedness criterion considered in [27]), one can apply the MSA to this problem, which yields the boundary conditions

$$P(\vec{r},\rho) = 1 \quad \text{for} \quad r < 1 \ , \tag{3.28}$$
$$C^{\dagger}(\vec{r},\rho) = 0 \quad \text{for} \quad r > 1 \ . \tag{3.29}$$

Except for the change in sign in (3.28), this is the pair of conditions (3.24) and (3.25), respectively. The sign change maps solutions of (3.1) at $-\rho$ into solutions at ρ. Thus in the MSA, percolation occurs at a positive (and hence physically realizable) ρ_c of $3/\pi$ for overlapping spheres. The solution of this problem (off the percolation threshold as well as on it) and some of its immediate generalizations were first obtained by Chiew and Glandt [18], who consider in the same work the percolation properties (in the Percus-Yevick approximation) of the permeable-sphere and adhesive-sphere models mentioned at the end of Section 2.

Acknowledgements

We thank the members of the Institute for Mathematics and its Applications for the opportunity to attend this Workshop and for their generous hospitality during our visit. We gratefully acknowledge the National Science Foundation and the Office of Basic Energy Sciences, U.S. Department of Energy, for support of research discussed in Sections 2 and 3, respectively. We are also indebted to many associates and co-workers for useful discussions and interactions. In particular we thank Joe Salacuse and Bjϕrn Hafskjold for early input that helped us undertand the connections between random-walk and fluid problems. But most of all we are indebted to our collaborators Sal Torquato and Peter Cummings for the privilege of working with them on the problems discussed herein.

References

1. S. Torquato, Ph.D. Thesis in Mechanical Engineering, State University of New York at Stony Brook, December 1980.
2. S. Torquato and G. Stell, J. Chem. Phys. 77, 1017 (1982), ibid. 78, 3262 (1983); 78, in press (1983).
3. H.L. Weissberg and S. Prager, Phys. of Fluids, 5, 1390 (1962).
4. J.K. Percus and G. Yevick, Phys. Rev. 110, 1 (1958).

5. J.K. Percus and G. Yevick, Phys. Rev. 136, 290 (1964); J.L. Lebowitz and J.K. Percus, Phys. Rev. 144, 251 (1966).

6. L. Verlet and J.J. Weis, J. Chem. Phys. 5, 939 (1972).

7. Z. Hashin and S. Shtrikman, J. Applied Phys. 33, 3125 (1962).

8. S. Prager, Physica 29, 129 (1963).

9. W.F. Brown, Trans. Rheology Soc., Pt. 9, No. 1, 357 (1965).

10. M. Beran, Il Nuovo Cimento 38, 771 (1965).

11. A.L. DeVera and W. Streider, J. Phys. Chem. 81, 1783 (1977).

12. D.W. Sundstrom and Y. Lee, J. Appl. Polym. Soc. 16, 3159 (1972).

13. a) J.E. Mayer and E. Montroll, J. Chem. Phys., 9, 2 (1941); J.E. Mayer, J. Chem. Phys. 15, 187 (1947) Eqs. (50) and (53).

    b) J.G. Kirkwood and Z.W. Salsburg, Discussions Faraday Soc. 15, 28 (1953). The KS set of equations is actually a special case of one due to Mayer [J. Chem. Phys. 15, 187 (1947), Eq. (54')]. See also S. Baer and J.L. Lebowitz, J. Chem. Phys., 40, (1964).

14. This was already noted in ref. [2] in regard to (2.1); the author has subsequently verified that it is equally true for (2.2).

15. We use an approach initiated by Boltzmann and developed by the author in "Boltzmann's Method of Evaluating and Using Molecular Distribution Functions", G. Stell, PIB Report (Aug. 1966).

16. See L. Blum and G. Stell, J. Chem. Phys. 71, 42 (1979), 72, 2212 (1980); J.J. Salacuse and G. Stell, J. Chem. Phys. 77, 3714 (1982).

17. R.J. Baxter, Aust. J. Phys. 21, 563 (1968).

18. Y.C. Chiew and E.D. Glandt, J. Phys. A, 1983 (in press).

19. P.T. Cummings and G. Stell, "Random Flights in Euclidean Space I. General Analysis and Results for Flights with Prescribed Hit Expectance Density about the Origin", SUSB CEAS Report #415, (February 1983).

20. T.H. Berlin and M. Kac, Phys. Rev. 86, 821 (1952).

21. H.W. Lewis and G.H. Wannier, Phys. Rev. 88, 682 (1952); Phys. Rev. 90, 1131 (1953).

22. G. Stell, J.L. Lebowtiz, S. Baer and W. Theumann, J. Math. Phys. 7, 1532 (1966).

23. See, for example, A.A. Maradudin, E.W. Montroll, G.H. Weiss, R. Herman, and H.W. Milnes, "Green's Functions for Monatomic Cubic Lattices" (Academie Royale de Belgique, Bruxelles, 1960) and G. Joyce, in "Phase Transitions and Critical Phenomena", Volume 2, ed. by C. Domb and M.S. Green (Academic Press, London, 1972). A number of other excellent surveys of results on random walks and related problems on periodic lattices have been written by E.W. Montroll. See, for example, E.W. Montroll, in "Applied Combinatorial Mathematics", ed. by E.F. Beckenback (Wiley, New York, 1964) and E.W. Montroll, J. Soc. Indust. Appl. Math. 4, 241 (1956). In this connection, see also the beautiful article by P.W. Kasteleyn, in "Graph Theory and

Theoretical Physics", ed. by F. Harary (Academic Press, London, 1967). We should also note the following two classic general studies of lattice walks: F. Spitzer, "Principles of Random Walk" (Van Nostrand, Princeton, 1964); M.N. Barber and B.W. Ninham, "Random and Restricted Walks" (Gordon and Breach, New York, 1970).

24. G. Stell, Phys. Rev. 184, 135 (1969).

25. L.S. Ornstein and F. Zernike, Proc. Acad. Sci. Amsterdam 17, 793 (1914). Equation (3.1) can be thought of as defining $c(r,\rho)$ in terms of $h(r,\rho)$. In the statistical theory of fluids one has an exact (but intractable) second independent relation among c, h, and the pair potential that has no counterpart in random flight theory. For quantitative fluid results, the second relation is typically replaced by a relatively simple approximation, such as Eq. (3.23).

26. B. Noble, "Methods Based on the Wiener-Hopf Technique" (Pergamon Press, London, 1958).

27. A. Coniglio, U. De Angelis, and A. Forlani, J. Phys. A 10, 1123 (1977).

VARIATIONAL BOUNDS ON THE DIFFUSIVE AND HYDRODYNAMIC PERMEABILITIES OF RANDOMLY PERFORATED SHEETS

George H. Malone

Department of Chemical Engineering and Materials Science

Suk Youn Suh and Stephen Prager*

Department of Chemistry
University of Minnesota
Minneapolis, Minnesota 55455

ABSTRACT

We define steady state permeabilities for a randomly perforated sheet of zero thickness with respect to both diffusive and hydrodynamic flow, and derive upper and lower variational bounds on them. For sparsely perforated sheets the bounds are sufficiently close to establish rigorously the existence of a logarithmic term in the expansion of the diffusive permeability in the perforation density; the hydrodynamic permeability shows no such behavior at low perforation densities. This contrast is one consequence of a quite general relationship between the diffusive permeability corresponding to a given perforation geometry and the hydrodynamic permeability for the complementary geometry generated by interchanging closed and open areas: logarithmic terms appear in the hydrodynamic permeability of a membrane constructed of randomly overlapping impermeable disks at the high porosity limit.

Variational bounds on the diffusive permeability can also be obtained if the perforation geometry is subject to temporal fluctuations, such as might be generated by allowing perforations to perform independent Brownian motions of their own in the plane of the membrane. Brownian motion of penetration sites can considerably enhance the diffusive permeability over what would have been obtained for stationary sites present at the same density.

---

* Correspondence concerning this article should be addressed to Stephen Prager.

1. Introduction

The work to be reported here has its origin in our efforts to model transport through very thin membranes, such as the lipid bilayers encountered in biological systems. When the thickness of a membrane becomes comparable to molecular dimensions, it can no longer be represented as a homogeneous slab, nor can its interaction with the environment be treated in terms of local equilibria at its surface. Such models [1] have been very successful for transport through thick films, where the permeation process can be clearly separated into an absorption or desorption occurring at the surfaces coupled to diffusional transport within the membrane. However this type of subdivision becomes questionable when the membrane thickness is only a few times the diameter of the penetrant molecules.

Models for thin membranes must take into account both spatial and temporal fluctuations in local structure which generate domains of high and low permeability that may move about the membrane in more or less random fashion. These considerations have given rise to a variety of pore or, more generally, penetration site models [2]; the former usually include hydrodynamic flow in addition to diffusion as possible transport mechanisms. A more detailed representation of the membrane structure requires a correspondingly more detailed treatment of the transport processes occurring in the medium surrounding the membrane (this point is often ignored, for instance by confining the discussion to events occurring within a single pore).

With this in mind, we investigate here a relatively simple model that exhibits in some form all of the features listed in the preceding paragraph: the representation of the membrane as an infinite plane sheet of zero thickness, with penetration sites introduced as randomly placed perforations (Fig. 1). The sheet is immersed in a continuum fluid which may itself flow through the perforations under the influence of a pressure differential, or simply act as host to a penetrant species diffusing in response to a concentration gradient (Fig. 2); we shall not consider the much more difficult case of the two transport mode operating simultaneously. In the hydrodynamic mode, we treat the flow in the zero

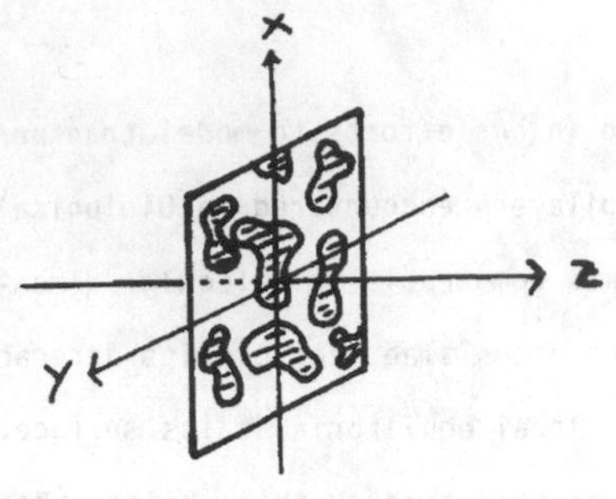

Figure 1. The randomly perforated sheet. Shaded areas (Ω) are open, unshaded areas ($\bar{\Omega}$) are closed to permeation.

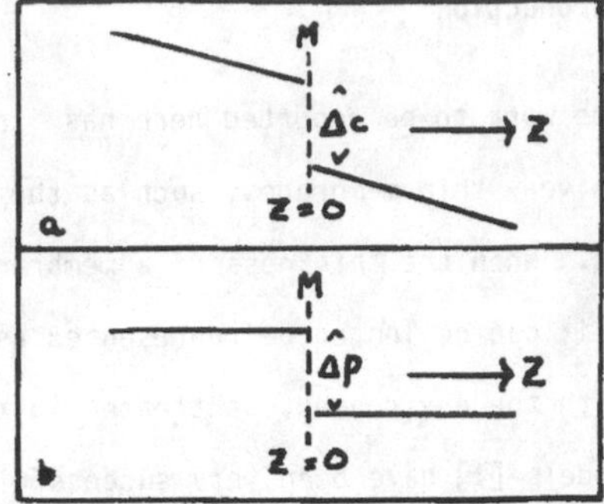

Figure 2. a) Steady state concentration profile of diffusing species for diffusive transport across a membrane (M). b) Steady state pressure profile for hydrodynamic flow through a membrane (M).

Reynolds number limit, and in the diffusive mode we assume that penetrant transport is governed by the usual laws of continuum diffusion.

The perforated sheet is a generalization of a problem long familiar in both hydrodynamics and diffusion theory: transport through a circular orifice [3]. Whatever its limitations as a model for membrane transport (even in bilayers, membrane pore diameters are small compared to the membrane thickness), it possesses several features of theoretical interest, which will be taken up in the following sections. Already in the course of formulating the transport equations, we shall see that there is an unexpected relationship between the hydrodynamic and diffusive permeabilities of perforated sheet membranes (Section 2). This type of model is particularly suited to application of the variational methods that have been used to obtain bounds on effective transport coefficients in random media [4]: almost alone among two-phase systems with impermeable domains, it permits bounding of transport coefficients both from above and from below (Sections 3 and 4). There is even a percolation problem of sorts associated with perforated sheets - not the percolation limit itself, of course, which is indeed trivial, but the manner in which it is approached as the perforation density goes to zero (Sections 4 and 5). Finally, temporal fluctuations in the membrane structure can be simulated by allowing the perforations to execute Brownian motions of their own (Section 6); this extension of the model generates a genuine many-particle problem, which can however still be treated by a corresponding extension of the variational methods of Section 3.

2. Definition of Two Permeabilities, and a Relation Between Them

Let the membrane extend over the entire xy-plane at $z = 0$, and let its open areas be denoted collectively by Ω, its impermeable areas by $\overline{\Omega}$. In the case of a solute diffusing through the membrane, the steady state situation is as sketched in Fig. 2a: the average solute flux $\langle q_d \rangle$ through any plane perpendicular to the z-direction is independent of z, and the average $\langle c \rangle$ of the concentration over the plane varies linearly with z, with a discontinuous drop $\langle \Delta c \rangle$ across the membrane itself*. The diffusive permeability of the membrane is defined as the flux per unit concentration drop:

$$\kappa_d \equiv \langle q_d \rangle / \langle \Delta c \rangle . \tag{1}$$

The case of transport by creeping viscous flow is shown in Fig. 2b: once again the mean flux $\langle q_h \rangle$ is independent of z, as is the mean pressure $\langle p \rangle$ on either side of the membrane, but there is a pressure drop $\langle \Delta p \rangle$ across the membrane itself. The hydrodynamic permeability is defined to be

$$\kappa_h \equiv \langle q_h \rangle / \langle \Delta p \rangle . \tag{2}$$

Consider now the steady concentration distribution $c(x,y,z) \equiv c(\vec{\rho},z) \equiv c(\vec{r})$ generated by a given distribution $q_d(x,y) = q_d(\vec{\rho})$ of solute flux across the plane $z = 0$ [3a]:

$$c(\vec{r}) = c_0 - \frac{\langle q_d \rangle}{D} z - \frac{\operatorname{sgn} z}{2\pi D} \int \frac{q_d(\vec{\rho}\,') - \langle q_d \rangle}{|\vec{r} - \vec{\rho}\,'|} d^2\vec{\rho}\,' , \tag{3}$$

---

* Throughout this article the angular bracket notation $\langle f \rangle$ is used to indicate the average $\frac{1}{A} \int_A f(x,y,0)\,dx\,dy$ of f over the area A of the membrane, in the limit $A \to \infty$. The symbol Δf refers to the limit of the difference $f(x,y,-z) - f(x,y,z)$ as $z \to 0$ from above.

where D is the solute diffusion coefficient, and the constant $c_0 \equiv \lim_{z\to 0} \langle c(\vec{\rho},-z) + c(\vec{\rho},z)\rangle/2$ merely sets the level of solute concentration. The first two terms in (3) represent the concentration distribution generated by a uniform flux $\langle q_d\rangle$, while the integral* is the summed contribution of the fluctuations $q_d(\vec{\rho}\,') - \langle q_d\rangle$. The concentration drop across the membrane at a point $\vec{\rho}$ in the plane $z = 0$ is therefore

$$\Delta c(\vec{\rho}) = \frac{1}{\pi D} \int \frac{q_d(\vec{\rho}\,') - \langle q_d\rangle}{|\vec{\rho} - \vec{\rho}\,'|} d^2\vec{\rho}\,' . \tag{4}$$

Now $\Delta c(\vec{\rho})$ must vanish in the open regions Ω of the membrane, whereas q_d must vanish in the closed areas $\bar{\Omega}$. The flux distribution must therefore satisfy the integral equation

$$\int \frac{q_d(\vec{\rho}\,') - \langle q_d\rangle}{|\vec{\rho} - \vec{\rho}\,'|} d^2\vec{\rho}\,' = 0 \qquad (\vec{\rho} \text{ in } \Omega)$$

$$q_d(\vec{\rho}) = 0 \qquad (\vec{\rho} \text{ in } \bar{\Omega}) . \tag{5}$$

The hydrodynamic transport of a fluid through the membrane can also be formulated as an integral equation, for we know that in creeping flow a unit point force in the z-direction exerted at a point $\vec{\rho}\,'$ in the $z = 0$ plane induces a velocity at $\vec{r}$ whose z-component is [3b]

$$u_z(\vec{r}) = \frac{1}{8\pi\eta} \frac{|\vec{r} - \vec{\rho}\,'|^2 + z^2}{|\vec{r} - \vec{\rho}\,'|^3} , \tag{6}$$

where η is the fluid viscosity. A distributed point force or pressure drop $\Delta p(\vec{\rho}\,')$ across the membrane will therefore give rise to a flux distribution

$$u_z\,(x,y,0) = q_h(\vec{\rho}) = -\frac{1}{8\pi\eta} \int \frac{\Delta p(\vec{\rho}\,') - \langle\Delta\vec{\rho}\rangle}{|\vec{\rho} - \vec{\rho}\,'|} d^2\vec{\rho}\,' \tag{7}$$

in the membrane plane**. Since q_h must be zero if ρ lies in $\bar{\Omega}$, and Δp must

---

* The integral extends over the entire membrane area (the notation $d^2\rho'$ signifies $dx'dy'$). It exists if the average of $q_d - \langle q_d\rangle$ over any circle of radius ρ vanishes $1/\rho^{1+\varepsilon}$ for some $\varepsilon > 0$ as $\rho \to \infty$.

** See footnote following equation (3).

vanish in Ω , we arrive at the integral equation

$$\int \frac{\Delta p(\vec{\rho}\,') - \langle\Delta p\rangle}{|\vec{\rho} - \vec{\rho}\,'|} d^2\vec{\rho}\,' = 0 \qquad (\vec{\rho} \text{ in } \overline{\Omega})$$

$$\Delta p(\vec{\rho}) = 0 \qquad (\vec{\rho} \text{ in } \Omega) \tag{8}$$

for the pressure drop distribution over the membrane.

Equations (5) and (8) present a very similar appearance, the only difference between them (apart from notation) being the interchange of Ω and $\overline{\Omega}$. We now consider a membrane M^* having a complementary perforation geometry to the original membrane M , such that the closed areas of M^* are the open areas of M and vice versa (i.e., $\Omega^* = \overline{\Omega}$ and $\overline{\Omega}^* = \Omega$). It follows that $(\Delta p - \langle\Delta p\rangle)$ for M^* will be proportional to $(q_d - \langle q_d\rangle)$ for M , and we can see from (4) and (7) that the same proportionality exists between q_h and Δc^* ; the permeabilities κ_h and κ_d are therefore related by the equation

$$\kappa_d^* \kappa_h = D/8\eta \tag{9}$$

It should be pointed out that equation (9) does not follow from a direct mathematical equivalence between the diffusion and hydrodynamic problems - the Laplace equation is quite different from the creeping flow Navier-Stokes equation, and, except on the membrane plane itself, the variations of q_d^* and Δp will not be simply related.

Since (9) is valid for any pair of complementary perforation geometries, it may in particular be applied to the class of geometries which are auto-complementary in the sense that M^* and M are statistically equivalent (a necessary, but not sufficient, condition for this is of course that the area fraction occupied by the perforations be ½). For such geometries $\kappa_d^* = \kappa_d$, and (9) becomes

$$\kappa_d \kappa_h = \frac{D}{8\eta} , \tag{10}$$

so that in these special circumstances we have a relation between the two permeabilities for the same membrane. It is interesting to note that if the geometry is altered while still maintaining its auto-complementarity (for example by a change in scale), κ_d and κ_h will change in opposite directions, one increasing while the other decreases.

So far as we are aware, the only other exact relation between transport coefficients in media of complementary geometry has been found by Keller [5a] for the electrical conductivities κ_1 and κ_2 in the two principal directions of an isotropic two-dimensional, two-phase medium. Taking advantage of the symmetry existing between the potential and stream functions in this case, Keller was able to show that the product $\kappa_1\kappa_2^*$, where κ_2^* is the conductivity in the 2-direction of the complementary medium, is always equal to the product of the conductivities of the individual phases. More recently Mendelson [5b] has pointed out that in two dimensions the conductivity of an isotropic medium with an auto-complementary geometry must therefore be just the geometric mean of the individual phase conductivities.

3. Variational Inequalities

Equation (9) does not of course provide information about the diffusive and hydrodynamic permeabilities separately; for that we return to equations (4) and (5), or (7) and (8). Exact solutions are out of the question if, as is usually the case, the perforation geometry is random, so that we can expect to have available only partial specifications in terms of appropriate spatial averages and correlations. Fortunately, equations (5) and (8) can be replaced by equivalent variational formulations, which ultimately lead to rigorous bounds on κ_d and κ_h.

In the case of solute diffusion, it can be shown [6] that (5) is the Euler-Lagrange equation corresponding to minimization of the integral

$$I\{q_d\} = \frac{1}{\pi D} \lim_{A\to\infty} \frac{1}{A} \iint_{AA} \frac{(\tilde{q}_d(\vec{\rho}) - \langle\tilde{q}_d\rangle)(\tilde{q}_d(\vec{\rho}\,') - \langle\tilde{q}_d\rangle)}{|\vec{\rho} - \vec{\rho}\,'|} d^2\vec{\rho}\, d^2\vec{\rho}\,' \qquad (11)$$

(A is again the membrane area) with respect to the trial flux distribution $\tilde{q}_d$, subject to the requirements that $\tilde{q}_d$ vanish in $\bar{\Omega}$ and $\langle\tilde{q}_d\rangle = \langle q_d\rangle$. Furthermore, multiplication of (4) by $q_d(\vec{\rho}) - \langle q_d \rangle$ and integration over $\vec{\rho}$ shows that, for the minimizing distribution $\tilde{q}_d = q_d$,

$$I\{q_d\} = \langle \Delta c \rangle \langle q_d \rangle$$

or (12)

$$\frac{1}{\kappa_d} = I\{q_d\}/\langle q_d \rangle^2$$

For $\tilde{q}_d \neq q_d$, we have therefore the inequality

$$\frac{1}{\kappa_d} < I\{\tilde{q}_d\}/\langle \tilde{q}_d \rangle^2 \quad , \tag{13}$$

which, in the case of a random, statistically homogeneous perforation geometry that is also isotropic in the membrane plane, can be written

$$\frac{1}{\kappa_d} < \frac{1}{\pi D} \int \frac{f(\xi)}{\xi} d^2\vec{\xi} \tag{14}$$

where $\vec{\xi} = \vec{\rho} - \vec{\rho}'$, $\xi = |\vec{\xi}|$, and $f(\xi)$ is the autocorrelation

$$\langle(\tilde{q}_d(\vec{\rho}) - \langle\tilde{q}_d\rangle)(\tilde{q}_d(\vec{\rho} + \vec{\xi}) - \langle\tilde{q}_d\rangle)\rangle/\langle q_d\rangle^2 \ .$$

A lower bound on $1/\kappa_d$ is also obtainable. We first note that a concentration drop $\Delta c(\vec{\rho}')$, extending over an area element $d^2\vec{\rho}'$ of the membrane at a location $\vec{\rho}'$ on the plane $z = 0$, will generate a flux perturbation whose z-component at a point $\vec{r} \equiv (\vec{\rho},z)$ in the surrounding medium is given by

$$\delta(q_d(\vec{r}) - \langle q_d \rangle) = \frac{D}{4\pi} (\Delta c(\vec{\rho}')d^2\vec{\rho}') \left[\frac{1}{|\vec{r} - \vec{\rho}'|^3} - \frac{3z^2}{|\vec{r} - \vec{\rho}'|^5} \right]$$

The inverse of equation (4) is therefore

$$q_d(\vec{\rho}) - \langle q_d \rangle = \frac{D}{4\pi} \lim_{z\to 0} \int \Delta c(\vec{\rho}') \left[\frac{1}{|\vec{r} - \vec{\rho}'|^3} - \frac{3z^2}{|\vec{r} - \vec{\rho}'|^5} \right] d^2\vec{\rho}'$$

(15)

$$= \frac{D}{4\pi} \int \frac{\Delta c(\vec{\rho}') - \Delta c(\vec{\rho})}{|\vec{\rho} - \vec{\rho}'|^3} d^2\vec{\rho}' \ .$$

Recalling that q_d must vanish in closed areas, we arrive at an integral equation for $\Delta c(\vec{\rho})$ which is fully equivalent to (5):

$$\int \frac{\Delta c(\vec{\rho}) - \Delta c(\vec{\rho}\,')}{|\vec{\rho} - \vec{\rho}\,'|^3} d^2\vec{\rho}\,' = \frac{4\pi}{D} \langle q_d \rangle \qquad (\vec{\rho} \text{ in } \overline{\Omega})$$

$$\Delta c(\vec{\rho}) = 0 \qquad (\vec{\rho} \text{ in } \Omega) \tag{16}$$

Equation (16) may also be formulated as a variational problem: we ask for the trial function $\Delta\tilde{c}(\vec{\rho})$ which minimizes the functional

$$J\{\Delta\tilde{c}\} \equiv \frac{D}{4\pi} \lim_{A\to\infty} \frac{1}{A} \iint_{AA} \frac{(\Delta\tilde{c}(\rho))^2 - \Delta\tilde{c}(\rho)\Delta\tilde{c}(\rho'))}{|\vec{\rho} - \vec{\rho}\,'|^3} d^2\vec{\rho}\,' \, d^2\vec{\rho} \tag{17}$$

subject to the subsidiary conditions $\Delta\tilde{c} = 0$ in Ω and $\langle\Delta\tilde{c}\rangle = \langle\Delta c\rangle$. Multiplication of (15) by $\Delta c(\vec{\rho})$ and integration over the plane shows that, for the correct flux and concentration distributions, I and J are equal:

$$J\{\Delta c\} = I\{q_d\} = \langle\Delta c\rangle\langle q_d\rangle = \kappa_d\langle\Delta c\rangle^2$$

so that

$$\frac{1}{\kappa_d} > \langle\Delta\tilde{c}\rangle^2 / J\{\Delta\tilde{c}\} \,. \tag{18}$$

For the random, statistically homogeneous, isotropic geometries with which we are mainly concerned, (18) becomes

$$\frac{1}{\kappa_d} > \frac{4\pi}{D} \left[\int \frac{h(0) - h(\xi)}{\xi^3} d^2\vec{\xi}\,\right]^{-1} \,, \tag{19}$$

where

$$h(\xi) \equiv \langle\Delta c(\vec{\rho})\Delta c(\vec{\rho} + \vec{\xi})\rangle / \langle\Delta c\rangle^2$$

Corresponding inequalities for the hydrodynamic permeability can of course be written at once through the use of the relation (9), upper and lower bounds on D/κ_d becoming respectively upper and lower bounds on $8\eta\kappa_h{}^*$.

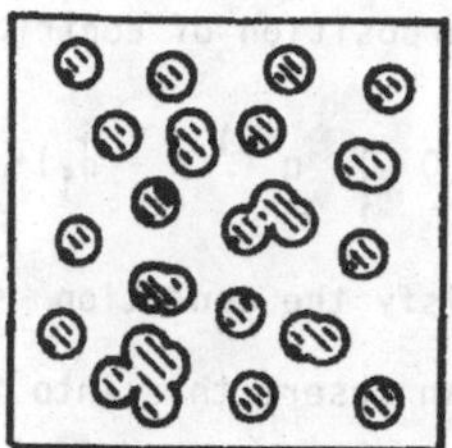

Figure 3. Perforation geometry generated by randomly overlapping circular holes (shaded areas are open, unshaded areas are closed).

4. Application to Random Circular Perforations

A membrane geometry with particularly simple statistics can be generated by distributing circular perforations of radius a at random over the membrane plane, allowing different perforations to overlap one another to form holes of more complex shape (Fig. 3). Although exact evaluation of the permeabilities is not, so far as we are aware, possible even in this case, the bounds of the preceding section can be calculated to provide a test of the variational approach over a wide range of perforation densities.

When the number n of perforations per unit area is small, different perforations will act independently of one another, and κ_d will be just n times the solute flow (per unit concentration difference) through a single hole in an otherwise impermeable sheet [3a]

$$\kappa_d^{(0)} = 2Dan \quad . \tag{20}$$

At a point $\vec{\rho}$ near an isolated hole of radius a centered at the origin, the flux is

$$q^{(0)}(\vec{\rho}) = \begin{cases} \dfrac{1}{2\pi a^2}\left(1 - \dfrac{\rho^2}{a^2}\right)^{-1/2} & (\rho < a) \\ 0 & (\rho > a) \end{cases} \tag{21}$$

for unit total solute flow through the hole.

A reasonable trial function q_d, particularly if n is not too high, may now be constructed by linear superposition of contributions of the form (21):

$$q_d(\vec{\rho}) = Q \sum_i q^{(0)}(\vec{\rho} - \vec{\rho}_i) \tag{22}$$

where Q is chosen so as to satisfy the condition $\langle \tilde{q}_d \rangle = \langle q_d \rangle$, and $\vec{\rho}_i$ is the position of the i'th hole. We can insert this into the inequality (14), to obtain

$$1/\kappa_d < 1/\kappa_d^{(0)} \tag{23}$$

i.e., the diffusive permeability is always greater than what one would estimate by simply adding contributions from individual perforations without taking into account their interactions through overlap of diffusion zones in the surrounding fluid.

At higher perforation densities we can obtain a better bound from (14) by using a trial function that weights each point in Ω equally, regardless of the number of perforations covering it:

$$q_d(\rho) = \begin{cases} q_0 & (\vec{\rho} \text{ in } \Omega) \\ 0 & (\vec{\rho} \text{ in } \overline{\Omega}) \end{cases} \tag{24}$$

The autocorrelation function f corresponding to (24) is

$$f(\rho) = (e^{n\omega(\rho)} - 1)/(e^{\pi n a^2} - 1)^2 \tag{25}$$

where

$$\omega \equiv \begin{cases} 2na^2 \text{ arc cos } (\frac{\rho}{2a}) - na\rho(1 - \frac{\rho^2}{4a^2})^{1/2} & (\rho < 2a) \\ 0 & (\rho > 2a) \end{cases}$$

is the overlap area between two holes separated by a distance ρ. Substitution into (14) gives, after rearrangement,

$$\frac{\kappa_d^{(0)}}{\kappa_d} < \frac{8na^2}{(e^{\pi n a^2}-1)^2} \int_0^{\pi/2} (e^{na^2(2\theta - \sin 2\theta)}-1)\sin\theta \, d\theta \tag{26}$$

The right hand side of (26) is plotted against the porosity

$$\sigma \equiv 1 - e^{-\pi n a^2}$$

in Fig. 4 (curve 1); the result is an improvement over (23) for $\sigma > 0.107$, but is some 8% higher in the limit $\sigma \to 0$.

To improve on (23) for small σ, we modify the trial function (22) by weighting the contribution of the ith perforation at a point $\vec{\rho}$ by a factor depending on the distribution of other perforations in the vicinity of $\vec{\rho}$:

$$\tilde{q}_d(\vec{\rho}) = Q \sum_i \left[q^{(0)}(\vec{\rho} - \vec{\rho}_i) \prod_{j \neq i} (1 + s(\vec{\rho} - \vec{\rho}_j))\right] \tag{27}$$

where $s(\vec{\rho}) \to 0$ as $\vec{\rho}$ goes to infinity, but is otherwise left unrestricted. The autocorrelation function f now becomes

$$\tilde{f}(\vec{\rho}) = -1 + e^{n\chi(\rho)}\left[\frac{1}{n}\phi(\rho) + (1 - \mu(\rho))^2\right] \tag{28}$$

where χ, ϕ, and μ are the convolutions

$$\chi(\rho) \equiv \int s(\vec{\xi})s(\vec{\xi} + \vec{\rho})d^2\vec{\xi}$$

$$\phi(\rho) \equiv \int q^{(0)}(\vec{\xi})q^{(0)}(\vec{\xi} + \vec{\rho})d^2\vec{\xi}$$

$$= \begin{cases} \dfrac{1}{\pi a^2} \ln\left(\dfrac{2a + \sqrt{4a^2 - \rho^2}}{\rho}\right) & (\rho < 2a) \\ 0 & (\rho > 2a) \end{cases}$$

$$\mu(\rho) = \int q^{(0)}(\vec{\xi})s(\vec{\xi} + \vec{\rho})d^2\vec{\xi}.$$

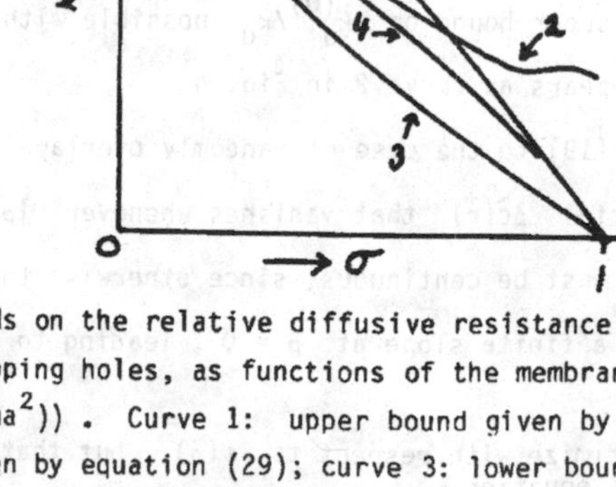

Figure 4. Bounds on the relative diffusive resistance $\kappa_d^{(0)}/\kappa_d$ for stationary randomly overlapping holes, as functions of the membrane porosity $\sigma \equiv (1 - \exp(-\pi n a^2))$. Curve 1: upper bound given by equation (26); curve 2: upper bound given by equation (29); curve 3: lower bound given by equations (34) and (36); curve 4: improved upper bound given by the trial function (37).

Upon substitution into (14) we obtain the inequality

$$\frac{\kappa_d(0)}{\kappa_d} < 4an \int_0^\infty \{ e^{n\chi(\rho)}[\frac{1}{n} \phi(\rho) + (1 + \mu(\rho))^2] - 1\} d\rho \tag{29}$$

The function $s(\vec{\rho} - \vec{\rho}_j)$ has the physical significance of a shielding term: it represents the fractional change in solute flux at $\vec{\rho}$ produced by a perforation at $\vec{\rho}_j$. A self-consistent field approach based on such reasoning has been used by Prager and Frisch [7a] (as well as by Malone, Hutchinson, and Prager [7b] for the corresponding hydrodynamic problem). The results strongly suggest that, at least for low n , $s(\rho)$ should behave as $1/\rho$ in the range $a \ll \rho \ll 1/na$, but as $1/\rho^3$ for $\rho \gg 1/na$. A reasonably tractable choice capable of satisfying these requirements is the two-parameter form*

$$s(\rho) = \alpha [\frac{a}{b} (1 + \frac{\rho^2}{b^2})^{-1/2} - (1 + \frac{\rho^2}{a^2})^{-1/2}] \qquad (b > a) \tag{30}$$

for which the functions χ and μ can be explicitly evaluated by means of Fourier transforms:

$$\chi(\rho) = 2\pi a^2 \alpha^2 \ \ell n \ \frac{[a + b + ((a + b)^2 + \rho^2)^{1/2}]^2}{[2a + (4a^2 + \rho^2)^{1/2}][2b + (4b^2 + \rho^2)^{1/2}]}$$

$$\mu(\rho) = \alpha \{ \text{arc csc} [\frac{1}{2a} ((b^2 + (a + \rho)^2)^{1/2} + (b^2 + (a - \rho)^2)^{1/2})] \tag{31}$$

$$- \text{arc csc} [\frac{1}{2a} ((a^2 + (a + \rho)^2)^{1/2} + (a^2 + (a - \rho)^2)^{1/2})]\}$$

With χ and μ given by (31), numerical evaluation of the integral in (29) is straightforward; we have, for each value of n , minimized it with respect to α and b to obtain the optimum upper bound on $\kappa_d^{(0)}/\kappa_d$ possible with the trial function (30). The result appears as curve 2 in Fig. 4.

To apply the lower bound (19) to the case of randomly overlapping circular perforations, we require a trial $\Delta\tilde{c}(\vec{r})$ that vanishes whenever $|\vec{\rho} - \vec{\rho}_i| < a$ for <u>any</u> i ; in addition $\Delta\tilde{c}(\vec{\rho})$ must be continuous, since otherwise the autocorrelation $h(\rho)$ will have a finite slope at $\rho = 0$, leading to an infinite

---

* It is also possible to optimize with respect to $s(\rho)$, but that requires solution of a non-linear integral equation.

value for the integral in (19) and a lower bound of zero. A trial function that satisfies these conditions is

$$\Delta\tilde{c}(\vec{\rho}) = \beta \prod_i (1 + v(|\vec{\rho} - \vec{\rho}_i|)) ,$$

$$v(\rho) = -1 \text{ if } \rho < a , \qquad (32)$$

$$v(\rho) \to 0 \text{ as } \rho \to \infty ,$$

corresponding to an autocorrelation

$$h(\rho) = e^{n\nu(\rho)} \qquad (33)$$
$$(\nu \equiv \int v(\vec{r})v(\vec{r} + \vec{\rho})d^2\vec{r})$$

and a lower bound

$$\frac{\kappa_d(0)}{\kappa_d} > 8na/\int_0^\infty \frac{e^{n\nu(0)} - e^{n\nu(\rho)}}{\rho^2} d\rho \qquad (34)$$

The obvious choice for $(1 + v(\rho))$ is the known $\Delta c(\rho)$ for a single isolated circular perforation [3a]; this would suggest the trial function

$$v(\rho) = \begin{cases} -1 & (\rho < a) \\ -\frac{2}{\pi} \text{ arc sec } \frac{\rho}{\omega} & (\rho > a) , \end{cases} \qquad (35)$$

but both $h(0)$ and $h(\infty)$ diverge in that case, since $v(\rho)$ vanishes at large ρ only as $1/\rho$. To remedy this deficiency we use instead

$$v(\rho) = \begin{cases} -1 & (\rho < a) \\ [\frac{a}{b'} - \frac{2}{\pi} \text{ arc csc } (\frac{\rho}{a})]/[1 - \frac{a}{b'}] & (a < \rho < b') \\ \frac{2}{\pi} [\frac{a}{b'} \text{arc csc}(\frac{\rho}{b'}) - \text{arc csc } (\frac{\rho}{a})]/[1 - \frac{a}{b'}] & (\rho > b') \end{cases} \qquad (36)$$

When $b' \gg a$ the trial function (36) is almost identical with (35) for $\rho < b'$, but vanishes as $1/\rho^3$ for $\rho \gg b'$. The $\nu(\rho)$ corresponding to (36) can be obtained in closed form:

$$\nu(\rho) = \frac{2a^2b'^2}{\pi(b'-a)^2} [\frac{1}{2} F(2, \frac{\rho}{a}) + \frac{a^2}{2b'^2} F(\frac{2b'}{a} , \frac{\rho}{a})$$

$$+ \frac{a}{b'} F(\frac{b'}{a} -1, \frac{\rho}{a}) - \frac{a}{b'} F(\frac{b'}{a} + 1, \frac{\rho}{a})]$$

where

$$F(s,w) \equiv \begin{cases} (2s^2 + w^2)\, \ell n\left(\frac{s}{w} + \sqrt{\frac{s^2}{w^2} - 1}\right) - 3s\sqrt{|s^2 - w^2|} & (w < s) \\ 0 & (w > s) \end{cases}$$

The lower bound (34) can now be evaluated numerically and optimized with respect to the single parameter b' to give the lower of the three plots in Fig. 4 (curve 3).

Fig. 4 also shows the result (curve 4) of a more elaborate upper bound trial function, arrived at by making the optimum linear combination of (27) and an improved version of (24), as follows:

$$\tilde{q}_d = Q_1 \tilde{q}_1 + Q_2 \tilde{q}_2 , \tag{37}$$

where

$$q_1(\vec{\rho}) = \sum_i \left[q^{(0)}(\vec{\rho} - \vec{\rho}_i) \prod_j (1 + s(\vec{\rho} - \vec{\rho}_j))\right]$$

$$q_2(\vec{\rho}) = 1 - \prod_j (1 + Q_3 q(\vec{\rho} - \vec{\rho}_j))$$

$$q(\vec{\rho}) = \begin{cases} q^{(0)}(\vec{\rho}) & \rho < a \\ 0 & \rho > a , \end{cases}$$

and Q_1 , Q_2 , Q_3 are variation parameters.

V. Limiting Behavior at Low Perforation Density

The diffusive resistance (or reciprocal permeability) of a sheet randomly perforated with circular holes has thus been shown to lie, at any perforation density, between curves 3 and 4 of Fig. 4, and corresponding bounds for the hydrodynamic permeability of a membrane composed of randomly overlapping disks (i.e. the complement to the randomly perforated sheet) can be obtained through equation (9). In choosing trial functions we have been guided in part by computational convenience, so that, even within the framework of the forms (32) and (37), tighter bounds could be obtained by improved selection of $s(\rho)$, $q(\rho)$ and $v(\rho)$. However, for the system of primary interest, the sparsely perforated membrane, upper and lower bounds on the diffusive resistance are already quite close: up to $\sigma = 0.1$, the difference between them is at most 5%.

Indeed, as the void fraction σ goes to zero, both (29) and (34) give the same limiting behavior of $\kappa_d - \kappa_d^{(0)}$:

$$\frac{\kappa_d - \kappa_d^{(0)}}{\kappa_d^{(0)}} = -\frac{8}{\pi^2}\,\sigma\,\ln\sigma + O(\sigma) \tag{38}$$

The logarithmic approach to the "percolation limit" is a result of the long range interaction between perforations, arising from the fact that the perturbation in the concentration of diffusing solute around an isolated perforation falls off only as the inverse first power of the distance, and has been suggested by the self-consistent field treatment of Prager and Frisch [7a], although with a coefficient of ½ rather than $8/\pi^2$. The coincidence of the upper and lower bounds to within terms of order σ rigorously establishes this somewhat unexpected behavior at low perforation densities, including the correctness of the $8/\pi^2$ coefficient.*

In contrast to the case of diffusive permeability, the hydrodynamic permeability through a randomly perforated membrane should not be expected to show a logarithmic dependence on σ at low perforation densities. Although this can be formally demonstrated [8] by devising upper and lower bounds on $(\kappa_h - \kappa_h^{(0)})/\kappa_h^{(0)}$ which vanish as the first power of σ as $\sigma \to 0$, it is already obvious from the inverse cube law decay of Δp in the vicinity of an isolated perforation [2] that there will be no surprises in $\kappa_h(\sigma)$; in the limit of small σ, it can be expanded in an ordinary Taylor series in σ, whose individual terms correspond to isolated holes, hole pairs, etc.

It is for the randomly overlapping disk geometry that κ_h will show logarithmic behavior when the number $\overline{n}$ of disks per unit area is small. From equations (9) and (38) we have

$$\frac{\kappa_h - \kappa_h^{(0)}}{\kappa_h^{(0)}} = \frac{8}{\pi^2}\,\overline{n}\pi a^2\,\ln(\overline{n}\pi a^2) + O(\overline{n}), \tag{39}$$

---

* The value of ½ obtained in Ref. 7a is the result of treating the perforations as point sources.

which may be compared to the result found by Malone, Hutchinson, and Prager [5] for a membrane of $\overline{n}$ spheres per unit area randomly distributed over a plane (a is now the sphere radius):

$$\frac{\kappa_{hs} - \kappa_{hs}^{(0)}}{\kappa_{hs}^{(0)}} = \frac{9}{\pi^2}\,\overline{n}\pi a^2\ \ln(\overline{n}\pi a^2) + O(\overline{n})\ . \qquad (40)$$

Although equations (39) and (40) have the same form, they differ in the numerical coefficient when written in terms of $\overline{n}$ and a ; when expressed in terms of $\overline{n}$ and the limiting permeabilities $\kappa_h^{(0)} = (16\ \eta a n)^{-1}$ and $\kappa_{hs}^{(0)} = (6\pi\eta a n)^{-1}$, however, they become identical:

$$\frac{\kappa - \kappa^{(0)}}{\kappa^{(0)}} = -\frac{1}{32\pi^2}\ \frac{1}{\eta^2\kappa^{(0)}\overline{n}}\ \ln(\eta^2\kappa^{(0)2}\overline{n}) + O(\overline{n})\ . \qquad (41)$$

One is led to speculate that (41) applies to creeping flow through random planar arrays of arbitrarily shaped obstacles in the $n \to 0$ limit.

6. Effect of Fluctuations in Perforation Geometry on Diffusive Permeation [9]

In the preceding sections we have obtained bounds on the permeabilities of randomly perforated sheets, assuming the perforations to be stationary. We conclude by showing how the variational methods employed for the purpose can be extended to membranes whose structures fluctuate with time because membrane elements are performing Brownian motions of their own.

The most straightforward way of introducing temporal fluctuations into the randomly perforated sheet model is to allow the perforations to perform independent Brownian motions on the membrane plane. The effect of this will be an increase in the diffusive permeability, since additional pathways are now available for a solute molecule to pass through the membrane. Some idea of what to expect can be obtained by considering first of all an isolated perforation. If the diffusion coefficient characterizing the Brownian motion of a perforation is D_p , then the joint motion of solute and perforation is described by the diffusion equation

$$\frac{\partial\psi}{\partial t} = D_p \frac{\partial^2\psi}{\partial\vec{\rho}^2} + D \frac{\partial^2\psi}{\partial\vec{r}^2} , \tag{42}$$

where $\vec{\rho}$ is the position of the perforation center on the membrane plane at $z = 0$, and $\psi(\vec{r},\vec{\rho},t)\, d^3\vec{r} d^2\vec{\rho}$ the probability of finding at time t a solute particle in the volume element $d^3\vec{r}$ at $\vec{r}$ and the perforation center in the area element $d^2\vec{\rho}$ at $\vec{\rho}$. Boundary conditions are

$$\frac{\partial\psi}{\partial z} = 0 \quad \text{at} \quad z = 0 \quad \text{if} \quad |\vec{r} - \vec{\rho}| > a$$

$$\psi \quad \text{continuous at} \quad z = 0 \quad \text{if} \quad |\vec{r} - \vec{\rho}| < a$$

$$\int (\psi(x,y,-z,\vec{\rho},t) - \psi(x,y,z,\vec{\rho},t)) d^2\vec{\rho} \rightarrow \Delta c \quad \text{as} \quad z \rightarrow \infty$$

The problem can be reduced to the stationary perforation (i.e. $D_p = 0$) case by transformation to scaled relative coordinates

$$\vec{\rho}\,' = (\vec{r}\,' - \vec{\rho}) , \qquad z' = (1 + \frac{D_p}{D})^{1/2} z ,$$

where $\vec{r}\,'$ is the projection of $\vec{r}$ on the membrane plane:

$$\frac{\partial\psi}{\partial t} = (D_p + D) \left[\frac{\partial^2\psi}{\partial\vec{\rho}\,'^2} + \frac{\partial^2\psi}{\partial z'^2} \right] ;$$

the boundary conditions remain unchanged. This is just the formulation of diffusion through a stationary hole of radius $\underline{a}$ if the solute diffusion coefficient is $D + D_p$. Transformation back to the original coordinates then gives, for the steady state permeability,

$$\kappa_d^{(0)}(D,D_p) = (1 + \frac{D_p}{D})^{1/2} \kappa_d^{(0)}(D,0) \tag{43}$$

To put it another way, the conversion from stationary to mobile perforations is, for a sparsely perforated membrane, accomplished by replacing the solute diffusion coefficient D with an effective coefficient $(D(D + D_p))^{1/2}$.

Unless D_p is small compared to D , the effect of perforation mobility is by no means negligible; if $D_p >> D$, the solute flux will be controlled neither by D nor by D_p , but by their geometric mean. Even at fixed D , a, and perforation density, considerable enhancement of the diffusive permeability is therefore possible through an increase in D_p.

At finite perforation densities, equation (42) must be replaced by the many-body diffuson equation, which reads, for a membrane with $N = nA$ perforations,

$$\frac{\partial\psi}{\partial t} = D_p \sum_{i=1}^{N} \frac{\partial^2\psi}{\partial\vec{\rho}_i^{\,2}} + D\,\frac{\partial^2\psi}{\partial\vec{r}^{\,2}} \tag{44}$$

The problem can no longer be reduced in terms of relative coordinates, and separation of variables is prevented by the boundary conditions:

$$\frac{\partial\psi}{\partial z} = 0 \quad \text{at} \quad z = 0 \quad \text{if} \quad |\vec{r} - \vec{\rho}_i| > a \quad \text{for } \underline{\text{all}} \ i$$

$$\psi \ \text{continuous at} \quad z = 0 \quad \text{if} \quad |\vec{r} - \vec{\rho}_i| < a \quad \text{for } \underline{\text{any}} \ i$$

$$-D\,\frac{\partial\psi}{\partial z} \to \langle q_d\rangle \quad \text{as} \quad z \to \pm\infty .$$

Although an exact treatment is thus not possible, variational bounds analogous to (14) and (19) can still be derived. Indeed it turns out that we need only replace the 3-dimensional steady-state Green's functions appearing under the integral signs in those two inequalities by their (2N + 3)-dimensional counterparts; thus in (14)

$$\frac{1}{4\pi\xi} \Longrightarrow \frac{\Gamma(N+1/2)}{2\pi^{N+(3/2)}}\,\frac{1}{\Xi^{2N+1}} \equiv \int_0^\infty \frac{e^{-\Xi^2/4\tau}}{(4\pi\tau)^{N+(3/2)}}\,d\tau , \tag{45a}$$

and in (19)

$$\frac{1}{4\pi\xi^3} \Longrightarrow \frac{\Gamma(N+(3/2))}{\pi^{N+(3/2)}}\,\frac{1}{\Xi^{2n+3}} \equiv \int_0^\infty \frac{e^{-\Xi^2/4\tau}}{(4\pi\tau)^{N+(3/2)}}\,\frac{d\tau}{\tau} ; \tag{45b}$$

in both inequalities the vector $\vec{\xi}$ must be replaced by the (2N + 2) - dimensional vector $\vec{\Xi} \equiv (\vec{\xi},(D/D_p)^{1/2}\rho_1,\ldots(D/D_p)^{1/2}\rho_N)$. These substitutions give the inequalities

$$\left[\frac{\Gamma(N+1/2)}{\pi^{N+1/2}}\int \frac{F(\Xi)}{\Xi^{2N+1}}\,d^{2N+2}\vec{\Xi}\right] > \frac{4\pi D}{\kappa_d} > \left[4\frac{\Gamma(N+(3/2))}{\pi^{N+1/2}}\int \frac{H(0)-H(\Xi)}{\Xi^{2N+3}}\,d^{2N+2}\vec{\Xi}\right]^{-1} , \tag{46}$$

where F and H are correlations derived from trial concentration jump and flux distributions $\Delta\tilde{\psi}(\vec{\Xi})$ and $\tilde{Q}(\vec{\Xi})$:

$$F(\vec{\Xi}) \equiv \frac{1}{\langle\tilde{Q}\rangle^2} \langle(\tilde{Q}(\vec{\Xi}') - \langle\tilde{Q}\rangle)(\tilde{Q}(\vec{\Xi}' + \vec{\Xi}) - \langle\tilde{Q}\rangle)\rangle$$

$$H(\vec{\Xi}) \equiv \langle(\Delta\tilde{\psi}(\vec{\Xi}')\ \Delta\tilde{\psi}(\vec{\Xi}' + \vec{\Xi})\rangle/\langle\Delta\tilde{\psi}\rangle^2$$

The angular brackets must here be understood to indicate averages extending over all $\vec{\Xi}'$. In the case of randomly overlapping perforations executing uncorrelated Brownian motions, all configurations are equally likely in the steady state, but the averaging process continues to be non-trivial because of the requirements imposed on $\Delta\tilde{\psi}$ and $\tilde{Q}$:

$$\begin{aligned} \Delta\tilde{\psi} &= 0 \quad \text{if} \quad |\vec{\xi} - \rho_i| < a \quad \text{for } \underline{\text{any}}\ i \\ \tilde{Q} &= 0 \quad \text{if} \quad |\vec{\xi} - \rho_i| > a \quad \text{for } \underline{\text{all}}\ i \end{aligned} \tag{47}$$

Moreover, in contrast to the case of stationary perforations, the correlations F and H are not isotropic in Ξ-space; the multidimensional integrals in (46) will not in general simplify to single integrals.

However any valid trial function for the stationary problem is also a trial function for the case of mobile perforations, and we have carried out some calculation on this basis. For trial functions that can be written as products (or sums of products) of factors depending only on the position of a single perforation, the identity

$$\Gamma(N + 1/2) \int\!\ldots\!\int \frac{g(\vec{\xi},\vec{\rho}_1)\ g(\vec{\xi},\vec{\rho}_2)\ \ldots\ g(\vec{\xi},\vec{\rho}_N)}{\Xi^{2N+1}}\, d^2\vec{\rho}_1 \ldots d^2\vec{\rho}_N$$

$$\equiv \left(\frac{D_p}{D}\right)^{N+(3/2)} \int_0^\infty \left(\int e^{-D\rho^2/4D_p\tau}\ g(\vec{\xi},\vec{\rho})\,d^2\vec{\rho}\right)^N \frac{e^{-\xi^2/4\tau}}{(2\tau)^{N+(3/2)}}\, d\tau$$

can be used to make what would otherwise be 2N-fold integrals tractable. Fig. 5 shows the bounds obtained for the trial functions (36) and (37), with $D = D_p$. When plotted as $\kappa_d^{(0)}(D,D_p)/\kappa_d(D,D_p)$, the lower bound is virtually indistinguishable from that obtained in the stationary case, but the upper bound has decreased appreciably over the result for fixed perforations. This narrowing of the interval within which the true value of $\kappa_d^{(0)}/\kappa_d$ must lie is not unexpected: it is usually easier to obtain good variational estimates of permeability coefficients

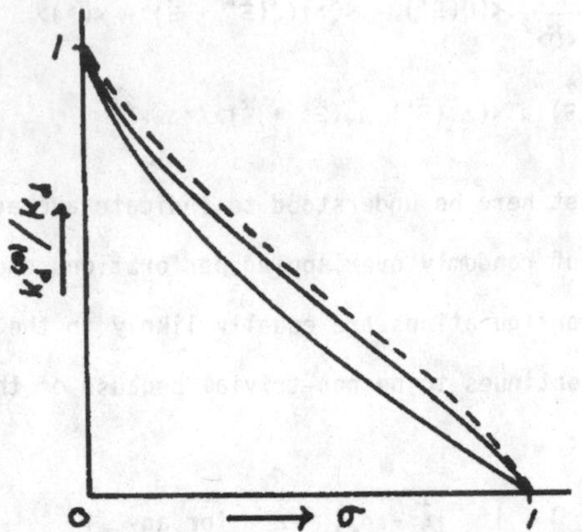

Figure 5. Bounds on the relative diffusive resistance $\kappa_d^{(0)}/\kappa_d$ for mobile randomly overlapping holes, as functions of σ for the case $D_p = D$. The dashed curve is the upper bound for $D_p = 0$ (curve 4 of Figure 4).

---

in systems that provide a larger selection of mechanisms by which the penetrant can traverse the membrane.

References

1. J. Crank and G.S. Park, "Diffusion in Polymers", Academic Press (NY), 1968.

2. W.D. Stein, "The Movement of Molecules across Cell Membranes", Academic Press (New York), 1967.

3. a) H.S. Carslaw and J.C. Jaeger, "Conduction of Heat in Solids", 2nd ed., Oxford Univ. Press (London), 1959.
   b) J. Happel and H. Brenner, "Low Reynolds Number Hydrodynamics", Prentice Hall (Englewood Cliffs), 1965.

4. R. Aris and W. Strieder, "Variational Methods Applied to Problems of Diffusion and Reaction", Springer (Berlin), 1973.

5. a) J.B. Keller, J. Math. Phys. 5, 548 (1964).
   b) K.S. Mendelson, J. Appl. Phys. 46, 917 (1975).

6. A.M. Arthurs, "Complementary Variation Principles", Clarendon Press (London), 1970.

7. a) S. Prager and H.L. Frisch, J. Chem. Phys. 62, 89 (1975).
   b) G.H. Malone, T.E. Hutchinson, and S. Prager, J. Fluid Mech. 65, 753 (1974).

8. G.H. Malone, Ph.D. Thesis, Univ. of Minnesota, 1975.

9. S.Y. Suh, MS Thesis, Univ. of Minnesota, 1976.

DIFFUSIONS AND RANDOM WALKS IN RANDOM MEDIA

George C. Papanicolaou

Courant Institute
New York University
New York, New York 10012

ABSTRACT

We formulate and analyze briefly some problems that arise when diffusion processes evolve in a random medium. We also consider some systems of interacting Brownian particles and examine their collective behavior.

1. Introduction

Physical problems lead very naturally to the formulation of partial differential equations (or difference equations) that have random coefficients because the processes in question take place in a random medium. Consider for example a conducting material of conductivity a_1 into which is imbedded at random identical spherical inclusions of material with different conductivity a_2. If we denote by $\omega = (x_1, x_2, \ldots)$ the positions (points in R^3) of the sphere centers and by $a(x,\omega)$ the conductivity of the composite material, then at any point x, $a(x,\omega)$ is equal to a_1 if x is outside the spheres and a_2 if it is inside any sphere. A typical problem to be analyzed is as follows: Compute the temperature distribution $u(x,\omega)$ of a large piece of the composite material D (large with respect to the size of the spheres) under given boundary conditions:

$$\nabla \cdot (a(x,\omega)\nabla u(x,\omega)) = 0 \quad \text{in} \quad D \,, \tag{1}$$

$$u(x,\omega) = \text{given function on} \quad \partial D \,.$$

Since the conductivity $a(x,\omega)$ depends on the configuration ω of the spherical inclusions, which we shall assume are randomly distributed, the solution $u(x,\omega)$ of (1) also depends on the configuration ω. The corresponding statistical properties of u are obtained by regarding it as a functional of the conductivity, a , via (1). In general (1) is an impossibly difficult problem to analyze, especially numerically, when the number of spherical inclusions contained in D is large, the radius of each sphere is small and the ratio of the conductivities a_2/a_1 is large or small.

The problem simplifies and becomes interesting if we look only for bulk information about the composite. If the number of spherical inclusions per unit volume is large and the radii are small, keeping the occupied volume fraction fixed, can we find a number a^* that does not depend on the configuration and describes the overall conducting properties of the composite? This number, or tensor, should naturally be called the effective conductivity and should not depend on the macroscopic shape D of the conducting material.

We formulate the definition of a^* in a more general setting as follows. Let $a(x,\omega)$ be a strictly stationary process such that

$$0 < a_1 \leq a(x,\omega) \leq a_2 < \infty \ , \quad x \in R^3 \ , \tag{2}$$

where $\omega \in \Omega$ and $(\Omega,\mathfrak{F},P)$ is a probability space. The set Ω is the set of all configurations, which need not be spherical inclusions. The microscopic conductivity $a(x,\omega)$ need not in general be two-valued but we require (2) to hold. Moreover we require stationarity, which means that the microscopic properties of the composite are statistically translation invariant. This is a necessary hypothesis if we want to be able to describe the material by bulk or effective properties.

Let $F(x,\omega)$ and $G(x,\omega)$ be two stationary random vector fields that represent heat flux and negative temperature gradient. One can easily show that these fields are uniquely defined by the requirements of stationarity, mean square integrability and

$$\nabla \cdot F = 0 \qquad \text{(no sources)} \tag{3}$$

$$\nabla \times G = 0 \qquad \text{(G is a gradient)} \tag{4}$$

$$F(x,\omega) = a(x,\omega)G(x,\omega) \qquad \text{(Fourier's law of heat conduction)} \tag{5}$$

$$\langle G \rangle = e \quad , \qquad e \text{ is a unit vector .} \tag{6}$$

Here $\langle G\rangle$ denotes the average of G over configurations ω and, by stationarity, is a constant

$$\langle G\rangle = \int_{\Omega} G(x,\omega)\, P(d\omega) \ . \qquad (7)$$

The problem defined by Equations (3)-(6) is an infinite-region version of (1); stationarity and (6) are the boundary conditions. We now define the effective conductivity tensor a^* by

$$a^*(e,\ell) = \langle F \cdot \ell\rangle \ , \quad \text{with } \ell \text{ a unit vector.} \qquad (8)$$

This definition simply means that the effective conductivity in the direction ℓ is the average heat flux flowing in this direction when unit average temperature gradient is prescribed in the direction e.

In the following sections we shall give the connections between the above definition of a^* and (i) an infinite volume limit definition and (ii) a diffusion process in a random medium definition. We shall then discuss some qualitative properties of a^*, namely bounds. In the last section we shall discuss notions of bulk quantities for interacting Brownian particles and the relationship of such problems to the ones that arise in random media.

2. Infinite Volume Limit.

The detailed mathematical analysis of this limit is given in [1] by the methods that were used in [2].

We formulate the problem in two dimensions for simplicity. Let $u^{(N)}(x_2,x_2,\omega)$ be the solution of the boundary value problem

$$\nabla \cdot (a(x_1,x_2,\omega)\nabla u^{(N)}(x_1,x_2,\omega)) = 0 \quad \text{in } -N < x_1 < N \ , \ -N < x_2 < N \ , \qquad (9)$$

$$\frac{\partial u^{(N)}}{\partial x_2} = 0 \quad \text{on } x_2 = \pm N \ , \quad -N < x_1 < N \ , \qquad (10)$$

$$u^{(N)} = N \quad \text{on } x_1 = -N \ , \ -N < x_2 < N$$

and

$$u^{(N)} = -N \quad \text{on } x_1 = +N \ , \ -N < x_2 < N \ . \qquad (11)$$

Define

$$a^{*,N}(\omega) = \frac{1}{2N} \int_{-N}^{N} a(x_1,x_2,\omega) \frac{\partial u^{(N)}(x_1,x_2,\omega)}{\partial x_1} dx_2 \quad . \qquad (12)$$

This is just the heat flux flowing down the temperature gradient. Note that it depends on the sample size N and the configuration ω of the random medium. Assuming that the random conductivity is ergodic, in addition to being stationary, then one can show that

$$\langle (a^{*,N} - a^*)^2 \rangle \to 0 \quad \text{as} \quad N \to \infty \qquad (13)$$

where $a^* = a^*(e_1,e_1)$ is given by (8) and e_1 is the unit vector in the x_1 direction. Clearly (13) is what one expects to be true and ergodicity is a natural hypothesis to impose. It would be interesting to show that $a^{*,N}(\omega) \to a^*$ almost surely as $N \to \infty$. This has not been proved yet but it is expected to be true.

3. Diffusions in a Random Medium

For each configuration $\omega \in \Omega$ let

$$L_\omega = \nabla \cdot (a(x,\omega)\nabla) \ . \qquad (14)$$

This differential operator generates a diffusion Markov process on R^3 which is reversible relative to Lebesgue measure since L_ω is self-adjoint. Let $Q_{x,\omega}$ denote the probability measure on paths (t) in R^3 (or in R^d in just the same way) starting from $x = (0)$. Note that this Markovian probability measure depends on the configuration $\omega \in \Omega$. It is the probability law of a reversible diffusion process in a random medium.

It is shown in [3] that

$$\langle (\frac{1}{t} E^{Q_{x,\omega}}\{((t) - x)^2\} - a^*)^2 \rangle \to 0 \quad \text{as} \quad t \to \infty \ . \qquad (15)$$

This means that the mean square displacement of the random diffusion $E^{Q_{x,\omega}}\{((t) - x)^2\}$ behaves like $a^* t$, for t large, in the mean square sense relative to the distribution of the configurations ω . Here we have assumed for simplicity that $a^*(e_j,e_j) = a^*$ independently of the direction e_j. Again a^* is given by (8).

On the basis of this result one can calculate a^* by a Monte Carlo simulation which is a convenient and relatively efficient way to get the effective conductivity.

Let us make some remarks about related problems.

The result analogous to (15) for the random diffusion operator

$$L_\omega = a(x,\omega)\Delta \quad \left(\text{or } \sum_{i,j} a_{ij}(x,\omega) \frac{\partial^2}{\partial x_i \partial x_j}\right), \tag{16}$$

which is not selfadjoint, is also valid [3] but it is obtained by a somewhat different method. In the selfadjoint case special tools (Hilbert space methods) are available that simplify and streamline the analysis considerably.

The result analogous to (15) for the operator

$$L_\omega = \Delta + b(x,\omega) \cdot \nabla \tag{17}$$

for $b(x,\omega)$ a stationary, zero mean random vector field is not expected to hold in general. In the one dimensional discrete analog of this problem Sinai [4] has shown that the root mean square displacement is of order $(\log t)^2$ rather than t, when $b(x,\omega)$ is white noise. For any dimension, when

$$b(x,\omega) = \nabla a(x,\omega) \tag{18}$$

where $a(x,\omega)$ is a stationary random process bounded above and below uniformly in x and ω, the analysis of (17) reduces to that of (14) almost word for word and the mean square displacement is linear in t (diffusive behavior of the random diffusion).

For dimensions $d \geq 3$, when

$$\nabla \cdot b(x,\omega) = 0 \tag{19}$$

and the power spectrum of b has a density, one can easily show [5], [6] that the analog of (15) is valid.

The lattice version of (14) which corresponds to a continuous time random walk on a lattice with random transition rates assigned to the bonds (so that the generator is selfadjoint) is analogous to (14); the detailed analysis is given in [7]. The lattice version of (16) which corresponds to a continuous time random walk on a lattice with transition rates assigned to sites (so that the generator is not selfadjoint) is carried out in [8].

It should be noted that the analysis of (17) (diffusion with random drift) encounters a "small divisor" problem which hypotheses such as (18) and (19) help simplify. Hilbert space methods, when applicable, also help neutralize small divisor difficulties.

4. Bounds for a^*.

In all problems that have effective diffusive behavior (i.e. (15) holds) one wants to know properties of a^*. How does a^* depend on specific properties of $a(x,\omega)$, the microscopic conductivity (or diffusivity)? It is very convenient for this purpose to have a characterization for a^* which does not involve limits, such as (3)-(8).

From the characterization (3)-(8) one can easily show that

$$\langle a^{-1}\rangle^{-1} \leq a^* \leq \langle a\rangle , \tag{20}$$

which are bounds that have been known from the beginning of this century. There are many other interesting bounds that can be obtained by recognizing that (3)-(6) admits a variational formulation [9, and references therein]. It is of considerable interest in electromagnetic problems to find estimates for the range of values of a^* when $a(x,\omega)$ is complex-valued in (3)-(6). In the complex case (3)-(6) does not admit a variational formulation so another method is required. Such a method, the analytic continuation method, was introduced by D. Bergman [10] but it is so far restricted to two-component materials. A mathematical analysis of this method is given in [1].

For many theoretical and applied problems it is important to be able to characterize the <u>exact</u> range of values of a^* when $a(x,\omega)$ belongs to a certain class. Moreover it is important to be able to display particular functions (nontrivial ones) $a(x,\omega)$ that realize each possible effective medium. One reason why this is interesting is the following. Suppose we want to built out of two types of conductors a material with specified bulk properties: maximum heat flux for some temperature distribution at the boundary and domain geometry. The optimal conducting material may be a composite i.e., a limit of mixed materials with finer and finer structure. Such problems were considered by Tartar [11], Lurie [12,13] and Kohn and Strang [14].

5. Interacting Brownian Particles.

A number of interesting asymptotic problems associated with the macroscopic behavior of interacting particle systems are analyzed in [15], [16], [17]. We shall formally describe a problem that we have recently analyzed in detail by the adapting the the methods we used in [2].

Let $\Phi(x)$ be a pair potential function that is repulsive and finite range: $\Phi > 0$, Φ smooth and $\Phi(x) = \Phi(-x)$, $x \in R^d$, $d > 2$. Consider the infinite system of interacting Brownian motions

$$dx_{k(t)} = - \sum_{j \neq k} \nabla\Phi(x_k(t) - x_j(t))dt + \sqrt{2}\, dw_k(t) \tag{21}$$

$$x_k(0) = x_k \ , \qquad k = 1,2,\ldots$$

Here $w_k(t)$, $k = 1,2,\ldots$, are standard independent Brownian motions. The existence and uniqueness of (21) was studied in [18] and its Gibbs equilibrium states in [19]. Formally the Gibbs equilibrium measures are given by the "density"

$$a(x_1,x_2,\ldots) = \exp\{- \sum_{i \neq j} \Phi(x_i - x_j)\}$$

and the generator by

$$L = \frac{1}{a} \sum \nabla_j \cdot (a\, \nabla_j) \tag{22}$$

where ∇_j is the gradient operator with respect to the j-th coordinate.

We single out a particle $x_1(t)$, say, as the one that is closest to the origin at time $t = 0$ and set

$$\xi(t) = x_1(t) \ , \ \ y_k(t) = x_k(t) - x_1(t) \ , \ k \geq 2 \ . \qquad (23)$$

The relative motion process $\{y_1(t), y_2(t), \ldots,\}$ turns out to be a Markov process with formal generator

$$L = \frac{1}{a} \ \{ \sum_{j \geq 2} \nabla_j \cdot (a \nabla_j) + (\sum_{k \geq 2} \nabla_k) \cdot (a \sum_{k \geq 2} \nabla_k)\} \qquad (24)$$

where

$$a(y_2, y_3, \ldots,) = \exp\{-1/2 \sum_{i \neq j \geq 2} \Phi(y_i - y_j) - \sum_{j \geq 2} \Phi(y_j)\} \qquad (25)$$

We show that the relative motion process is well defined when (24) is interpreted correctly.

Assume that $\{y_k\}$ is initially distributed with its equilibrium law formally given by (25). Then we show that

$$\frac{1}{t} \langle (\xi(t) - x)^2 \rangle \to a_s^* \quad \text{as} \quad t \to \infty \qquad (26)$$

which says that the tagged particle has diffusive behavior in the long run. The bracket in (26) is average relative to measure with formal density (25). The diffusion coefficient a_s^* is called the selfdiffusion coefficient of the tagged particle. It is a functional of the pair potential Φ and the mean particle density at equilibrium. From a variational characterization of a^* one can extract properties of a_s^* at various concentrations. For example, at low concentrations a_s^* decreases as the concentration increases.

Acknowledgement

This research was supported by the Air Force Office of Scientific Research under grant number AFOSR 80-0228.

References

1. K. Golden and G. Papanicolaou, Bounds for effective parameters of heterogeneous media by analytic continuation, Comm. Math. Phys. to appear.

2. G. Papanicolaou and S.R.S. Varadhan, Boundary value problems with rapidly oscillating random coefficients, in "Random Fields" edited by J. Fritz, J. Lebowitz, D. Szaxa, Janos Bolyai series, North-Holland, 1981, 835-873.

3. G. Papanicolaou and S.R.S. Varadhan, Diffusion with random coefficients, in Essays in Stat. & Probability, edited by G. Kallianpur, P. Krishnaiah and J. Ghosh, North Holland 1982, 253-262.

4. Ya. Sinai, Limit behavior of one-dimensional random walks in random environments, Theory Prob. Appl. 27 (1982), 247-258.

5. G. Papanicolaou and O. Pironneau, On the asymptotic behavior of motion in random flows, in Stoch. Nonl. Systems, edited by L. Arnold and R. Lefever, Springer 1981, 36-41.

6. R. Figari, E. Orlandi and G. Papanicolaou, Proceedings of Kyoto Conference, July 1982, to appear.

7. R. Kuhnemann, Comm. Math. Phys. to appear.

8. G. Lawler, Comm. Math. Phys. to appear.

9. W. Kohler and G. Papanicolaou, Bounds for effective conductivity of random media, in Macroscopic properties of disordered media, R. Burridge, S. Childress and G. Papanicolaou editors, Springer Lect. Notes Phys. #154 (1982), 111-130.

10. D. Bergman, The dielectric constant of a composite material - a problem in classical physics, Physics Reports C, 43 (1978) 377-407.

11. L. Tartar, Estimation de coefficients homogenises, Springer Lect. Notes Math. #704 (1977) 364-373.

12. K.A. Lurie, On the optimal distribution tensor of a specific resistence of the working fluid in a channel of an MHD-generator, PMM 34 (1970)#7.

13. K.A. Lurie, A.V. Cherkaev and A.V. Fedorov, Regularization of optimal design problems for bars and plates I, II, J. Optim. Theory Appl. 37 (1982) 499, 523.

14. R. Kohn and G. Strang, Structural design optimization, homogenization and relaxation of variational problems, Springer Lect. Notes Phys. #154 (1982) 131.

15. J.L. Lebowitz and H. Spohn, Steady State Self-Diffusion at Low Density, J. Stat. Phys. 29 (1982) 39.

16. J.L. Lebowitz and H. Spohn, Microscopic Basis for Fick's Law for Self-Diffusion, J. Stat. Phys. 28 (1982) 539.

17. C. Kipnis, J.L. Lebowitz, E. Presutti and H. Spohn, Self-Diffusion for Particles with Stochastic Collisions in One Dimension, to appear.

18. R. Lang, Unendlich-dimensionale Wienerprozesse mit Wechselwirkung, Z. Wahr. u. Verw. Geb. 38 (1977) 55-72 and 39, 277-299.

19. D. Ruelle, Superstable interactions in classical statistical mechanics, Comm. Math. Phys. 18 (1970) 127-159.

WAVES IN DISORDERED MEDIA

Michael J. Stephen
Department of Physics
Rutgers University
Piscataway, New Jersey 08544

ABSTRACT

The propagation of waves in disordered media is considered. Examples are provided by sound waves in materials of randomly varying composition or in turbulent fluids or electromagnetic waves in materials with randomly varying dielectric constant. The nature of the states at a given frequency is determined by studying the diffusion of the energy associated with the wave. In one and two dimensions diffusion vanishes at large distances leading to the conclusion that all states are localized. Above two dimensions a mobility edge in the diffusivity exists showing the existence of extended states at low frequencies and localized states at high frequencies. Phonon localization lengths are derived.

1. Introduction

In this article we discuss waves in disordered media [1]. This is a subject that arises in a number of areas of physics and a considerable body of literature exists on the problem [2]. This study was stimulated by recent work [3] on electrons in disordered solids, the existence of localized and extended states for electrons in such materials and the appearance of mobility edges in the conductivity. An extended state is one which extends across the whole material in contrast to a localized state which is confined to some region. The mobility edge separates (in energy) the localized and extended states. The d.c. conductivity at zero temperature is only non-zero if the states of the electrons at the Fermi energy are extended.

We have attempted to apply some of the methods developed to deal with the electron problem to the problem of waves in disordered media in order to understand when localized and extended states occur and the appearance of mobility edges in the diffusivity (of energy).

As an example of this type of problem we may consider a longitudinal sound wave propagating in a disordered medium (e.g. turbulent water or a solid with varying composition). The linearized wave equation for the pressure is

$$[1/c^2(\vec{r})][\partial^2/\partial t^2]P(\vec{r},t) = \nabla^2 P(\vec{r},t), \qquad (1.1)$$

where $c(\vec{r}) = (\partial P/\partial\rho)^{1/2}$ is the sound speed and depends on position within the medium. A second example is provided by phonons (quantized vibrational excitations) in a disordered crystal. If $\vec{u}$ is the small displacement of the medium, in a continuum approximation it satisfies the equation

$$m(\vec{r})[\partial^2/\partial t^2]\vec{u} = [\mu(\vec{r}) + \lambda(\vec{r})]\, \vec{\nabla}\vec{\nabla}\cdot u + \vec{\mu}(r)\nabla^2\vec{u}, \qquad (1.2)$$

where μ and λ are the Lame coefficients and m is the mass density. In general both the elastic constants μ and λ and the mass density (isotopic disorder) can depend on position. Eq. (1.2) contains both longitudinal and transverse waves but scattering will mix these waves and the distinction between them will be lost. A third example is provided by a light wave propagating in a medium with spatially varying dielectric constant $\varepsilon(\vec{r})$ (e.g., a light wave passing through a turbulent atmosphere). After elimination of the magnetic field, Maxwell's equations for the electric field $\vec{E}$ can be written in the form

$$[\varepsilon(\vec{r})/c^2][\partial^2/\partial t^2]\vec{E} = \nabla^2\vec{E} + \vec{\nabla}(\vec{E}\cdot\nabla \ln \varepsilon), \qquad (1.3)$$

where c is the vacuum velocity of light. The last term in (1.3) comes from writing $\vec{\nabla}\cdot(\varepsilon\vec{E}) = 0$ as $\vec{\nabla}\cdot\vec{E} = -\vec{E}\cdot\vec{\nabla} \ln \varepsilon$. Eq. (1.3) takes the same form as the previous equations if we neglect the second term on the right. This requires that $a \gg \lambda$, where a is the length scale associated with the disorder (i.e., $\nabla \ln \varepsilon \approx a^{-1}$) and λ is the wavelength of the wave.

The above examples lead us to consider the scalar wave equation

$$m(r)[\partial^2/\partial t^2]\phi(\vec{r},t) = \nabla^2\phi(\vec{r},t). \qquad (1.4)$$

As the equation is linear and the disorder is assumed time independent, we can take the Fourier transform with respect to time, so that

$$-m(\vec{r})\omega^2\phi = \nabla^2\phi \tag{1.5}$$

(with ϕ now a function of position $\vec{r}$ and frequency ω). This equation then describes the simplified problem that we will consider. We have then neglected the possible longitudinal and transverse nature of the waves. The disorder is assumed to be in the mass density. From the usual ideas of universality in critical phenomena we expect that the detailed nature of the disorder is not important. We have examined some cases in which the elastic constants are random functions of position and have obtained results qualitatively the same as those described below.

We emphasize several important features of the problem (1.5) being considered: (a) The disorder is assumed independent of the time. This is not always the case. This assumption has the consequence that the scattering is purely elastic, the frequency of the wave is constant and energy is conserved. (b) We have made the harmonic approximation i.e. for phonons the phonon-phonon interactions are neglected and the inelastic scattering arising from such interactions is neglected. This is discussed further below when we discuss thermal diffusivity.

We will characterize the disorder in the following way. The mass density $m(\vec{r})$ (sound velocity or dielectric constant) is taken in the form $m(\vec{r}) = 1 + m'(\vec{r})$ where m' is a Gaussian random variable with

$$\langle m'(\vec{r})\rangle = 0 \tag{1.6}$$

$$\langle m'(\vec{r})m'(\vec{r}')\rangle = B(\vec{r} - \vec{r}') .$$

The assumption that m' is Gaussian, i.e., is characterized by its first two moments (1.6), is unnecessary. It can be shown that only the second moment of m' is relevant. Any possible higher order cumulant moments are not important in the theory developed below. Higher moments of m' must actually exist because it is necessary that $m'(r) > -1$. The correlation function B characterizing the disorder may have the following forms:

$$B(\vec{r} - \vec{r}\,') = \gamma^2 \delta(\vec{r} - \vec{r}\,') \qquad (1.7a)$$

$$= \gamma^2 e^{-|\vec{r}-\vec{r}\,'|/a} \qquad (1.7b)$$

$$= \gamma^2 |\vec{r} - \vec{r}\,'|^{-\sigma} . \qquad (1.7c)$$

Thus in case (a) the disorder has zero correlation length, in (b) the correlation length is a and in (c) the disorder is characterized by a power law decay. We will confine the discussion mainly to case (a). Case (b) differs only qualitatively from (a). Case (c) is interesting because it is appropriate in the case of a turbulent medium, at least for some range of $|\vec{r} - \vec{r}\,'|$, but it has not been analyzed.

In the present problem described by Eq. (1.5) the energy is a conserved quantity and the energy density $E(\vec{r},t)$ of the wave satisfies a conservation relation

$$(\partial/\partial t)E(\vec{r},t) + \vec{\nabla}\cdot\vec{J}_E(\vec{r},t) = 0 , \qquad (1.8)$$

where $\vec{J}_E$ is the energy current. We propose to study the nature of the normal modes at frequency ω of (1.5) by introducing a velocity $\phi(\vec{r},0)$ of the medium localized around $r = 0$ at $t = 0$ and then examining how the energy (at frequency ω) associated with this disturbance spreads out. In a disordered medium of the type considered here the energy should diffuse so that

$$\vec{J}_E = -D\vec{\nabla}E(\vec{r},t) \qquad (1.9)$$

where D is the diffusion constant. If the states at frequency ω are extended, then the diffusion constant at frequency ω , $D(\omega)$ (defined below) will be finite and if the states are localized, $D(\omega)$ will vanish. It is assumed that the scattering due to the disorder is sufficient to give a finite i.e. not infinite, diffusion constant.

The diffusive motion of the energy is like a random walk problem and we can write the diffusion constant as

$$D = \lim_{t \to \infty} \frac{1}{t} \left[\int r^2 E(\vec{r},t)d\vec{r} / \int E(\vec{r},t)d\vec{r} \right] . \qquad (1.10)$$

Using the expression for the energy density $E(\vec{r},t) = m(\vec{r})\dot{\phi}^2(\vec{r},t) + (\vec{\nabla}\phi)^2$ and expressing the energy at $\vec{r},t$ in terms of the initial disturbance by means of the Green's function of (1.5), one can rewrite Eq. (1.10) in the form

$$2D = \lim_{\eta\to 0} 2\eta \left[\int d\omega\, \omega^2 \int dr\, r^2 \overline{|G(r,0,\omega_+)|^2} \Big/ \int d\omega\, \omega^2 \int dr\, \overline{|G(r,0,\omega_+)|^2}\right] , \quad (1.11)$$

where G is the Green's function of (1.5), $\omega_+ = \omega + i\eta$ and η is a small, positive, real quantity. The horizontal bar indicates an average over all realizations of the random mass field. The integral over all frequencies appearing in (1.11) leads to a spectral decomposition of the diffusivity

$$D = \int d\omega\, E(\omega)D(\omega) / \int d\omega\, E(\omega) , \quad (1.12)$$

where $E(\omega)$ is the energy density associated with the mode of frequency ω defined by

$$E(\omega) = \eta\omega^2 \int d\vec{r}\, \overline{|G(\vec{r},u,\omega_+)|^2}$$

Eq. (1.12) defines the <u>frequency dependent diffusivity</u> $D(\omega)$. Equations of the form of (1.12) are familiar in the theory of thermal diffusivity, in which case $E(\omega) = \hbar\omega\, \rho(\omega)[\partial\, n(\omega)/\partial T]$ where $\rho(\omega)$ is the density of states [4] and $n(\omega) = (e^{\beta\hbar\omega} - 1)$ is the Bose distribution (here $2\pi\hbar$ is Planck's constant and $\beta = 1/kT$) .

The expression for the diffusivity (1.11) involves the average of the product of an advanced and retarded Green's function [5]. This shows that we are dealing with an interference phenomenon, and the diffusion constant depends on the long range behavior of this product. It is important that the average of the product of Green's functions not be replaced by the product of the averages. The average Green's function is always short ranged, i.e., $G(r,\omega) \approx e^{-r/\ell}$ where ℓ is the mean free path. The Green's function in the disordered medium has a random phase and the averaging over this phase, leads to this short range behavior. In the average of the product of advanced and retarded Green's functions the random phase cancels, leading to long range behavior.

If we let

$$F(\vec{r}) = |\overline{G(r,0,\omega_+)}|^2 , \quad (1.13)$$

then we expect that the spatial Fourier transform of this function for weak scattering is

$$F(K) = 1/(\eta\, C_0 + C_2 K^2) . \quad (1.14)$$

$F(K)$ is essentially the Green's function for the diffusion equation (with η playing the role of frequency). $F(K)$ must satisfy the sum rule $\lim_{K\to 0} \eta\, F(K=0) = [\pi\rho(\omega)/4\omega^2]$ where $\rho(\omega)$ is the density of states, which gives $C_0^{-1} = \pi\rho/4\omega^2$. The frequency dependent diffusion constant is $D(\omega) = 2dC_2/C_0$. The form (1.14) is similar to the perpendicular susceptibility of the Heisenberg magnet, in which η plays the role of the magnetic field and C_2 is proportional to spin wave stiffness [6]. This analogy will prove useful below.

2. Localization as a Phase Transition

The electron localization problem has been approached mainly in two equivalent ways. The first begins by treating the scattering by diagrammatic methods [7]. In order to get localization, it is necessary to go beyond the usual ladder approximation and sum the so called crossed graphs. These graphs were fist considered by Langer and Neal [8]. In the electron problem this procedure has been followed by Abrahams and Ramakrishan [7] and others. The second method, introduced by Wegner [9], is to make this problem look like a phase transition problem as closely as possible and then to apply some of the techniques familiar from that field. This is the method employed. Briefly it begins with the introduction of a Lagrangian and the identification of an order parameter. The mean field theory is carried out by finding the stationary points of the Lagrangian. In the present case the mean field theory is identical with the coherent potential approximation [10]. The fluctuations around the mean field theory lead to the identification of the diffusivity as the relevant coupling constant in a model called the non-linear σ model [9]. A recursion relation for the diffusivity is obtained by integrating out the short-wavelength fluctuations, and from this recursion relation the mobility edge, correlation length and critical exponents are obtained.

3. Lagrangian and Mean Field Theory

To obtain the Lagrangian for the model (1.5) we introduce a generating function Z which allows us to calculate the average Green's functions and products as integrals over replica fields. As we require both advanced and retarded Green's functions, we introduce two sets of replica fields ϕ and ψ, each of which is an n component vector with components ϕ_α and ψ_α $(\alpha = 1,\ldots n)$. Then

$$Z = \langle \int e^{iL(\phi)-iL(\psi)}(d\phi)(d\psi)\rangle_c , \tag{3.1}$$

where

$$L(\phi) = (1/2)\int d\vec{r}[\omega_+^2 m(\vec{r})\phi^2(\vec{r}) - (\vec{\nabla}\phi)^2] \tag{3.2}$$

$$L(\psi) = (1/2)\int d\vec{r}[\omega_-^2 m(\vec{r})\psi^2(\vec{r}) - (\vec{\nabla}\psi)^2] ,$$

and the angular brackets in (3.1) indicate an average over the disorder. The average Green's functions are then given by

$$G(\vec{r},0,\omega_+) = -i\langle\phi_\alpha(\vec{r})\phi_\alpha(0)\rangle \tag{3.3}$$

$$G(\vec{r},0,\omega_-) = i\langle\psi_\alpha(\vec{r})\psi_\alpha(0)\rangle ,$$

with similar results for the average of products of Green's functions. In (3.3) and formulae below, it is understood that the $\eta \to 0$ limit is to be taken. This limit then removes the unwanted factor resulting from the integrals over the fields in (3.1). The replica method was first applied to this problem by Aharony and Imry [10] where the generating function (3.1) is discussed.

It is important to note the following symmetry properties of Eq. (3.1):
(a) Each Lagrangian $L(\phi)$ and $L(\psi)$ is a scalar function of the vector fields ϕ and ψ and is thus invariant under rotations of these vectors.
(b) If $\eta = 0$ the total Lagrangian $L(\phi) - L(\psi)$ is also invariant under the

transformation $\phi = \alpha\phi + \beta\psi$ and $\psi = \alpha\psi + \beta\phi$ with $\alpha^2 - \beta^2 = 1$. Thus the variable η breaks this symmetry and plays a similar role to that of a magnetic field in magnetic materials.

We can take advantage of the replica method and average over the disorder in (3.1). For simplicity in presentation we assume it is a Gaussian variable with zero correlation length, i.e., (1.7a). The relevant part of (3.1) is

$$\langle \exp[(i/2)\omega^2 \int d\vec{r}(\phi^2 - \psi^2)m'(\vec{r})]\rangle_c \tag{3.4}$$

$$= \exp[-(\gamma^2\omega^4/8) \int dr(\phi^2 - \psi^2)^2] ,$$

We thus see that averaging introduces quartic terms into the Lagrangian and mixes the two sets of replicas.

We can handle the quartic term in (3.4) by taking it apart by means of a Gaussian transform

$$\exp[-(\gamma^2\omega^4/8)\int d\vec{r}(\phi^2-\psi^2)^2] = \int dQ e^{-(1/2)\int[Q_{++}{}^2(\vec{r})+Q_{--}{}^2(\vec{r})+2Q_{+-}{}^2(\vec{r})]d\vec{r}}$$

$$e^{-(i\omega^2\phi/2)\int(Q_{++}(\vec{r})\phi\phi - Q_{--}\psi\psi + iQ_{+-}\phi\psi)d\vec{r}} . \tag{3.5}$$

We refer to the Q variables introduced in this way as the order parameter. Q is quite a complicated object: it is a tensor in the replica space $\alpha = 1,\ldots,n$ and also contains indices $i,j = +$ or $-$ referring to the advanced and retarded functions. What is important to us is that from the Q field we can generate the Green's functions

$$\langle Q_{++}\rangle = (\omega^2/2)\gamma G_+(\vec{r},\vec{r}) , \tag{3.6}$$

$$\langle Q_{--}\rangle = (\omega^2/2)\gamma G_-(\vec{r},\vec{r}) ,$$

$$\langle Q_{+-}(\vec{r})Q_{+-}(0)\rangle = (\omega^4/4)\gamma^2\ G_+(\vec{r},0)G_-(\vec{r},0) . \tag{3.7}$$

In particular the correlation function (3.7) of the Q-field gives the correlation function required in (1.11) for the determination of the diffusion constant.

The mean field theory is obtained by substituting (3.5) for (3.4) in (3.3) and evaluating the Q integral by the method of stationary phase. This stationary point has $Q^{\alpha\beta}_{++} = Q\delta_{\alpha\beta} = Q^{\alpha\beta*}_{--}$ and $Q_{+-} = 0$ and the self consistent equation for Q is

$$Q = (\omega^2/2V) \sum_k G_+(k,\omega_+) \ , \tag{3.8}$$

where $G^{-1}(k,\omega) = \omega^2 - k^2 - \gamma\omega^2 Q$. Eq. (3.8) is the usual coherent potential approximation equation [11]. At low frequencies it leads to a density of states $\rho(\omega) = \omega^{d-1}$ in d dimensions, and to a scattering rate

$$\tau^{-1}(\omega) = (\omega\gamma/2)Q_i(\omega) = (\pi\omega^2/8)\rho(\omega)\gamma^2 = \omega^{d+1} \ , \tag{3.9}$$

where Q_i is the imaginary part of Q. Eq. (3.9) is a well known result and in three dimensions leads to a scattering rate $\approx \omega^4$, which is equivalent to Rayleigh scattering [12].

4. Fluctuations

To determine the diffusion constant we examine the fluctuations around the mean field theory or stationary phase point found in the previous section. The important fluctuations are those in the Q_{+-} field and it is not difficult to show that those fluctuations are determined by a Lagrangian

$$L_F = \int d\vec{r}_1 d\vec{r}_2 Q_{+-}(\vec{r}_1) C(\vec{r}_1 - \vec{r}_2) Q_{+-}(\vec{r}_2) \ , \tag{4.1}$$

where repeated indices are to be summed and

$$C(\vec{r}_1 - \vec{r}_2) = \delta(\vec{r}_1 - \vec{r}_2) - (\omega^4/2)\gamma^2 |G_+(\vec{r}_1 - \vec{r}_2)|^2 \ . \tag{4.2}$$

The Fourier transform of this function can be shown from (3.8) to have the form

$$C(K) = (4/\omega^4\gamma^2)(\eta C_0 + C_2 K^2) \ , \tag{4.3}$$

where $C_0 = 4\omega^2/\pi\rho(\omega)$; in two dimensions, $C_2 = \omega^2/2\pi^2\rho^2\gamma Q_i$ and in three dimensions, $C_2 = \omega^3/3\pi^2\rho^2\gamma Q_i$. In each case we obtain a diffusion constant

$$D(\omega) = 2dC_2/C_0 \approx \tau(\omega) = \omega^{-(d+1)} \tag{4.4}$$

(with d the dimensionality), which is equivalent to the value obtained from kinetic theory.

5. Renormalization

We wish to determine the nature of the normal modes of frequency ω by considering how the diffusion constant $D(\omega)$ renormalizes as we integrate over short wavelength fluctuations of the field theory and rescale to longer lengths [13]. Following Wegner [9] we define an effective Lagrangian based on (4.1) and (4.3). In the spirit of the previous mean field approximation we restrict Q to the vicinity of the previously determined saddle point (3.6). However, in doing so, we note that the exact Lagrangian (3.1) with (3.5) has the symmetry properties discussed in section 3, i.e., for $n = 0$, $L(Q)$ is invariant under a rotation of both sets of replicas as discussed above Eq. (3.4) Thus from the saddle point (3.6) we can generate new saddle points by such rotations and instead of restricting Q to the neighborhood of the original saddle point we should extend it to a neighborhood that encompasses the entire manifold of saddle points. This leads us to generalize (4.1) to

$$L_F = (1/2)\int d\vec{r}_1 d\vec{r}_2 Q_{ij}(\vec{r}_1) C(\vec{r}_1 - \vec{r}_2) Q_{ij}(\vec{r}_2) \qquad (5.1)$$
$$= (1/2) \sum_K C_2 K^2 \, |Q_{ij}(K)|^2 ,$$

where repeated indices are to be summed over. The restriction of (5.1) to the neighborhood of the manifold of saddle points is accomplished by requiring that

$$(1/n)|Q_{ij}(\vec{r})|^2 = 2Q_i^2 , \qquad (5.2)$$

when Q_i is given in (3.9). It is convenient to introduce a normalized order parameter $q_{ij} = Q_{ij}/Q_i$ and the fluctuation Lagrangian is

$$L_F = (1/g^2) \sum_K K^2 |q_{ij}(K)|^2 , \qquad (5.3)$$

where $1/g^2 = (\pi/4d)\rho(\omega)D(\omega) \simeq \omega^{-2}$ is the frequency dependent diffusivity.

The physical picture that emerges is that of a phase transition in a fixed length spin system $(|q_{ij}|^2 = 2n)$. Straightforward momentum shell integration and rescaling yields the following differential recursion relation for the resistance in $d = 2 + \varepsilon$ dimensions (for $n = 0$):

$$dg^2(L)/d \ln L = -\varepsilon g^2 + 2g^4 . \tag{5.4}$$

This is identical in form to the scaling relation for the resistance of an electron system in a random potential [3,9]. In two dimensions and below there is a single unstable fixed point at $g^2 = 0$. The flow is towards infinite resistance on longer length scales so that all states are localized. For $d = 1$ (5.4) can be approximated by

$$dg^2/d \ln L = g^2 , \tag{5.5}$$

which has the solution $g^2/g_0^2 = L/L_0$. The length scale L at which the resistance g^2 reaches some fixed value, say $g^2 = 1$, defines the <u>localization length</u> ξ. Thus

$$\xi \approx L_0/\omega^2 \quad (d = 1) , \tag{5.6}$$

where we have made use of the asymptotic behavior of the bare conductance g given below (5.3). A similar analysis for $d = 2$ yields an even more rapidly diverging localization length at low frequency

$$\xi \approx L_0 e^{1/\omega^2} \quad (d = 2) . \tag{5.7}$$

Above two dimensions $(\varepsilon > 0)$ a new unstable fixed point at $g^{*2} = \varepsilon/2$ appears which represents a <u>mobility edge</u> between localized and extended phonon states. The solution of (5.4) is

$$L/L_0 = \{[g_0^2/g^2] [(g^2 - g^{*2})/(g_0^2 - g^{*2})]\}^{1/\varepsilon} . \tag{5.8}$$

Since $g^{*2} = \varepsilon/2$ corresponds to the mobility edge ω_c, we may write $g^2 - g^{*2} \approx \omega - \omega_c$. Then setting $g^2 = 1$ in (5.8), we see that the localization length diverges at the mobility edge as

$$\xi \approx L_0/(\omega - \omega_c)^{1/\varepsilon} , \quad (d = 2 + \varepsilon) . \tag{5.9}$$

This is the same localization length exponent that occurs in the problem of electron localization. In the region of extended states the conductivity is

$$1/g^2 \approx (\omega_c - \omega)^t , \quad t = (d - 2)\nu = 1 . \tag{5.10}$$

The feature that distinguishes phonon localization from that of electrons is the asymptotic form of the low resistance $g_0^2 \approx \omega^2$. For sufficiently low frequencies g^2 will always lie below the nontrivial fixed point and hence flow towards zero resistance.

6. Discussion

We have shown that the transition from extended to localized states of a disordered elastic medium can be described as a phase transition in a generalized fixed length spin system. The "spin" is a complex matrix and the resistance g^2 plays the role of temperature. The theory leads to the appearance of a mobility edge with extended states at low frequencies.

In dimension $d = 2$ and below we have shown that the energy diffusion constant vanishes at all frequencies. This means that any initial localized distrubance will remain localized. It also implies that the solutions of (1.5) are localized in space for typical realizations of the disorder $m(\vec{r})$. In three dimensions the diffusion constant vanishes above a certain frequency ω^* (the mobility edge) and is finite for $\omega < \omega^*$. We thus expect localized states for $\omega > \omega^*$ and extended states for $\omega < \omega^*$.

We have presented only the case of site-diagonal disorder [11]. We have studied the effects of introducting disorder into the Lamé coefficients in a simpler scalar theory described by a Lagrangian

$$L = \int dx \left[\frac{1}{2} M \dot{\phi}^2 - \frac{1}{2} V(x)(\nabla\phi)^2 \right] , \tag{6.1}$$

where the fluctuating part of the elastic constant $V(x) = V_0 + V'(x)$ satisfies $\langle V'(x)V'(x')\rangle = \gamma^2\delta(x - x')$. The localization properties of this model are the same as those described above.

The low temperature thermal transport properties are determined by the thermal conductivity

$$K(T) = \int d\omega\rho(\omega)\hbar\omega[\partial n(\omega)/\partial T]\, D_o(\omega) \qquad (6.2)$$

where $n(\omega) = (e^{\beta\hbar\omega} - 1)$ is the Bose factor. The mean field theory predicts a low frequency ω^{-2} behavior for $\rho(\omega)D_o(\omega)$, and as a result the thermal conductivity diverges. This is the usual result for a model that does not take into account inelastic scattering processes arising from lattice anharmonicity. Phonon-phonon interactions introduce a new length scale into the problem, the inelastic mean free path, which must be longer than the localization length ξ if localization of phonons in the harmonic approximation is to be important. The competing effects of localization and of elastic scattering have been discussed recently be Jackle [14]. Another way to observe phonon localization might be to study phonons injected into solids at sufficiently low temperatures where inelastic scattering effects are small.

Acknowledgements

This work was supported in part by the National Science Foundation under Grant No. DMR-81-06151. I am grateful to the Physics Department at Boston University for hospitality while this work was being carried out.

References

1. S. John, H. Sompolinsky and M.J. Stephen, Phys. Rev. B (to be published).
2. Tatarski, V. Wave Propagation in a Turbulent Medium, McGraw Hill (1961).
3. E. Abrahams, P.W. Anderson, D.C. Licciardello and T.V. Ramakrishnan, Phys. Rev. Lett. 42, 673 (1979), L. Schafer and F. Wegner, Z. Physik B38, 113 (1980); A.J. McKane and M. Stone, Annals of Physics 131, 36 (1981).
4. C. Kittel, Introduction to Solid State Physics Chap. 6, J. Wiley, NY (1971).
5. J.D. Jackson, Classical Electrodynamics, 2nd Ed. p. 225, J. Wiley, NY (1975).

6. See Ref. 9.

7. E. Abrahams and T.V. Ramakrishnan, J. Non-crystalline Solids 35, 15 (1980).

8. J. Langer and T. Neal, Phys. Rev. 169, 508 (1966).

9. F.J. Wegner, Z. Phys. 25, 327 (1976); F.J. Wegner, Phys. Reports. 67, 15 (1980).

10. A. Aharony and Y. Imry, J. Phys. C L487 (1977).

11. R.J. Elliott, J.A. Krumhansl and P.L. Leath, Rev. Mod. Phys. 46, 465 (1974).

12. J. Ziman, Electrons and Phonons, Chap. 19, Oxford Clarendon Press (1960).

13. D.R. Nelson and R.A. Pelcovits, Phys. Rev. B16 2191 (1977).

14. J. Jackle, Solid State Commun. 39, 1261 (1981).

MICROSCOPIC ORIGINS OF STOCHASTIC CRACK GROWTH

W.J. Pardee, W.L. Morris and B.N. Cox

Rockwell International Science Center
Thousand Oaks, CA 91360

ABSTRACT

Physical arguments are made to obtain a mathematical model of the stochastic growth of surface fatigue cracks in a ductile metal alloy. The model is a set of coupled partial differential equations for the expected statistical density of cracks per unit area. The differential equations describe the smooth, deterministic local evolution of crack states, with the stochastic effects of abrupt local changes of material in the crack path appearing as transitions between distinct subspaces of single crack state space. Results are related to observables such as statistical distributions of crack growth rate and of time for at least one crack to reach macroscopic length.

1. Introduction

Procedures to design structures and components (bridges, airplanes, rocket engine turbine pump blades) to withstand prescribed maximum loads are well established, even if the material has a statistical variability in yield stress (the stress at which the material exhibits appreciable irreversible deformation) or in ultimate strength (the stress at which fracture occurs). As a result, structures normally do not fail because of a single overload, but because of damage that accumulates over many cycles of loading and unloading at stresses less than the nominal yield stress. Several microscopic processes are involved in this accumulation of fatigue damage, but their net contribution to failure is through the formation and growth of cracks. Most metal components and structures are fabricated from polycrystalline alloys, and the size, shape, and orientation of the individual crystals (grains) profoundly influence the rate of crack formation and early growth under cyclic loading. This paper describes a theory to predict the important lifetime related statistical distributions that result from the locally deterministic growth of cracks in a random microstructure. This fatigue crack growth has an abstract similarity to the percolation processes of other papers in this volume, but the description presented here is rather different. As one would expect from the paper of Montroll and Shlesinger [1], the lifetime distributions are approximately log-normal. Typical lifetime distributions in the number N of constant amplitude load cycles for samples from three different, nominally identical, heats of a common aluminum alloy are shown in Fig. 1. These samples were fatigued at rather high stress amplitude (peak amplitude of about 80% of nominal macroscopic yield stress) and failed quite early. At lower applied stress amplitudes lifetimes are much longer, and the coefficient of variation (square root of the variance/mean) can exceed 100%.

Long cracks grow very predictably (deterministically) for given load and environment; intrinsic variability in lifetime arises from the often long period required for a crack to form and reach macroscopic size. This paper deals only with short cracks, cracks no more than 10 mean grain diameters in length. For typical alloys this means a crack is "long" when it reaches about 0.5 mm. The "state" of such a short crack evolves [2] deterministically except when one of the

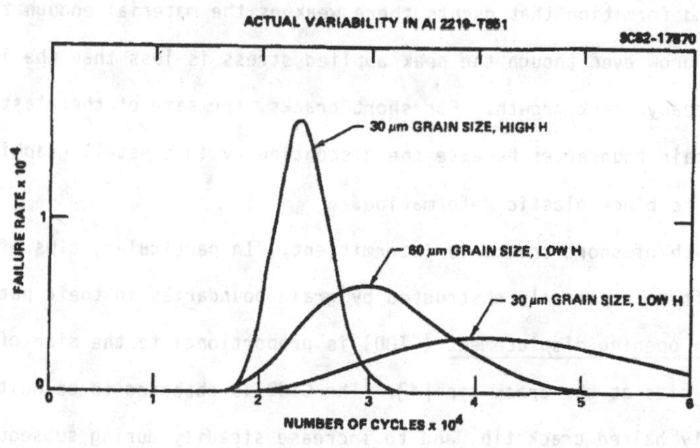

Figure 1

crack's tips arrives at or leaves a grain boundary. At such points the laws governing the continued evolution of the crack's state change stochastically. In the following, the state of a crack is defined and a set of partial differential equations is obtained to describe the continuous portions of crack evolution. The stochastic role of the microstructure appears in the transfer of cracks between subspaces of the single crack state space. Such transfers appear either as boundary conditions or as distributed "sources". The solution of these differential equations is an expected statistical (number, not probability) density of cracks per unit area. Section 4 of this paper relates these densities to several quantities of more direct experimental and engineering significance.

2. Essential Physical Phenomena

Under the simple tensile loads considered here, the force for crack extension is largest in the direction perpendicular to the applied stress. No serious loss in usefulness is incurred by considering only those cracks that are oriented parallel to this force. The presence of a crack disturbs the stress field in a way that produces stress concentrations near the crack tips. Very near the crack tip the resulting stress will exceed the yield stress and small regions of crystal are deformed irreversibly. This deformed region is called the plastic zone. The irreversible deformation that occurs there weakens the material enough to permit the crack to grow even though the peak applied stress is less than the level needed for steady crack growth. For short cracks, the size of the plastic zone is limited by grain boundaries because the discontinuity in crystallographic orientation tends to block plastic deformation.

The growth of short cracks is intermittent. In particular, tips of surface cracks are often temporarily obstructed by grain boundaries in their path [3]. The <u>crack tip opening displacement</u> (CTOD) is proportional to the size of the plastic zone size at the crack tip [4]. The CTOD is observed to be quite small for a recently halted crack tip, and to increase steadily during subsequent fatigue until, at some sufficiently large value, the crack tip breaks through the grain boundary and resumes motion [2]. This process is evidently the incubation of a new plastic zone in the next grain. The embryonic plastic zone's maturity is conveniently described by a normalized damage accumulation parameter β, with $\beta = 0$ at a just stopped crack tip and $\beta = 1$ at a tip about to resume growth.

The parameter β is thought to be a measure of the plastic deformation at the crack tip. The shear stress driving this deformation is largest at 45° to the applied stress axis. It is found empirically that the rate v_β ($v_\beta = d\beta/dN$) at which β increases is proportional to the larger (D) of the two possible 45° distances to the next grain boundary. Though no completely satisfactory model exists for v_β, it is known that it must be a function of crack length (2c) as well as D, and it is explicitly dependent on N because of microscipic hardening mechanisms. Thus, any blocked crack tip has associated with it two variables, β and D.

A growing crack tip's motion is governed [2] by the effective local cyclic stress intensity range. That local stress intensity is a function of crack length and of the size of the grain at the crack tip. We take the appropriate size to be the distance z to the next grain boundary along a line parallel to the crack. Then, if the crack has length 2c and one tip is moving toward the next grain boundary at a rate $-v_z = -dz/dN$ and the other at $-v_z' = -dz'/dN$, the length increases as $2v_c = -v_z - v_z'$. There are two tacit assumptions here: the two tips grow essentially independently, and surface constraints are the primary determinants of growth rate. These assumptions are known to provide a good description of short crack growth in alloys where plastic slip is much easier near the surface.

Thus, we distinguish three types of cracks, those with two surface tips growing (type 2), one tip growing (type 1), or neither tip growing (type 0, zero growth rate). The independent variables for a type 2 crack are its length (2c), and the distances z and z' between the tips and the next grain boundaries. Type 1 cracks are described by length and, at the growing tip, the distance z to the next grain boundary, and, at the blocked tip, by β and D. Type 0 cracks are described by five independent variables: length (2c), D, β, D', and β'. The geometrical interpretation of these variables is illustrated for type 1 cracks in

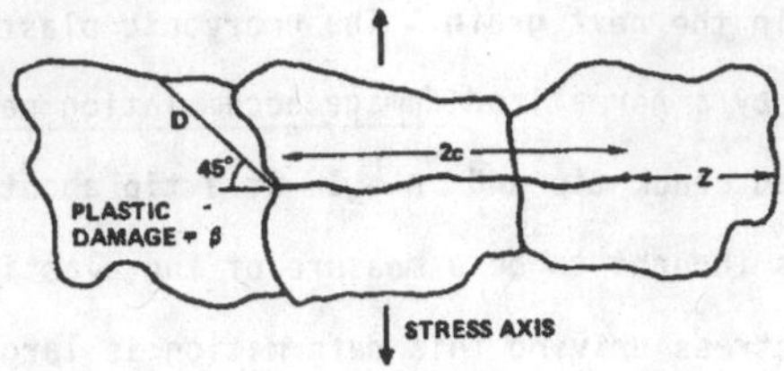

Figure 2

Fig. 2. Experience with crack initiation and early growth in aluminum alloys has shown that no smaller set of state space variables is sufficient to determine a crack's state after N + 1 cycles from its state after N cycles. Let a representative point in type j state space $\Omega^{(j)}$ be denoted $\underset{\sim}{\xi}^{(j)}$, with

$$\underset{\sim}{\xi}^{(0)} = (c,\beta,D,\beta',D') , \tag{1}$$

$$\underset{\sim}{\xi}^{(1)} = (c,z,\beta,D) , \tag{2}$$

and

$$\underset{\sim}{\xi}^{(2)} = (c,z,z') . \tag{3}$$

Each coordinate vector has a corresponding velocity $\underset{\sim}{v}^{(j)}$,

$$\underset{\sim}{v}^{(j)} = d\underset{\sim}{\xi}^{(j)}/dN . \tag{4}$$

The growth rates of crack length (2c) and of normalized CTOD (β) are non-negative; the rates of change of the z variables are always negative or zero; the rates of change of the D variables are always zero.

The fundamental dependent variables of our theory are the state space crack densities $f^{(j)}$. These are most simply defined in terms of the cumulative state space crack distributions $F^{(j)}$, with $F^{(j)}(N,\underset{\sim}{\xi}^{(j)})$ the expected number of cracks per unit specimen area with each state vector component less than the corresponding component of $\underset{\sim}{\xi}^{(j)}$. Then the densities are defined by differentiation, with $f^{(j)}$ the result of differentiating $F^{(j)}$ once with respect to each of its arguments except N. For example, the statistical density $f^{(1)}$ is related to $F^{(1)}$ by

$$f^{(1)} = \partial^4 F^{(1)}/\partial c\ \partial z\ \partial\beta\ \partial D\ . \tag{5}$$

Its interpretation is that the quantity Δn defined by

$$\Delta n = f^{(1)}\ (N,c,z,\beta,D)\ A\ dc\ dz\ d\beta\ dD \tag{6}$$

is the expected number of type 1 cracks in area A with: length between $2c$ and $2(c + dc)$; with the growing tip no more than $z + dz$ nor less than z from the next grain boundary; with accumulated damage at the blocked tip between β and $\beta + d\beta$; and with the parameters of the grain in front of the blocked tip in the interval $(D, D + dD)$. The other two densities have similar interpretations. The expected number of cracks per unit specimen area can be thought of as the limit of a sequence of observations on samples of increasing area as the area becomes arbitrarily large. Let the rate of formation of cracks of type j , that is, the state space density of cracks of type j formed per unit specimen area per cycle be denoted $s^{(j)}\ (N,\underset{\sim}{\xi}^{(j)})$. The most common physical mechanisms for crack formation in aluminum alloys are by brittle fracture of constituent particles, grain boundary separation, and formation at persistent slip bands. An example of a source function $s^{(j)}$ is given in Appendix A .

One other empirical fact will be important if the theory is to describe long crack growth correctly. Long cracks can cross grain boundaries without stopping. It appears [5] that a crack tip will not be blocked, even temporarily, by a grain boundary if the elastic CTOD without fatigue exceeds some threshold value. Since CTOD increases with crack length there will, therefore, exist a long crack threshold $2c_L$ above which all cracks are type 2 .

3. Evolution equations and boundary conditions

The evolution equations for the densities $f^{(j)}$ are obtained by considering a small region $d\Omega^{(j)}$ at $\underset{\sim}{\xi}^{(j)}$ in the state space $\Omega^{(j)}$ and letting it evolve in time. The number of cracks in a neighborhood $d\Omega^{(j)}$ of $\underset{\sim}{\xi}^{(j)}$ at N cycles is equal to the number in the region $d\Omega^{(j)'}$ of $\underset{\sim}{\xi}^{(j)} + \underset{\sim}{v}^{(j)}\ \Delta N$, plus the new cracks created by $s^{(j)}$, plus any cracks added from the other distributions, minus any cracks lost to the other distributions. It is useful to define three gradient operators $\nabla^{(j)}$ by

$$\nabla^{(j)} = (\partial/\partial c, \partial/\partial \xi_2^{(j)} \ , \ \partial/\partial \xi_3^{(j)}) \ , \ j = 0,1,2 \ , \qquad (7)$$

with $\xi_2^{(j)} = \beta, z, z$ and $\xi_3^{(j)} = \beta', \beta, z'$ for $j = 0,1,2$ respectively. Only three components of each $\xi^{(j)}$ are dynamic variables, changing continuously with fatigue. The other variables are stochastic parameters that change only when a crack's trajectory reaches a state space boundary in one of the dynamic variables. At such boundaries a crack changes type and begins a trajectory in one of the other state spaces.

The state space volume magnification factor along a crack's trajectory is simply the Jacobian

$$\frac{d\Omega^{(j)'}}{d\Omega^{(j)}} = 1 + \nabla^{(j)} \cdot \underset{\sim}{v}^{(j)} \Delta N + O(\Delta N^2) \ . \qquad (8)$$

Thus, the density at the transformed point $\xi^{(j)'} = \xi^{(j)} + \underset{\sim}{v}^{(j)} \Delta N$ is given by

$$f^{(j)} \ (N', \xi^{(j)'}) \ d\Omega^{(j)'} = f^{(j)} \ (N, \xi^{(j)}) \ d\Omega^{(j)} + \text{transfers} + \text{sources.} \qquad (9)$$

Transfer of cracks from type 0 to type 1 occurs when a crack crosses a $\beta = 1$ boundary. At this instant the crack acquires a value of z distributed according to the conditional probability density $g(z|D)$, where D is the maximum 45° slip distance from the crack tip to the remote boundary of the newly penetrated grain. The z in this distribution is the width of the grain perpendicular to the applied uniaxial load at the point where the crack tip entered the grain. Transfers from higher j to lower always occur from a $z = 0$ boundary to a $\beta = 0$ boundary, and thus are boundary conditions, not source terms in the right hand side of Eq. (9). By expanding the left side of Eq. (9) in a Taylor's series and using the volume dilation factor Eq. (8) one obtains

$$\frac{\partial f^{(j)}}{\partial N} + \nabla^{(j)} \cdot (\underset{\sim}{v}^{(j)} f^{(j)}) = s^{(j)} + \text{transfer } (j - 1 \rightarrow j) \ . \qquad (10)$$

The rate $J_e^{(j)}$ at which cracks of type j cross any surface with unit normal $\underset{\sim}{e}^{(j)}$ is given by $\underset{\sim}{e}^{(j)} \cdot \underset{\sim}{v}^{(j)} f^{(j)}$, and, in particular, the rate at which type 0 cracks cross the $\beta = 1$ boundary to become type 1 is $J_\beta^{(0)}$ evaluated at $\beta = 1$, with $J_\beta^{(0)}$ given by

$$J_\beta^{(0)} = v_\beta^{(0)} f^{(0)} . \tag{11}$$

The vector flux is simply $\underset{\sim}{J}^{(j)} = \underset{\sim}{v}^{(j)} f^{(j)}$ where $\underset{\sim}{v}^{(j)}$ has only three non-zero components. The explicit equations for the evolution of the $f^{(j)}$ for $N > 0$ are now easily stated. The density $f^{(0)}$ of type 0 cracks is defined on the region $0 < c < c_L$, $0 < \beta$, $\beta' < 1$, $0 < D'$, D and satisfies there the first order partial differential equation

$$\frac{\partial f^{(0)}}{\partial N} + \nabla^{(0)} \cdot \underset{\sim}{J}^{(0)} = s^{(0)} . \tag{12}$$

The corresponding equation for the evolution of type 1 cracks is

$$\frac{\partial f^{(1)}}{\partial N} + \nabla^{(1)} \cdot \underset{\sim}{J}^{(1)} = s^{(1)} + T^{1,0} , \tag{13}$$

with the transfer rate $T^{1,0}$ of cracks from type 0 into type 1 space given by

$$T^{1,0} = \int dD'[J_\beta^{(0)}(N,c,1,\beta,D',D) + J_\beta^{(0)}(N,c,\beta,1,D,D')]g(x|D') . \tag{14}$$

The density $f^{(2)}$ is defined for c,z and z' positive. For $c < c_L$, $f^{(2)}$ satisfies

$$\frac{\partial f^{(2)}}{\partial N} + \nabla^{(2)} \cdot \underset{\sim}{J}^{(2)} = s^{(2)} + T^{2,1} \tag{15}$$

with the transfer rate $T^{2,1}$ given by

$$T^{2,1} = \int dD[J_\beta^{(1)}(N,c,z,1,D)g(z'|D) + J_\beta^{(1)}(N,c,z',1,D)g(z|D)]/2 . \tag{16}$$

There are no type 0 or type 1 cracks for $c > c_L$, so type 2 cracks reaching a $z = 0$ or $z' = 0$ boundary re-enter type 2 space with a new value of z or z'. The equation for $f^{(2)}$ for $c > c_L$ is

$$\frac{\partial f^{(2)}}{\partial N} + \nabla^{(2)} \cdot \underset{\sim}{J}^{(2)} = s^{(2)} + T^{2,2} \tag{17}$$

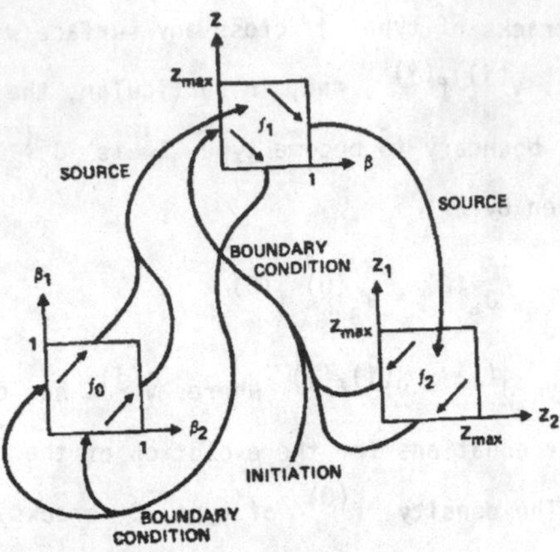

Figure 3

with $T^{2,2}$ given by

$$T^{2,2} = J_{z'}^{(2)}(N,c,z,0)\ g(z') + J_{z}^{(2)}(N,c,0,z')\ g(z)\ . \tag{18}$$

The probability density g in Eq. (18) is related to the previously defined conditional density by $g(z) = \int dD\ g(z|D)\phi(D)$, where ϕ is the probability density of slip distances.

The boundary conditions on the densities $f^{(j)}$ can be understood most easily by reference to the flow of cracks between the three spaces as shown in Fig. 3. The flux of cracks leaving type 2 state space at the $z = 0$ boundary is equally divided between the $\beta = 0$ and $\beta' = 0$ boundaries, satisfying

$$J_{\beta}^{(0)}(N,c,0,\beta',D,D') = -(1/2)J_{z}^{(1)}(N,c,0,\beta',D')\phi(D)\ , \tag{19}$$

$$J_{\beta}^{(0)}(N,c,\beta,0,D,D') = -(1/2)J_{z}^{(1)}(N,c,0,\beta,D)\phi(D')\ . \tag{20}$$

The cracks arriving at the $\beta = 0$ boundary of type 1 state space are the ones that left the $z = 0$ or $z' = 0$ boundary of type 2 state space, and thereby satisfy the boundary condition

$$J_{\beta}^{(1)}(N,c,z,0,D) = -2\ J_{z}^{(2)}(N,c,z,0)\phi(D)\ . \tag{21}$$

In this last equation the symmetry of $\underset{\sim}{J}^{(2)}$ under interchange of the 1 and

2 variables has been used.

A set of three functions $f^{(j)}$ which satisfy the above differential equations and boundary conditions, and which have everywhere prescribed initial values

$$f^{(j)}(0,\xi^{(j)}) \equiv f_0^{(j)}(\xi^{(j)}) , \tag{22}$$

is unique. To see this suppose a second solution $g^{(j)}$ existed; then the difference $h^{(j)} = g^{(j)} - f^{(j)}$ is a solution of the corresponding homogeneous system ($s^{(j)} = 0$) and satisfies

$$h^{(j)}(0,\xi^{(j)}) = 0 \tag{23}$$

and

$$\frac{\partial h^{(j)}}{\partial N}(0,\xi^{(j)}) = 0 \tag{24}$$

and is therefore always everywhere zero.

One can verify that these boundary conditions and the evolution equations imply that the total number of cracks $Q_T(N)$ satisfies

$$\frac{dQ_T}{dN} = \sum_{j=0}^{2} \int d\Omega^{(j)}\, s^{(j)}(N,\xi^{(j)}) . \tag{25}$$

The constancy of Q_T in the absence of sources is a sensitive test of the accuracy of any numerical solution.

4. Stochastic Crack Growth and Lifetime Distributions

Two complementary kinds of statistical information can be obtained from the solutions $f^{(j)}$ to the differential equations of Section 3. The first of these is the description of various kinds of crack growth rate experiments. Such experiments are of fundamental value in determining the microscopic growth mechanisms and of practical value in enabling an engineer to design more accurately for damage tolerance. The second major class of applications is the prediction of the probability $P(N,2c)$ that a crack longer than $2c$ exists after N fatigue cycles. This family of distributions provides lifetimes and time-to-(engineering) crack-initiation, and, broadly speaking, permits an engineer to modify structures and alloys for greater durability.

Most experimental measurements of crack growth rate record only length (2c), not z or β, so it is useful to define the reduced distributions $\rho^{(j)}(N,c)$ by integration of the corresponding $f^{(j)}(N,\underset{\sim}{\xi}^{(j)})$ over all variables except N and c. For example, $\rho^{(2)}$ is defined by

$$\rho^{(2)}(N,c) = \int dz\, dz'\, f^{(2)}(N,c,z,z') . \tag{26}$$

Denote by $R^{(j)}(N,c)$ the total number of cracks per unit area longer than 2c after N cycles,

$$R^{(j)}(N,c) = \int_c^\infty dc'\, \rho^{(j)}(N,c') . \tag{27}$$

Note that the expected total number of cracks per unit area $Q_T(N)$ is related to the $R^{(j)}$ by the equation

$$Q_T(N) = \sum_{j=0}^{2} R^{(j)}(N,0) . \tag{28}$$

A simple but useful crack growth experiment is the empirical determination as a function of crack length of the fraction of crack tips stopped at grain boundaries. Since the number of stopped type 2 crack tips is zero, the expected number of type 1 stopped tips is $\rho^{(1)}(N,c)dc$, and the expected number of type 0 stopped tips is $2\rho^{(0)}(N,c)dc$, the expected value of the fraction, $f_B(N,c)$, is given by

$$f_B(N,c) = (2\rho^{(0)} + \rho^{(1)})/[2(\rho^{(0)} + \rho^{(1)} + \rho^{(2)})] . \tag{29}$$

This relatively simple experiment can be quite sensitive to the models assumed for incubation (v_β) and growth rate (v_c).

The measurement of crack growth rates $v_c = dc/dN$ of an ensemble of cracks provides considerably more information. The distribution in growth rates can be obtained from the $f^{(j)}$. Denote by ρ the sum over j of the $\rho^{(j)}$,

$$\rho(N,c) = \sum_{j=0}^{2} \rho^{(j)}(N,c) . \tag{30}$$

The average growth rate $\nu(N,c)$ of cracks of length $2c$ can then be calculated from the equation

$$\nu(N,c) = [\rho(N,c)]^{-1} \sum_{j=0}^{2} \int \frac{d\Omega^{(j)}}{dc} v^{(j)} f^{(j)} (N,\xi^{(j)}) \qquad (31)$$

where the measure $d\Omega^{(j)}/dc$ denotes integration over all components of $\xi^{(j)}$ except c . By differentiating ρ one obtains (Appendix B) the very simple result that

$$\frac{\partial \rho}{\partial N} + \frac{\partial}{\partial c} (\nu\rho) = S , \qquad (32)$$

where the source S is the sum

$$S(N,c) = \sum_{j=0}^{2} \int \frac{d\Omega^{(j)}}{dc} s^{(j)} (N,\xi^{(j)}) . \qquad (33)$$

The appealing simplicity of Eq. (32) is deceptive. The average growth rate $\nu(N,c)$ does not follow (for short cracks) standard long crack empirical formulae such as the common relationship $dc/dN = A(\Delta K)^m$, where ΔK is the stress intensity range (essentially stress range multiplied by the square root of the crack length). In fact, it cannot be obtained, in general, except by solution of the full system of equations. Equation (32) can be a numerically convenient technique for determining ν from either theoretical or experimental distributions $\rho(N,c)$.

Experimental comparison of stochastic growth rates is probably best done with the cumulative probability distribution $\Psi(N,v,\Delta K;\sigma_M)$ that a crack fatigued under uniform surface stress amplitude σ_M and at cyclic stress intensity range ΔK is instantaneously growing at a rate less than v . This is obtained from the $f^{(j)}$ by a change of variables from c to ΔK , then dividing by the number $\rho(N,c)$ of cracks per unit area. The corresponding density $\psi = \partial\Psi/\partial v$ is, in general, singular because of the presence, even in $\rho^{(2)}$, of zero growth rate cracks.

The calculation of lifetime distributions from the $f^{(j)}$ is somewhat simpler. Let $Q_T(N)$ denote the total number of cracks per unit area after N cycles. Suppose that $Q_T(N) > 0$. By appropriate choice of the unit of area, one can suppose that $Q_T(N) >> 1$. First consider an area A large enough to contain an expected number of cracks $n = Q_T A >> 1$. The probability $P(N,2c^*)$ that this area has at least one crack longer than $2c^*$ is given by

$$P = 1 - q^n , \tag{34}$$

where q is the probability that any one crack is shorter than $2c^*$,

$$q = 1 - R(N,c^*)/Q_T(N) , \tag{35}$$

and R is the sum of the $R^{(j)}$. Since Q_T is large P can be rewritten as

$$P = 1 - (1 - R/Q_T)^{AQ_T} = 1 - e^{-RA} . \tag{36}$$

Equation (36) is valid even if AQ_T is not large, as can be seen by considering a large ensemble of m little areas A , such that $mAQ_T >> 1$. Let q denote the probability that one of the m little regions has no crack longer than $2c^*$. Then, since mA is a large region, the probability q^m that none of them has such a crack is given by

$$q^m = (1 - R/Q)^{mQ_T A} = e^{-mRA} , \tag{37}$$

and therefore, q is given by

$$q = e^{-RA} \tag{38}$$

The distribution P is, for appropriate choice of c^* , the probability of engineering crack initiation or the probability of failure. The corresponding density $\partial P/\partial N \equiv L$ can be expressed as

$$L(N,c^*) = \nu(N,c^*)\ \rho\ (N,c^*)\ A \exp(-RA) , \tag{39}$$

if c^* is large enough that $s^{(j)}\ (N,c^*) = 0$.

5. Summary and Discussion

A theory relating local microstructural effects on crack initiation and growth to the statistics of crack growth and of fatigue has been stated as a coupled set of first order partial differential equations and their boundary conditions. The solution of those equations is the expected number of cracks per unit area with certain properties. The calculation of growth rate and failure distributions from the theory is described. Solution of these equations for specific models for the local dynamic laws v_β, v_c, and v_z can be an effective method for testing those laws by experiment. Numerical techniques for obtaining such solutions will be presented elsewhere.

Two important classes of phenomena have been omitted, crystallographic crack growth and crack coalescence. Growth along a crystallographic plane can be much more rapid than the ductile growth mode described here. Though in some alloys it is unusual for grains to be suitably oriented to permit this growth mode, it does lead to earlier failure when it occurs. Though too uncommon to influence average lifetime, crystallographic crack growth does result in the important early failure part of the lifetime distribution departing from log-normal behavior. Similar considerations apply to crack coalescence. Choices of material and load range can significantly alter the importance of these phenomena. They are not always negligible.

The theory presented here differs from other pictures of stochastic fatigue failure [6,7] in two respects: (i) our theory attributes stochastic fatigue failure to statistically distributed microscopic geometrical constraints on continuum fracture mechanics rather than statistical distributions of microscopic variations in mechanical properties like fracture toughness, and (ii) our theory predicts the results of both microscopic (growth rate, incubation) and macroscopic (lifetime) fatigue experiments. These microscopic geometrical properties, such as the distributions of grain size and shape, are measurable by realistic experiments, permitting application of the theory to interpretation of fatigue experiments, and eventually to alloy and structural design. Note, in particular, that the theory presented here treats local growth and incubation as deterministic processes. Arone [6], in contrast, assumes that the incremental growth $2\Delta c$ is a normally distributed random variable. That distribution is not assumed here, but can be predicted. McCartney [7] ascribes variation in lifetime to variation in initial defect size. That assumption is false for small initial defects; scatter in growth is then much larger than scatter in nucleation of the crack. The kinetic theory presented here embodies much the same assumptions as the computer simulation of Morris, James and Buck [8]. This theory differs in the omission of an assumed dependence on grain orientation, and in that the computer simulation of [8] really is a simulation of a finite number of fatigue experiments, while the distributions predicted from this paper are limits of expectation values as total sample area and total number of cracks become infinite. Including grain orientation as a parameter is formally easy, but computationally prohibitive because of the increase in number of dimensions. A more interesting and physically significant omission is the neglect of crack-crack interactions, in particular, the nonlinear process of crack coalescence. That problem poses substantial new problems, both formally and computationally, in this formalism.

Acknowledgement

This work was supported by the Defense Advanced Research Projects Agency under DSS Contract NDA903-80-C-0641. The formalism presented here is a concrete application of an abstract kinetic theory of cracks first suggested to the authors by John Richardson (unpublished).

Appendix A

Fracture of brittle constituent particles is the dominant initiation mechanism in Al 2219-T851 at cyclic stress amplitudes above about 60% of the bulk yield stress. Such brittle fracture produces cracks without a mature plastic zone at either tip. Such a crack is type 0, but several aspects of the formalism are better behaved if one defines the initiation event to include the incubation time necessary to start the first tip moving. Then the source is pure type 1 ($S^{(0)} = S^{(2)} = 0$). The damage accumulation mechanism that fractures the particle is similar to the incubation process. The geometry is defined in Fig. 4.

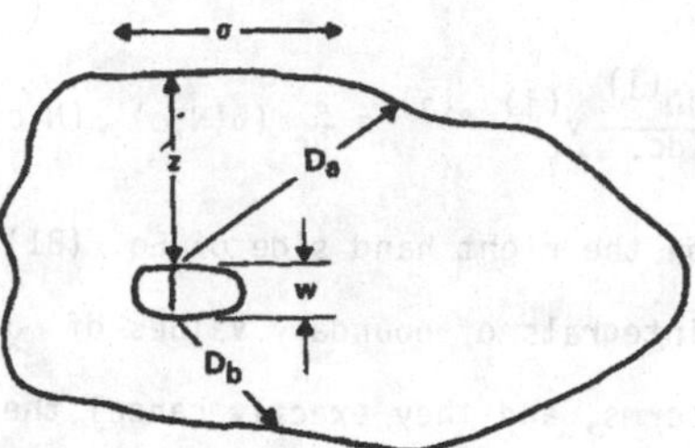

Figure 4

A number of cycles N_o are needed to fracture the particle and an additional number N_a , depending on D_a and w , are needed to start the first tip moving. At that time ($N = N_o + N_a$) the crack has length w and the damage β_b , at the other tip is given by

$$\beta_b(N,c,D_a,D_b) = \int_{N_o}^{N} dN' \; v_\beta(N',c,D_b) \; . \qquad (A1)$$

Then if ϕ_a is the probability density for the variable D_a , ϕ_w for particle size, ϕ_b for D_b , and ϕ_b the conditional probability density for z given D_a , the source $s^{(1)}$ is given by

$$s^{(1)}(N,c,z,\beta,D) = \phi_b(D)\;\phi_w(c)\int dD'[\phi_a(D')\;\phi_3(z|D') \qquad (A2)$$

$$\theta(N - N_o(D',c) - N_a(D',c))\;\delta(\beta - \beta_b(N,c,D',D))].$$

With these definitions $s^{(1)}$ is continuous in all variables.

Appendix B

The evolution equation for the density $\rho(N,c)$ is obtained by first differentiating ρ to obtain

$$\frac{\partial\rho}{\partial N} = \sum_j \int \frac{d\Omega^{(j)}}{dc}\;\frac{\partial f^{(j)}}{\partial N} = \sum_j \int d\Omega^{(j)}\;[\nabla\cdot(\underset{\sim}{v}^{(j)} f^{(j)}) - s^{(j)} - T^{(j)}] \quad (B1)$$

where $T^{(j)}$ represents the transfer terms $T^{1,0}$, $T^{2,1}$, $T^{2,2}$ in Eq. (13)-(18). The derivative with respect to c on the right side of Eq. (32) can be written in the form

$$\frac{\partial}{\partial c} \sum_j \int \frac{d\Omega^{(j)}}{dc} v_c^{(j)} f^{(j)} = \frac{\partial}{\partial c} (\nu(N,c) \rho(N,c)) . \tag{B2}$$

The other derivatives on the right hand side of Eq. (B1) can be integrated and expressed in terms of integrals of boundary values of $\underset{\sim}{J}^{(j)}$. These terms are the boundary transfer terms, and they exactly cancel the integrals of the $T^{i,j}$. This can best be shown by first simplifying the $T^{i,j}$ integrals by noting that

$$\begin{aligned} \int dzd\beta dD\ T^{1,0} &= \int dzd\beta dDdD'[J_\beta^{(0)}(N,c,1,\beta,D',D)\ g(z|D') \\ &\quad + J_\beta^{(0)}(N,c,\beta,1,D',D)\ g(z|D)] \\ &= \int d\beta dDdD'\ [J_\beta^{(0)}(N,c,1,\beta,D',D) \\ &\quad + J_\beta^{(0)}(N,c,\beta,1,D',D)] \end{aligned} \tag{B3}$$

and

$$\begin{aligned} \int dzdz'\ T^{2,1} &= \int dzdz'\ dD[J_\beta^{(1)}(N,c,z,1,D)\ g(z'|D) \\ &\quad + J_\beta^{(1)}(N,c,z',1,D)\ g(z|D)]/2 \\ &= \int dzdD\ J_\beta^{(1)}(N,c,z,1,D) \end{aligned} \tag{B4}$$

and

$$\int dzdz'\ T^{2,2} = 2 \int dz\ J^{(2)}(N,c,z,0) . \tag{B5}$$

The derivatives with respect to variables other than c can be simply evaluated. The results for type 0 are

$$\begin{aligned} &\int d\beta\ d\beta' dDdD'\ [\frac{\partial}{\partial\beta} J_\beta^{(0)} + \frac{\partial}{\partial\beta'} J_{\beta'}^{(0)}] \\ &= \int d\beta'\ dDdD'\ [J_\beta^{(0)}(N,c,1,\beta',D,D') + J_\beta^{(0)}(N,c,\beta',1,D,D') \\ &\quad - J_\beta^{(0)}(N,c,0,\beta',D,D') - J_\beta^{(0)}(N,c,\beta',0,D,D')] \\ &= \int dzd\beta\ dD\ T^{1,0} + \int d\beta'\ dDdD'[J_z^{(1)}(N,c,0,\beta',D')\ \phi(D) \\ &\quad + J^{(1)}(N,c,0,\beta',D)\ \phi(D')]/2 \\ &= \int dzd\beta dD\ T^{1,0} + \int d\beta\ dD\ J_z^{(1)}(N,c,0,\beta,D) . \end{aligned} \tag{B6}$$

The contribution for type 1 is similar,

$$\int d\beta dz dD \, [\frac{\partial}{\partial\beta} \, J_\beta^{(1)} + \frac{\partial}{\partial z} \, J_z^{(1)}]$$

$$+ \int dz dD [J_\beta^{(1)}(N,c,z,1,) - J_\beta^{(1)}(N,c,z,0,D)]$$

$$- \int d\beta dD \, J_z^{(1)}(N,c,0,\beta,D)$$

$$= \int dz dz' \, T^{2,1} + 2 \int dz dD \, J_z^{(2)}(N,c,z,0) \, \phi(D)$$

$$- \int d\beta dD \, J_z^{(1)}(N,c,0,\beta,D) \; . \tag{B7}$$

The contribution from type 2 is simply given by the expression

$$\int dz dz' \, [\frac{\partial}{\partial z} \, J_{z'}^{(2)} + \frac{\partial}{\partial z'} J_z^{(2)} \,] = -2 \int dz \, J_z^{(2)}(N,c,z,0) \; . \tag{B8}$$

Collecting terms produces the simple continuity equation given in Eq. (32).

References

1. E.W. Montroll and M.F. Shlesinger, this volume.

2. Morris, W.L., James, M.R. and Buck, O. "Growth Rate Models for Short Surface Cracks in Al2219-T851," Met. Trans. A, Vol. 12A, 1981, pp. 57-64.

3. Zurek, A.K., James, M.R. and Morris, W.L., "The Effect of Grain Size on Fatigue Growth of Short Cracks", submitted to Met. Trans. A, 1982.

4. Rice, J.R., "A Path Independent Integral and the Approximate Analysis of Strain Concentration by Notches and Cracks", J. Appl. Mech., Vol. 35, 1968, pp. 379-386.

5. James, M.R., Morris, W.L. and Zurek, A.K., "On the Transition from Near-Threshold to Intermediate Growth Rates in Fatigue, in preparation.

6. Arone, R., "A Statistical Model for Fatigue Fracture Under Constant Amplitude Cyclic Loading", Eng. Fract. Mech., Vol. 14, 1981, pp. 189-194.

7. McCartney, L.N., "Extensions of a Statistical Approach to Fracture", Int. J. Fract., Vol. 15, 1979, pp. 477-487.

8. Morris, W.L., James, M.R. and Buck, O., "Computer Simulation of Fatigue Crack Initiation", Eng. Fract. Mech., Vol. 13, 1980, pp. 213-221.